THERMODYNAMIC PROPERTIES
OF ORGANIC COMPOUNDS

PHYSICAL CHEMISTRY
A Series of Monographs

Edited by

ERNEST M. LOEBL

*Department of Chemistry, Polytechnic Institute of Brooklyn
Brooklyn, New York*

1. W. Jost: Diffusion in Solids, Liquids, Gases, 1952
2. S. Mizushima: Structure of Molecules and Internal Rotation, 1954
3. H. H. G. Jellinek: Degradation of Vinyl Polymers, 1955
4. M. E. L. McBain and E. Hutchinson: Solubilization and Related Phenomena, 1955
5. C. H. Bamford, A. Elliott, and W. E. Hanby: Synthetic Polypeptides, 1956
6. George J. Janz: Thermodynamic Properties of Organic Compounds — Estimation Methods, Principles and Practice, revised edition, 1967
7. G. K. T. Conn and D. G. Avery: Infrared Methods, 1960
8. C. B. Monk: Electrolytic Dissociation, 1961
9. P. Leighton: Photochemistry of Air Pollution, 1961
10. P. J. Holmes: Electrochemistry of Semiconductors, 1962
11. H. Fujita: The Mathematical Theory of Sedimentation Analysis, 1962
12. K. Shinoda, T. Nakagawa, B. Tamamushi, and T. Isemura: Colloidal Surfactants, 1963
13. J. E. Wollrab: Rotational Spectra and Molecular Structure, 1967
14. A. Nelson Wright and C. A. Winkler: Active Nitrogen, 1968
15. R. B. Anderson: Experimental Methods in Catalytic Research, 1968

Thermodynamic Properties of Organic Compounds

ESTIMATION METHODS, PRINCIPLES AND PRACTICE

George J. Janz
Department of Chemistry
Rensselaer Polytechnic Institute
Troy, New York

REVISED EDITION

ACADEMIC PRESS · New York and London · 1967

Copyright © 1967, by Academic Press Inc.
ALL RIGHTS RESERVED.
NO PART OF THIS BOOK MAY BE REPRODUCED IN ANY FORM,
BY PHOTOSTAT, MICROFILM, OR ANY OTHER MEANS, WITHOUT
WRITTEN PERMISSION FROM THE PUBLISHERS.

ACADEMIC PRESS INC.
111 Fifth Avenue, New York, New York 10003

United Kingdom Edition published by
ACADEMIC PRESS INC. (LONDON) LTD.
Berkeley Square House, London W.1

Library of Congress Catalog Card Number: 67-27838

Second Printing, 1968

PRINTED IN THE UNITED STATES OF AMERICA

Preface to the Revised Edition

A number of significant contributions to the estimation techniques for organic compounds have appeared since this book was first published. To retain the original plan of the book essentially unchanged, the new material is found as Addenda to the respective chapters in Part I, and as additional tables in Part II. It has thus been possible to update the *Revised Edition* without undue increase in size of this volume, a feature that is useful from the practical viewpoint. The recommendations of the International Union of Pure and Applied Chemistry on notation for the thermodynamic free energy have been adopted in this edition, and the preparation of the *Revised Edition* has also provided the opportunity for corrections of errata. The author wishes to acknowledge the generous cooperation of various colleagues and publishers for permission to draw on new material for this edition. It is also a pleasure to acknowledge correspondence with various colleagues since this book first appeared; their thoughtful comments and suggestions were a welcome assistance to the preparation of this revised edition.

GEORGE J. JANZ

Rensselaer Polytechnic Institute
May, 1967

Preface to the First Edition

The preparation of the present work was undertaken to fulfill two basic needs in the field of chemical thermodynamics. The first need, experienced both in teaching at the senior and graduate levels and in research activities, was for a text bringing up to date the advances in the practical methods for computing thermodynamic properties based on the long recognized principle that the regularity and systemization found in organic compounds applies to thermodynamic properties as it does to chemical and physical properties. The second need, of great importance to the application of the various methods, was for compilation of the data and correlations in one work for ready reference in practical calculations.

The development of the subject matter proceeds from an introductory account of the calculation of free energy changes to the use of the results as a criterion of the thermodynamic stability and feasibility of the processes considered. The theoretical calculation of the thermodynamic properties from molecular parameters, spectroscopic data, and statistical thermodynamics is considered, the discussion of the basic principles being limited as required for comparison with the empirical methods. The greater part of the book is devoted to the topic of estimation methods, which, simultaneously with the theoretical treatments, have been developed in the search for procedures to calculate the numerical values of the thermodynamic properties in a simple manner, and with the help of as few data as possible. The constructions of the correlations for each empirical estimation method are examined to enable extensions where necessary. Numerical examples are given to guide the application of the procedures in practice.

The thermodynamic approach to various problems in the field of chemical reactions is discussed, with special interest in systems for which a portion of the requisite data is missing. The methods developed to date make possible a preliminary thermodynamic analysis of a wide variety of problems. Although, in many instances, the predictions are admittedly approximate, the results give a deeper insight into the fundamental principles and provide a guide for the experimental studies.

While the empirical methods of estimation essentially relate to organic compounds, the fundamental principles must be recognized as quite general in nature. The extension of these techniques to all types of polyatomic compounds, organic, inorganic, or mixed in structure, can be anticipated, the progress in this direction being dependent on the amount of data available in each field, and to some extent, on the experience and skill of the investigator.

In the second part of the book the tables of data and correlations of the many contributors to the field of estimation of thermodynamic properties have been consolidated so that convenient application of each method is possible. The tables are reproduced as compiled in the original work of each investigator. For permission to draw on these data, it is a pleasure to express thanks to the following: Dr. G. S. Parks, Dr. F. D. Rossini, Dr. K. S. Pitzer, Dr. W. B. Person, Dr. G. C. Pimentel, Dr. K. M. Watson, Dr. J. L. Franklin, Dr. M. Souders, Dr. C. S. Matthews, Dr. C. O. Hurd, Dr. J. Sherman, Dr. D. W. Van Krevelen, Dr. H. A. G. Chermin, Dr. E. J. Prosen, the late Dr. M. S. Kharasch, Dr. T. L. Cottrell, Dr. P. Fugassi, Dr. C. E. Rudi, Dr. D. R. Stull, Dr. F. D. Mayfield, Sir Hugh S. Taylor, Dr. J. Turkevich, and Dr. M. L. Huggins.

The author is indebted to his colleagues, Professors J. B. Cloke and W. H. Bauer, at Rensselaer Polytechnic Institute, and in particular to Dr. Frederick C. Nachod of the Sterling Winthrop Research Institute, for stimulating suggestions and encouragement in the course of this work. The preparation of the manuscript by Marguerite L. Janz and assistance in the task of proofreading by S. S. Danyluk and M. De Crescente are gratefully acknowledged.

The author will be very grateful to have his attention directed to any errors found in this work, and will be pleased to receive criticism and suggestions that may increase the usefulness of the book.

G. J. J.

Troy, New York, November 1957

Contents

Preface to the Revised Edition v
Preface to the First Edition vii

PART I

METHODS OF ESTIMATION

CHAPTER 1. The Free Energy Change in a Chemical Reaction 3

    1. Chemical Thermodynamics in Research 3
    2. Standard Free Energy Change 4
    3. Modified van't Hoff Isochore 5
    4. Statistical Thermodynamic Functions 6
    5. Combined Thermodynamic Functions 7
    6. Free Energy Change and Equilibrium Conversions 7
    7. Effect of Pressure 9
    8. Estimation of Thermodynamic Properties 11

CHAPTER 2. Thermodynamic Properties of Simple Polyatomic Systems by Statistical Thermodynamic Methods 13

    1. Introduction 13
    2. Molecular Energy of an Ideal Gas 13
    3. Partition Function and Thermodynamic Properties 15
    4. Rigid Rotator-Simple Vibrator 17
    5. Calculation of Statistical Thermodynamic Functions 19
    6. Internal Rotation 26

CHAPTER 3. Thermodynamic Properties of Long Chain Hydrocarbons . . . 35

    1. Introduction 35
    2. Construction of the Correlations 36
    3. Calculation of Thermodynamic Properties 39
    4. Branch Chain Hydrocarbons 44
    5. Unsaturated Hydrocarbons 44
    6. The Steric Factor 45
    7. Random Kinking and Ball-Like Molecules 45

CHAPTER 4. The Method of Structural Similarity 50

    1. Introduction 50
    2. Construction of the Correlation 50
    3. Paraffins 50

CONTENTS

    4. Unsaturated Hydrocarbons 51
    5. Cyclic Hydrocarbons 51
    6. Organic Compounds Containing Oxygen 53
    7. Nitrogen, Halogen, and Sulfur-Containing Organic Compounds . 54
    8. Entropy and Free Energy Regularities 55

CHAPTER 5. The Methods of Group Contributions 58

    1. Introduction 58
    2. Methods 58
    3. Construction of the Correlations 60
    4. Calculation of Thermodynamic Properties 68
    5. Nonhydrocarbon Groups 75
       Addendum 76

CHAPTER 6. The Method of Group Equations 86

    1. Introduction 86
    2. Symmetry Number 86
    3. Internal Rotation 88
    4. Calculation of Thermodynamic Properties 90
    5. Extension of Thermodynamic Values from the Aliphatic to the
       Aromatic Series 95

CHAPTER 7. Heat of Formation and Heat Capacity 99

    1. Introduction 99
    2. Heat of Formation 99
    3. Bond Energies and Binding Energies 99
    4. Heat of Formation from Group Increments 103
    5. Gaseous Free Radicals and Ions 106
    6. Heat of Combustion by Group Increments 108
    7. Heat of Vaporization 113
    8. Heat Capacity 119
    9. Method of Generalized Vibrational Assignments 119
   10. Temperature Dependence of Heat Capacity 123
       Addendum 125

CHAPTER 8. Applications of the Thermodynamic Method 129

    1. Introduction 129
    2. Comparison of Comprehensive Estimation Methods 129
    3. Hydrogenation of Benzene 133
    4. The Thermal Dimerization of Butadiene 136
    5. Thermodynamic or Kinetic Control 139
    6. Ring Closure 141
    7. Cyclic Additions 144
    8. Other Applications 146
       Addendum 147

PART II

NUMERICAL DATA

TABLE 1.	Bond Energies	151
TABLE 2.	Atomic Covalent Radii and Bond Angles	152
TABLE 3.	A Six-Place Table of the Einstein Functions	153
TABLE 4.	Restricted Internal Rotational Free Energy $(-G/T)$ Contributions	169
TABLE 5.	Free Energy Increase from Free Rotation $(G - G_f)/T$	170
TABLE 6.	Restricted Internal Rotational Entropy Contribution $(S_{r'})$	171
TABLE 7.	Entropy Decrease from Free Rotation $(S_f - S_{r'})$	172
TABLE 8.	Restricted Internal Rotational Heat Content Contribution, $(H_T - H_0)/T$	173
TABLE 9.	Restricted Internal Rotational Heat Capacity Contribution $(C_p°)$	174
TABLE 10.	Potential Barriers Hindering Internal Rotation	175
TABLE 11.	Additive Increments in Free Energy Function, $-(G° - H_0°)/T$, of Normal Paraffins	176
TABLE 12.	Additive Increments in Heat Content, $(H° - H_0°)/T$, of Normal Paraffins	176
TABLE 13.	Additive Increments in Heat Capacity, $C_p°$, of Normal Paraffins	177
TABLE 14.	Thermodynamic Properties of Normal Heptane	177
TABLE 15.	[CH$_2$] Increment for Normal Paraffins	178
TABLE 16.	Additional Increments Extending the Infinite Chain Method to Branch Chain Paraffins	178
TABLE 17.	Parent Group Properties	179
TABLE 18.	Contributions of Primary CH$_3$ Substitution	179
TABLE 19.	Secondary Methyl Substitutions	181
TABLE 20.	Multiple Bond Contributions and Additional Corrections	182
TABLE 21.	Nonhydrocarbon Group Contributions Replacing [CH$_3$] Group	183
TABLE 22.	$(H° - H_0°)$ Increments for Hydrocarbon Groups	184
TABLE 23.	$\Delta H_f°$ Increments for Hydrocarbon Groups	185
TABLE 24.	$(G° - H_0°)$ Increments for Hydrocarbon Groups	186
TABLE 25.	$\Delta G_f°$ Increments for Hydrocarbon Groups	187
TABLE 26.	$\Delta G_f°$ and $\Delta H_f°$ Increments for Nonhydrocarbon Groups	188
TABLE 27.	Vibrational Group Contributions to Heat Content (zero pressure)	189
TABLE 28.	Characteristic Internal Rotational Contributions to Heat Content (zero pressure)	190
TABLE 29.	Vibrational Group Contributions to Heat Capacity (zero pressure)	191
TABLE 30.	Characteristic Internal Rotational Contributions to Heat Capacity (zero pressure)	192
TABLE 31.	Group Contributions to $\Delta S_f°$ and $\Delta H_f°$ (Type I groups)	193
TABLE 32.	Conjugation and Adjacency Contributions to $\Delta S_f°$ and $\Delta H_f°$ (Type I groups)	194
TABLE 33.	Vibrational Group Contributions to $(S_{fT}° - S_{f298}°)$ (Type II groups)	195
TABLE 34.	Characteristic Internal Rotational Contributions to $(S_{fT}° - S_{f298}°)$	196
TABLE 35.	Vibrational Group Contributions to $(H_{fT}° - H_{f298}°)$ (Type II groups)	197
TABLE 36.	Characteristic Internal Rotational Contributions to $(H_{fT}° - H_{f298}°)$	198
TABLE 37.	$\Delta G_f°$ Increments for Hydrocarbon Groups	199
TABLE 38.	$\Delta G_f°$ Corrections for Ring Formation and Branching Effects	200

TABLE 39. $\Delta G_f°$ Increments for Nonhydrocarbon Groups 201
TABLE 40. $\Delta G_f°$ Parameters for Some Simple Organic and Inorganic Compounds 203
TABLE 41. $\Delta H_{f298}°$ Group Increments for Radicals 203
TABLE 42. $\Delta H_{f298}°$ Group Increments for Ions 203
TABLE 43. Atomic Contributions to $C_p°$ and $S°$ (25°C, 1 atmos.) 204
TABLE 44. Bond Contributions to $C_p°$, $S°$, and $\Delta H_f°$ at 25°C, 1 atmos. 204
TABLE 45. Group Contributions to $C_p°$, $S°$, and $\Delta H_f°$ for Ideal Gases at 25°C, 1 atmos. Hydrocarbons 205
TABLE 46. Group Contributions to $C_p°$, $S°$, and $\Delta H_f°$ for Ideal Gases at 25°C, 1 atmos. Halogen Compounds 206
TABLE 47. Group Contributions to $C_p°$, $S°$, and $\Delta H_f°$ for Ideal Gases at 25°C, 1 atmos. Miscellaneous 207
TABLE 48. Values of Group Contributions to the Heat of Formation Used in Bryant Calculations 208
TABLE 49. Values of Substitution Group Contributions to the Heat of Formation Involving the Halogens 208
TABLE 50. Log K_{fG} for Paraffinic Groups 209
TABLE 51. Log K_{fG} for Olefinic Groups 209
TABLE 52. Log K_{fG} for Acetylenic Groups 210
TABLE 53. Log K_{fG} for Conjugate Olefin Groups 210
TABLE 54. Log K_{fG} for Aromatic Groups 210
TABLE 55. Log K_{fG} for Oxygen-Containing Groups 211
TABLE 56. Log K_{fG} for Nitrogen-Containing Groups 211
TABLE 57. Log K_{fG} for Sulfur-Containing Groups 212
TABLE 58. Log K_{fG} for Halogen-Containing Groups 212
TABLE 59. Log K_{fG} for Some Organic Compounds 213
TABLE 60. Correction Factors for Cyclization 214
TABLE 61. Correction Factors Necessitated by Introducing Lateral Chains . . 215
TABLE 62. Thermodynamic Properties for n-Propyl Halides 216
TABLE 63. Aliphatic Hydrocarbon Groups 216
TABLE 64. Aromatic Hydrocarbon Groups 218
TABLE 65. Branching in Paraffin Chains 218
TABLE 66. Branching in Cycloparaffins 219
TABLE 67. Branching in Aromatics 219
TABLE 68. Ring Corrections 220
TABLE 69. Oxygen-Containing Groups 220
TABLE 70. Nitrogen- and Sulfur-Containing Groups 221
TABLE 71. Halogen-Containing Groups 221
TABLE 72. Bond Contributions for Heats of Atomization, Formation, and Combustion of Gases and Liquids 223
TABLE 73. Empirical Equations for Heat of Combustion 224
TABLE 74. Structural Correlations for Calculation of Heat of Combustion . . . 225
TABLE 75. Atomic and Structural Parachor Contributions 228
TABLE 76. Generalized Bonding Frequencies and Constants to Evaluate Einstein Functions, $(C_p°)_i = a_i + b_iT + c_iT^2$ 228
TABLE 77. Contribution of Generalized Bond Vibrational Frequencies to Heat Capacity 229
TABLE 78. Heat Capacity Solutions to Einstein Functions (one degree of freedom) 230

Author Index 237

Subject Index 241

PART I

METHODS OF ESTIMATION

CHAPTER 1

The Free Energy Change in a Chemical Reaction

1. Chemical Thermodynamics in Research

In the field of organic chemical reactions the application of thermodynamics has proved particularly useful for the prediction of reaction equilibria and the evaluation of the thermodynamic feasibility of a given process. In the former case, where precisely determined thermodynamic data are available, the reaction equilibrium can frequently be predicted more accurately than could be established by direct measurements. In the latter case, estimates of the required thermodynamic properties by empirical or semiempirical methods can frequently be achieved to predict the free energy changes for the given process over a temperature range for use as a guide to the experimental investigation.

The thermodynamic criterion for reaction equilibrium and spontaneous processes can be summarized in terms of the free energy change, $\Delta G°$, as follows: if $\Delta G°$ is negative, the reaction is promising; if $\Delta G°$ is greater than zero but less than $+10$ kcal., it is of questionable value but warrants investigation; if $\Delta G°$ is larger than $+10$ kcal., the reaction is predicted to be not feasible except under unusual conditions. It should be recalled that such considerations predict only the possible *equilibrium* yields, i.e.,

$$\Delta G° = -RT \ln K \tag{1.1}$$

and shed no light on the actual products and yields. The latter are ultimately dependent on the relative rates and energies of activation in the chemical processes.

The desirability of having a complete set of thermodynamic properties for organic compounds is thus self-evident. Provided such data are sufficiently precise, the question as to whether a reaction may possibly proceed at all in the desired direction and to what extent under given conditions of temperature and pressure can be answered thermodynamically by simple calculation rather than by the empirical and time consuming experimental method of trial and error. Thus an important use of the thermodynamic method is to indicate the most favorable temperature and pressure conditions for the reaction. If the reaction does not have a convenient velocity under these conditions a catalyst must be sought to promote the reaction velocity. In this way chemical thermodynamics can be used to guide the experimental investigation of a reaction which can occur to only a limited extent under a given set of conditions. Where the picture is complicated by several types of decomposition reactions in addition to the desired process, the use of the thermodynamic characteristics of the individual reactions serves to express in a succinct manner the possibility of various reactions at different temperature intervals.

2. Standard Free Energy Change

For application to chemical reactions, tables of free energy data are assembled generally in terms of the standard free energy of formation, $\Delta F^\circ_{f(T)}$, so that the equilibrium constant of a reaction can be calculated directly from the expression:

$$\Delta G^\circ = -RT \ln K \tag{1.1}$$

The standard free energy change in a chemical reaction may be defined by the equation:

$$\Delta G^\circ = \sum_{\text{prod.}} \Delta G_f^\circ - \sum_{\text{react.}} \Delta G_f^\circ \tag{1.2}$$

i.e., it is the change in free energy accompanying the conversion of reactants to products, all being in their standard states. The standard states for a substance as a solid or a liquid are, respectively, the pure solid in its most stable form, and the pure liquid in its most stable form, each at one atmosphere pressure and at the specified temperature. In the case of a gas, the standard state is taken as the gas at unit fugacity. For ideal gases, the fugacity is unity when the pressure is one atmosphere. The standard free energy of formation, $\Delta G^\circ_{f(T)}$, is defined as the change in free energy corresponding to the formation of the substance in its standard state from its elements in their standard states. The standard free energy of formation for any element, accordingly, is zero.

The standard free energy change in a chemical reaction may be obtained by several procedures: (a) the use of known free energy changes for suitable chemical reactions which lead, on summation, to the free energy change for the desired process; (b) the use of experimental equilibrium data (i.e. Eq. 1.1); (c) from measurements of the electromotive force of an electrical cell in which the system undergoes the transformation reversibly; and (d) from thermal data alone and the basic equation:

$$\Delta G_T^\circ = \Delta H_T^\circ - T\Delta S_T^\circ \tag{1.3}$$

which is applicable to any isothermal reaction. The methods are all familiar from elementary work in physical chemistry. In the present case, the last method differs from the first three procedures in that the need for experiments under equilibrium conditions is avoided. It is of special interest since a primary objective of the thermodynamic approach in research is the prediction of the feasibility or spontaneity of a given reaction. In the latter case the standard free energy change is calculated generally by: (a) the modified form of the van't Hoff isochore, (b) the statistical thermodynamic functions, and (c) the combined thermodynamic functions.

3. Modified van't Hoff Isochore

The calculation of the standard free energy change in a chemical reaction as a function of temperature is possible if the heat of reaction and the entropy change at any one temperature (e.g., 298.1° K) are known, and the heat capacity data for the products and reactants over the temperature range in question can be obtained. Thus for the reaction:

$$bB + cC = xX + yY \tag{1.4}$$

if the heat capacities of B, C, X, and Y can be expressed in the form of the simple power series equation:

$$C_p = a + bT + cT^2 \tag{1.5}$$

where a, b, and c are constants characteristic for each substance, it can be readily shown that the heat of reaction and entropy change can be expressed by the equations:

$$\Delta H_T° = I_H + \Delta a T + \frac{\Delta b}{2} T^2 + \frac{\Delta c}{3} T^3 \tag{1.6}$$

and

$$\Delta S_T° = I_S + \Delta a \ln T + \Delta b T + \frac{\Delta c}{2} T^2 \tag{1.7}$$

where the Δ's refer to the sums of the coefficients of the products minus the sums of the coefficients of the reactants, i.e.,

$$\Delta a = [(xa_X + ya_Y) - (ba_B + ca_C)] \tag{1.8}$$

The integration constants, I_H and I_S, are evaluated by using the known heat of reaction and entropy change at any one temperature, e.g., $\Delta H°_{298}$ and $\Delta S°_{298}$.

From the fundamental expression for the free energy change of the chemical reaction, i.e.,

$$\Delta G_T° = \Delta H_T° - T \Delta S_T° \tag{1.3}$$

and the temperature dependent equations for $\Delta H_T°$ and $\Delta S_T°$ (i.e., Eqs. 1.6, 1.7), the following equations for $\Delta G°$ and $\ln K_p$ are obtained:

$$\Delta G_T° = I_H + (\Delta a - I_S)T - \Delta a T \ln T - \frac{\Delta b}{2} T^2 - \frac{\Delta c}{6} T^3 \tag{1.9}$$

$$\log K_p = \left[-\frac{I_H}{T} + (I_S - \Delta a) + \Delta a \ln T + \frac{\Delta b T}{2} + \frac{\Delta c T^2}{6} \right] \frac{1}{(2.303\,R)} \tag{1.10}$$

The latter can be recognized as the integrated van't Hoff isochore modified to take into account the variations of ΔH and ΔS with temperature.

The data required for the prediction of the free energy change at any temperature by this method are the heat capacities of the products and reactants over the temperature range in question, and the values of the heat of reaction and entropy change at one specific temperature. The latter two quantities can be calculated from a knowledge of the heats of formation, ΔH°_{f298}, and entropies, S°_{298}, of the products and reactants concerned.

4. Statistical Thermodynamic Functions

When the thermodynamic functions from statistical thermodynamic calculations:

Free energy function, $-(G^\circ - H_0^\circ)/T$; Entropy, S°:

Heat content function, $(H^\circ - H_0^\circ)/T$; Heat capacity, C_p°

are available for both reactants and products, the free energy change can be calculated without recourse to the van't Hoff isochore.

A knowledge of the heat of reaction at one temperature is essential to evaluate the zero-point energy, ΔH_0°, i.e.,

$$\Delta H_0^\circ = \Delta H^\circ_{298} - \Delta(H^\circ_{298} - H_0^\circ) \qquad (1.11)$$

where ΔH°_{298} is the heat of reaction at 25° C and $\Delta(H^\circ_{298} - H_0^\circ)$ is the difference of heat content for products and reactants. ΔH_0° is not purely an experimental quantity, but is calculated by using the values of $(H^\circ_{298} - H_0^\circ)$ and as such is subject to the errors in the latter. It follows from the free energy function that the free energy change at any temperature is given by:

$$\frac{\Delta G^\circ}{T} = \Delta\left[\frac{G^\circ - H_0^\circ}{T}\right] + \frac{\Delta H_0^\circ}{T} \qquad (1.12)$$

where the Δ signifies the summation of the coefficients of the products minus the summation of the coefficients of the reactants as defined earlier in this chapter.

Comprehensive tabulations of the separate statistical thermodynamic functions over a temperature range up to 1500° K have been published by Rossini and co-workers[1] as part of a program of research for the American Petroleum Institute. A survey of the entropies, statistical calculations, and thermochemistry covering the past fifteen years has also appeared recently[2]. A discussion of the methods used for calculating these functions from spectroscopic and molecular data follows in the next chapter. Some of the

---

[1] F. D. Rossini, K. S. Pitzer, R. L. Arnett, R. M. Braun, and G. C. Pimentel, eds., "Selected Values of Physical and Thermodynamic Properties of Hydrocarbons and Related Compounds," Carnegie Press, Pittsburgh, 1953.

[2] F. D. Rossini, D. D. Wagman, W. H. Evans and E. J. Prosen, *Ann. Rev. Phys. Chem.*, **1**, 1 (1950).

more recent semiempirical methods to be considered lead to estimates of the thermodynamic properties in the form of the statistical thermodynamic functions.

5. Combined Thermodynamic Functions

The combined thermodynamic functions, $*G°/T$, $*H_T°$, are defined by the equations:

$$-*G_T°/T = -(G° - H_0°)/T - \Delta H_0°/T \quad (1.13)$$

and

$$*H_T° = (H_T° - H_0°) + \Delta H_0° \quad (1.14)$$

thus giving the free energy and heat changes for the hybrid reaction:

$$\text{elements (standard state, 0°K)} = \text{compound (T°K)} \quad (1.15)$$

Comparison with Eqs. 1.11 and 1.12 of the preceding section shows that the calculation of an equilibrium constant or heat of reaction is reduced to a single summation using the tabulated combined functions. This scheme has been used by Rodebush in the International Critical Tables.[3] While the combined functions thus offer an advantage of some simplicity in calculations, the numerical values of the combined functions generally are larger than the separate functions and change more rapidly with temperature. Graphical treatment and interpolations are less difficult with the separate thermodynamic functions of the preceding section. It should also be noted that in any revision of tabulated data owing to more precise heat of combustion data, only the $\Delta H_0°$ values need be changed in the tables of the separate functions. With the combined functions the whole tables of $-*G°/T$ and $*H_T°$ would require change.

6. Free Energy Change and Equilibrium Conversions

Consider the free energy change for the generalized chemical reaction:

$$b\text{B} + c\text{C} = x\text{X} + y\text{Y} \quad (1.16)$$

where the reactants and products are at any arbitrary concentrations, not necessarily the equilibrium values. The free energy change for the reaction would be expressed by:

$$\Delta G = xG_\text{X} + yG_\text{Y} - bG_\text{B} - cG_\text{C} \quad (1.17)$$

The difference in the free energy change for the above process, ΔG, and that for the same reaction but with each reactant and product in its thermodynamic standard reference state is thus given by:

$$\Delta G - \Delta G° = x(G - G°)_\text{X} + y(G - G°)_\text{Y} - b(G - G°)_\text{B} - c(G - G°)_\text{C} \quad (1.18)$$

[3] "International Critical Tables," Vol. V. McGraw-Hill, New York, 1929.

From the thermodynamic definition of activity it follows that the difference between the molar free energy of each substance in the actual state and in the standard state is:

$$(G - G°) = RT \ln a \qquad (1.19)$$

Substitution of the latter in (1.18) readily leads to the relation:

$$\Delta G - \Delta G° = RT \ln \frac{(a_X)^x (a_Y)^y}{(a_B)^b (a_C)^c} \qquad (1.20)$$

which is the *reaction isotherm* first derived by van't Hoff (1886). Using the reaction isotherm the free energy change accompanying the transfer of reactants at any arbitrary concentrations (activities) to products likewise at arbitrary concentrations can be computed. When the change takes place under equilibrium, $\Delta G = 0$, and (1.20) simplifies to:

$$-\Delta G° = RT \ln \frac{(a_X)^x (a_Y)^y}{(a_B)^b (a_C)^c} = RT \ln K \qquad (1.21)$$

where now the activities are no longer general but refer to their values in the system at equilibrium. The standard free energy change, $\Delta G°$, is the free energy change in this case corresponding to the formation of the indicated number of moles of X and Y, in their standard states from the indicated number of moles of B and C, in their standard states (cf. Eq. 1.1).

The use of $\Delta G°$ as a criterion of the thermodynamic feasibility of a process has already been discussed. The numerical significance of a positive free energy change may be illustrated more definitely by the following numerical examples.

Example 1.1. Calculate the mole per cent conversion of A or B at equilibrium for the following reaction at 300° C and atmospheric pressure:

$$A + B = C$$

when starting with an equimolar mixture of A and B initially. Given that the standard free energy change for this reaction at 300° C is + 5240 cal.

SOLUTION:

(a) Since $\Delta G°_{573 °K} = +\ 5240$ cal, one has:

$$-5240 = 4.575\ (573) \log K \qquad \text{(cf. Eq. 1.1)}$$

i.e., the value of K is 1×10^{-2}.

(b) Comparison of the number of moles present initially and after the reaction has reached equilibrium is used to develop the expression for the equilibrium conversion. It is assumed that $100\,x$ is the mole per cent of A or B converted at equilibrium.

I. THE FREE ENERGY CHANGE IN A CHEMICAL REACTION

	Initial	Number of moles Change	At equilibrium	Mole fraction at equilibrium
A	n	$-nx$	$n(1-x)$	$\dfrac{1-x}{2-x}$
B	n	$-nx$	$n(1-x)$	$\dfrac{1-x}{2-x}$
C	0	$+nx$	nx	$\dfrac{x}{2-x}$
	$2n$	$-nx$	$n(2-x)$	

Therefore

$$P_A = P_B = \frac{1-x}{2-x} P \text{ and } P_C = \left(\frac{x}{2-x}\right) P$$

Accordingly at equilibrium,

$$K = \frac{x(2-x)}{(1-x)^2 P} = 1 \times 10^{-2}$$

i.e., $x = 0.005$.

Thus the maximum conversion attainable for the specified concentrations would be 0.5 mole per cent.

Example 1.2. Calculate the mole per cent conversion of A or B at equilibrium for the reaction:

$$A + B = 3C$$

starting at 300° C with an equimolar mixture of A and B initially. Assume that $\Delta G°_{573°K}$ is $+ 2620$. (ANSWER: 18 mole per cent).

7. Effect of Pressure

Since the standard states have been defined so as to be independent of the pressure of the system, it follows that $\Delta G°$ and consequently the equilibrium constant, K, will be independent of the external pressure. Thus differentiation of Eq. 1.1 with respect to total pressure gives:

$$\left[\frac{\partial (\Delta G°)}{\partial P}\right]_T = -RT \left(\frac{\partial \ln K}{\partial P}\right)_T = 0 \qquad (1.22)$$

It should be noted that although the true equilibrium constant of a gas reaction does not vary with the external pressure, the actual *position of equilibrium* will be altered if the reaction is one in which the number of molecules changes.

10　　I. METHODS OF ESTIMATION

Thus from Eq. 1.21 it follows that at higher pressures for a gaseous reaction, the equilibrium constant becomes:

$$K = \frac{(f_X)^x (f_Y)^y}{(f_B)^b (f_C)^c} \tag{1.23}$$

since under nonideal conditions the fugacities or true measure of the escaping tendency must be used. In a mixture of gases the fugacity of each component in the mixture is simply

$$f_i = f_i^\circ N_i \tag{1.24}$$

where f_i° is the fugacity of the i th component in its pure state at the total pressure, P, on the system. Accordingly the expression for the equilibrium constant (1.23) may be written as:

$$K = \left[\frac{(f_X^\circ)^x (f_Y^\circ)^y}{(f_B^\circ)^b (f_C^\circ)^c}\right] \left[\frac{(N_X)^x (N_Y)^y}{(N_B)^b (N_C)^c}\right] \tag{1.25}$$

or as:

$$K = \left[\frac{(\gamma_X^\circ)^x (\gamma_Y^\circ)^y}{(\gamma_B^\circ)^b (\gamma_C^\circ)^c}\right] \left[\frac{(N_X)^x (N_Y)^y}{(N_B)^b (N_C)^c}\right] P^{[(x+y)-(b+c)]} \tag{1.26}$$

where γ° is the activity coefficient of the gas, defined as $\gamma_i^\circ = f_i^\circ/P$. When the gases in the reaction are ideal the activity coefficient for each gas is unity and the pressure effect in the position of equilibrium is given by the expression:

$$\left[\frac{(N_X)^x (N_Y)^y}{(N_B)^b (N_C)^c}\right] = KP^{-[(x+y)-(b+c)]} \tag{1.27}$$

Thus only when there is no change in the number of molecules in the reaction, i.e., $[(x+y)-(b+c)]$ is zero, will the equilibrium conversions be independent of the pressure.

Example 1.3. Calculate the mole per cent conversion of A and B at equilibrium for the reaction as given in *Example 1.1* at 300° C but at 100 atmospheres pressure. Assume the gases involved in the reaction are ideal.

SOLUTION:

When the gases involved in the reaction are ideal, the activity coefficient for each gas is unity. The equilibrium concentrations accordingly would be given by:

$$\frac{x(2-x)}{(1-x)^2} = KP, \text{ since } [(x+y)-(b+c)] = -1$$

Since the equilibrium constant depends only on temperature and not on pressure, it follows that:

$$\frac{x(2-x)}{(1-x)^2} = 1.0 \text{ at 100 atmospheres.}$$

Solving gives $x = 0.29$, i.e., the conversion is 29 mole per cent at this high pressure.

Example 1.4. Calculate the molar ratio of octane to hydrogen at equilibrium for the one step synthesis:

$$8C_{(graphite)} + 9H_{2(g)} = C_8H_{18(g)}$$

at 600° K and 500 atmospheres pressure of hydrogen. At 600° K for this reaction the equilibrium constant is equal to 2.5×10^{-23} (ANSWER: 0.1).

The foregoing examples illustrate the importance of considering the influence of pressure as well as temperature on the equilibrium conversions in chemical processes. For real gases, Eq. 1.26 rather than Eq. 1.27 applies. Thus even where there is no change in the number of molecules, one may expect an additional effect due to departure of the gases from ideal behavior.

8. Estimation of Thermodynamic Properties

When adequate heat content and entropy data are available the calculation of the free energy change for a reaction is readily achieved by substitution into the equation:

$$\Delta G_T° = \Delta H_T° - T\Delta S_T° \tag{1.3}$$

The conversion of the free energy data from one temperature to another can be carried out by the several alternate methods oulined in the preceding section. The first extensive survey of the thermodynamic properties of organic compounds was the monograph by Parks and Huffman.[4] An empirical estimation method based on structural similarity was proposed in this work to meet the need for thermodynamic data for application of the thermodynamic method to organic compounds and reactions.

Theoretically it is possible to calculate these data according to statistical thermodynamics. However, the rigorous application of this method is limited to the more simple polyatomic systems, the mathematics rapidly becoming very difficult and laborious as the complexity of the system increases through lack of symmetry or in the number of constituent atoms in the molecule. In addition, the lack of spectroscopic data limits the use of this approach in the organic field. While precise thermodynamic data at present have been established for a great many compounds by the above theoretical method or by calorimetric and other experimental investigations, only a very small part of the overwhelming number of known organic compounds has been examined. Simultaneously with the above developments, estimation methods have been sought to calculate the numerical values of

[4] G. S Parks and H. M. Huffman, "The Free Energies of Some Organic Compounds," *Am. Chem. Soc.*, Monograph No. 60. Chemical Catalogue Co., New York, 1932.

the thermodynamic properties in a simple manner and with the help of as few data as possible.

The estimation methods may be divided into two categories, precise and approximate. The method of statistical thermodynamics, the treatment of long chain aliphatic hydrocarbons by Pitzer,[5] and the method of group equations of Rossini, Pitzer, and co-workers used in compiling the rather accurate data for gaseous hydrocarbons,[1] are the more precise estimation methods available at present. The several semiempirical methods based on the principle that, within certain recognized limits, molecular structural groups have the same contributions to the thermodynamic property no matter what may be the total molecular structure, which have appeared since the work of Parks and Huffman,[4] are somewhat more approximate in nature. With the theoretical development of the subject and the much larger amount of precisely established data, it has proved possible to develop quite accurate and specific relations between structure and thermodynamic properties of organic compounds. The construction and use of these correlations is the subject of this book. The tables of numerical values as compiled by the various investigators are given for ready application of the various thermodynamic methods in the situations where the lack of adequate information makes necessary the use of estimation procedures. It should be noted that the units of temperature, in thermodynamic computations, are on the Absolute or Kelvin scale unless otherwise indicated.

[5] K. S. Pitzer, *J. Chem. Phys.*, **8**, 711 (1940); *Chem. Revs.*, **27**, 39 (1940).

CHAPTER 2

Thermodynamic Properties of Simple Polyatomic Systems By Statistical Thermodynamic Methods

1. Introduction

The calculation of the thermodynamic properties of polyatomic systems from molecular and spectroscopic data and statistical thermodynamics is well established and described in most advanced text books in physical chemistry.[6,7,8] Attention is directed in this chapter to the results and equations in this field founded on theoretical treatment.

2. Molecular Energy of an Ideal Gas

In the development of the relationships between molecular energies and thermodynamic properties it is customary to treat the contributions resulting from the various forms of energy separately. From theoretical considerations in conjunction with experimental data on heat capacities of gases and molecular spectra, the energy of a polyatomic molecule may be expressed accordingly by the equation:

$$\varepsilon_{\text{total}} = \varepsilon_{\text{trans.}} + \varepsilon_{\text{rot.}} + \varepsilon_{\text{int. rot.}} + \varepsilon_{\text{vib.}} + \varepsilon_{\text{elec.}} + \varepsilon_{\text{n.s.}} \quad (2.1)$$

in which the subscripts refer to the translational, rotational, internal rotational, vibrational, electronic, and nuclear spin forms of energy. If the value of each energy level is determined by experimental means, and if the population of each level is known, it follows that the total energy of the system referred to the ground state as zero can be evaluated by the appropriate summations. Statistical mechanics applied to this problem led to the development of the equations making it possible to calculate from molecular parameters and spectroscopic data the thermodynamic properties of polyatomic molecules. The nature of this problem can be understood from a consideration of the factors contributing to the energy of the system.

According to modern concepts, the variation of the amount of energy in any degree of freedom of a molecule takes place in steps corresponding to the increments of energy termed quanta. The levels of translational energy are so closely spaced that it may be considered to vary continuously, and,

---

[6] G. Herzberg, "Infrared and Raman Spectra." Van Nostrand, New York, 1945.
[7] F. D. Rossini, "Chemical Thermodynamics," Wiley, New York, 1950.
[8] R. W. Gurney, "Introduction to Statistical Mechanics," McGraw-Hill, New York, 1951.

thus, can be calculated by classical mechanics. It can be resolved into three directions. Hence each molecule has three degrees of translational freedom. The external rotational energy relates to the rotation of the molecule as a whole. As in the rotation of large bodies, it may be resolved into components about each of three perpendicular axes passing through the center of the rotor. A nonlinear polyatomic system has, thus, three degrees of external rotational freedom in addition to the translational degrees of freedom. The rotational energy is a function of the moments of inertia about these principal axes of rotation. In a linear molecule the moment of inertia about the axis joining the atoms is negligible and the rotational energy, accordingly, results from rotation about the two remaining perpendicular axes.

For a nonlinear polyatomic system of n nuclei, thus, $3n-6$ coordinates remain for describing the relative motion of the nuclei with a fixed orientation of the system as a whole, i.e., there are $3n-6$ vibrational degrees of freedom corresponding to the number of different "fundamental" modes of vibration for the polyatomic system. From the vibrational frequency assignment, determined from molecular spectroscopic studies, the vibrational contribution can thus be determined. For complex polyatomic systems a fundamental vibrational frequency assignment on anything but empirical lines presents special problems.

The internal rotational energy results from the rotation of groups of atoms in the molecule with respect to other groups in the same molecule. In an ethane-like molecule, for example, the CH_3 group rotates relative to the rest of the system about the C—C single bond. The energy of internal rotation is a function of the moments of inertia of the rotating group about the axis of rotation and the magnitude and symmetry of the potential barrier hindering free internal rotation. Knowledge of the latter has been gained from the more simple polyatomic systems by comparison of the thermodynamic properties calculated from statistical thermodynamics and molecular spectra, assuming free internal rotation, with the experimental values achieved by calorimetric methods. In the more complex molecules this information is used to guide the calculation of the contributions of restricted internal rotational energy. The accuracy of the thermodynamic calculations depends on the correctness of the fundamental assignment of vibrational frequencies and the potential energy barriers hindering internal rotation.

The process of evaluating the rotational-vibrational contributions for complex polyatomic systems is both difficult theoretically and laborious requiring extensive molecular and spectroscopic data and details of structure. The thermodynamic properties established by statistical thermodynamic methods for polyatomic molecules are, accordingly, limited.

The contributions to the thermodynamic functions arising from electronic states of energy may be calculated in the normal manner by statistical mechanics.[7] Since generally the molecules are predominantly in the ground electronic state, and the separation of the energy levels is very great, a very high excitation would be required to raise a significantly large number of molecules to a higher electronic state. The contribution arising from excited electronic states of energy for most polyatomic systems is, accordingly, negligible at conventional temperatures. For example, molecules which contain no odd electrons or unpaired electrons generally possess electronic levels which are apt to be 100,000 cal. above the nondegenerate ground level and thus are not appreciably occupied except at very high temperatures. The separation of rotational levels is of the order of 20 cal. per mole, and the separation of vibrational levels is of the order of 3000 cal. per mole. Consequently the contribution of the latter two to the partition function at room temperature, is appreciable.

In connection with the contributions from nuclear spin, unless the reaction is such that the atoms themselves are changed, the number of atoms, each with its characteristic nuclear spin, remains constant. The contributions due to nuclear spin, being the difference in the sums for all the atoms in the products and reactants respectively, thus, cancel in the conventional type of chemical process. Accordingly it is customary not to include the contribution of nuclear spin in calculation of the thermodynamic properties of polyatomic systems by statistical thermodynamics.

3. Partition Function and Thermodynamic Properties

The statistical thermodynamic equations are related to the properties of the individual molecules by the application of the Boltzmann statistics to a system in thermal equilibrium. According to the Maxwell-Boltzmann distribution law, the number of atoms or molecules, N_i, in a state for which ε_i is the energy in excess of the zero-point energy, and g_i the total statistical weight, is given by:

$$N_i = N_0 g_i e^{-\varepsilon_i/kT} \tag{2.2}$$

where N_0 is the number in the lowest energy level, k the Boltzmann constant and T the absolute temperature. The factor g_i represents the number of levels possessing energy differing by such small amounts from ε_i that, for practical purposes, these levels may be treated as g_i levels of identical energy, ε_i. For one mole the total number of atoms or molecules, N, is equal to the sum over all the energy states:

$$N = N_0 \sum g_i e^{-\varepsilon_i/kT} \tag{2.3}$$

where the summation, called the partition function, is usually denoted by the symbol Q. The total energy in excess of the zero-point energy is:

$$(E^\circ - E_0^\circ) = N_0 \sum \varepsilon_i g_i e^{-\varepsilon_i/kT} \tag{2.4}$$

Substituting for N_0 from Eqs. (2.3, 2.4), it follows that:

$$(E° - E_0°) = RT^2 \frac{d \ln Q}{dT} = -R \frac{d \ln Q}{d(1/T)} \qquad (2.5)$$

Since the heat capacity, heat content, entropy, and free energy all may be expressed as functions of the energy, it follows that from a knowledge of Q, the thermodynamic properties of the polyatomic system can be calculated. Further, for a perfect gas the translational and internal energies, $\varepsilon_{trans.}$ and $\varepsilon_{int.}$ are entirely independent of each other, and the partition function Q can be separated into a product:

$$Q = Q_{trans.} \cdot Q_{int.} \qquad (2.6)$$

where $Q_{trans.}$ can be evaluated from classical mechanics, and $Q_{int.}$ can be calculated if the total internal energy, $\varepsilon_{int.}$ and the statistical weights, g_i, have been determined from the spectrum. The internal partition function is frequently simply known as the *partition function* or the *state sum*.

The evaluation of the partition function is greatly simplified if it is assumed that the various types of internal energies are also independent of each other (Eq. 2.1). This is not strictly true, but the neglect of any interaction of one type of energy with another does not lead to any significant error in the thermodynamic function thus calculated. The partition function accordingly is simply expressed as the product of the individual partition functions for translational, rotational, vibrational, electronic, and nuclear spin energy contributions, i.e.,

$$Q = Q_{trans.} \cdot Q_{rot.} \cdot Q_{int.\ rot.} \cdot Q_{vib.} \cdot Q_{elec.} \cdot Q_{n.s.} \qquad (2.7)$$

Assuming the electronic contribution to be negligible and neglecting the contribution of nuclear spin energy, since this effect cancels in chemical reactions, the partition function method of obtaining thermodynamic functions of polyatomic molecules depends primarily on the evaluation of the distribution of molecules over available rotational and vibrational states at a given temperature. For the computation of the rotational-vibrational contributions, the further assumption is made that the polyatomic system behaves as a rigid rotator (no centrifugal stretching) possessing one simple harmonic oscillator for each vibrational degree of freedom. The contribution of hindered internal rotation is calculated separately, assuming rotation of the group with respect to other groups possible in this model.

According to statistical mechanics the thermodynamic functions for one mole of a perfect gas, in terms of the total partition function, are given by the expressions:

Energy Relative To Zero Point Energy:

$$(E° - E_0°) = RT^2 \frac{d \ln Q}{dT} = -R \frac{d \ln Q}{d(1/T)} \qquad (2.8)$$

Heat Content Function:

$$(H° - H_0°)/T = (E° - E_0°)/T + R \qquad (2.9)$$

Free Energy Content Function:

$$(G° - H_0°)/T = -R \ln Q + R \ln N \qquad (2.10)$$

Entropy:

$$S° = \left(\frac{H° - H_0°}{T}\right) - \left(\frac{G° - H_0°}{T}\right) = R(1 - \ln N) + RT \frac{d \ln Q}{dT} + R \ln Q \qquad (2.11)$$

Heat Capacity:

$$C_p = C_v + R = \left(\frac{\partial(E° - E_0°)}{\partial T}\right) + R = \frac{R}{T^2}\left[\frac{d^2 \ln Q}{d(1/T)^2}\right] + R \qquad (2.12)$$

The inherent relation of the thermodynamic properties of a polyatomic system with molecular structure is thus attributable to the nature of the partition function. It is of interest to examine the parameters of the individual partition functions to determine the factors bearing on the principle of additivity in these functions.

4. Rigid Rotator-Simple Vibrator

The evaluation of the translational, rotational, and vibrational partition functions for a polyatomic system, assuming a rigid rotator-simple vibrator model for the molecule, has been adequately described in most advanced texts and it is sufficient for the present discussion to examine the results. The expressions for the separate partition functions are summarized below:

(a) Translational

(a) $(2\pi m k T/h^2)^{3/2} V$ \qquad (2.13)

(b) $(2\pi m k T/h^2)^{3/2} \left(\frac{NkT}{P}\right)$ \qquad (2.14)

(b) Vibrational

$(1 - e^{-hc\omega/kT})^{-1}$ \quad (one degree of freedom) \qquad (2.15)

(c) Rotational

(a) $(8\pi^2 kT/h^2)(I/\sigma)$ \quad (linear molecules) \qquad (2.16)

(b) $(8\pi^2 kT/h^2)^{3/2} \pi^{1/2} (I_A I_B I_C)^{1/2} (\sigma)^{-1}$ \quad (nonlinear molecules) \qquad (2.17) where h, k, and c are the Planck and Boltzmann constants and the velocity of light, respectively, and T, P, and N, the temperature, pressure, and Avogadro's number for the system being considered.

In the translational partition function the only parameter specific for each polyatomic system is the mass, m or M, per molecule or mole, respectively. It follows that a contribution relating to a structural modification by a simple group or atom would thus be additive with reference to the translational contribution.

Similarly, the only parameters characteristic of the polyatomic system in the total vibrational partition function are the fundamental vibrational frequencies, ν, one for each degree of freedom. The vibrational motions of most polyatomic systems may be treated with sufficient accuracy in the simple harmonic-oscillator approximation. The neglect of anharmonicities is permissible safely only for the lower vibrational states, i.e., at the lower temperatures, where interaction of the vibration-rotational energy states is negligible. The fundamental vibrations thus are, in a first approximation, the modes of simple harmonic oscillators, one for each vibrational degree of freedom, in which the displacements of the atoms about any bond are governed by the force constant for that bond and the atomic masses. If the electronic structure for a bond between two atoms is the same in one molecule as in another, the restoring force for the vibration between the two atoms would be the same, i.e., the force constants would be the same. From elementary valence theory one would predict the C—H bond to have essentially the same electronic structure, and therefore the same force constant in different molecules. This is indeed observed, and is recognized in tables of characteristic bond frequencies compiled for polyatomic molecules. The point may be further illustrated in the case of the C—H vibrations, where the mass of the hydrogen nucleus is much less than that of the other nuclei in the molecule. In the first approximation, the H nucleus may be considered as oscillating against an infinitely large mass, and therefore the vibration frequency would depend practically only on the force by which the H atom is bonded to the rest of the molecule. It follows that the fundamental vibrational frequency would be nearly the same for different molecules with the same C—H force constant. Therefore the vibrational contribution to the thermodynamic property introduced by a simple structural change would be additive, especially if the bond or group contribution has been derived from an environment quite similar to that in which it is found in the final molecule. It must be realized that structural modifications such as the introduction of double bond conjugation or strong polar groups will influence the force constants of a particular group.

Inspection of the rotational partition function shows that its value is dependent on two parameters specific for each polyatomic system, the products of the principal moments of inertia, I, or $I_A I_B I_C$, and the symmetry number, σ. The external symmetry number of the molecule is defined as the number of indistinguishable positions into which the molecule can be turned by simple rigid rotations. Thus, σ is 2, 4, 6, and 12 for the systems, acetylene, ethylene, ethane and benzene, respectively. Since these rotations produce identical states which at sufficiently high temperatures contribute approximately equally to the partition function, the factor σ^{-1} is introduced to account for the effect of symmetry. Structural modifications in a polyatomic system by atomic or group increments would be essentially additive in nature with reference to the moments of inertia since the mass changes for such increments may form only a small part of the total mass of the parent molecule. The structural increments however, are obtained from lower members or related compounds in the class under consideration. The symmetry number of each polyatomic system under consideration must be taken into account separately in this calculation. The rotational contributions for the rigid rotator model of the polyatomic system to the thermodynamic properties are, thus, additive, providing cognizance is taken of the differences in the symmetry numbers. In practice, an additional factor arising in the potential barriers restricting internal rotation and the internal symmetry numbers must also be taken into account, as discussed in a later section of this work.

5. Calculation of Statistical Thermodynamic Functions

A summary of the equations of the statistical thermodynamic functions with numerical constants for calculations of the properties of molecules in the ideal gaseous state is given in Table 2.1. The calculation of the translational contributions presents little problem; the molecular or formula weight, in grams per mole, is the only parameter required specifically for each polyatomic system. Some comments on the vibrational and rotational contributions are in order to guide their application in practice.

The summations for the vibrational contributions are taken over all of the fundamental vibrational frequencies. For a molecule of n atoms the number of vibrational degrees of freedom is $3n - 5$ and $3n - 6$ for linear and nonlinear molecules, respectively. These are reduced in number, one for each hindered internal rotation, if restricted internal rotation is present in the polyatomic system. Calculation of the contributions for restricted internal rotation is discussed separately in the next section. A discussion of the methods concerned with the assignment of the vibrational frequencies from spectroscopic data to the fundamental vibrational modes of the molecule is outside the scope of the present work. This problem has been adequately treated in texts dealing with the theoretical and empirical aspects of vibrational spectra and molecular spectroscopy.[6,9] The functions, $x^2 e^x/(e^x - 1)^2$, $x/(e^x - 1)$, and $-\ln(1 - e^{-x})$ appearing in the formulas of vibra-

---

[9] E. Bright Wilson, Jr., J. C. Decius, and P. C. Cross, "Molecular Vibrations", McGraw-Hill, New York, 1955; see also L. J. Bellamy, "The Infrared Spectra of Complex Molecules". Methuen, London, 1964; K. S. Pitzer; *J. Chem. Phys.* **5**, 473 (1937).

TABLE 2.1

EQUATIONS FOR CALCULATING THE TRANSLATIONAL, ROTATIONAL (RIGID ROTATOR), AND VIBRATIONAL (HARMONIC OSCILLATOR) CONTRIBUTION TO THE THERMODYNAMIC FUNCTIONS[a] FOR LINEAR AND NONLINEAR POLYATOMIC MOLECULES IN THE IDEAL GAS STATE AT ATMOSPHERIC PRESSURE

(a) Translation contributions (all molecules)

$$(C_p°)_{\text{trans.}} = [(H° - H_0°)/T]_{\text{trans.}} = 4.9680$$

$$-[(G° - H_0°)/T]_{\text{trans.}} = 6.8635 \log M + 11.4392 \log T - 7.2820$$

(b) Rotation contributions (linear molecules)

$$(C_p°)_{\text{rot.}} = [H° - H_0°]/T]_{\text{rot.}} = 1.9872$$

$$-[(G° - H_0°)/T]_{\text{rot.}} = 4.5757 \log (I \times 10^{39}) + 4.5757 \log T - 4.5757 \log \sigma - 2.7676$$

(c) Rotation contributions (nonlinear molecules)

$$(C_p°)_{\text{rot.}} = [(H° - H_0°)/T]_{\text{rot.}} = 2.9808$$

$$-[(G° - H_0°)/T]_{\text{rot.}} = 2.2878 \log (I_A \times I_B \times I_C \times 10^{117})$$

$$+ 6.8635 \log T - 4.5757 \log \sigma - 3.0140$$

(d) Vibration contributions (all molecules)

$$[(H° - H_0°)/T]_{\text{vib.}} = 1.9872 \sum_i [(x_i/(e^{x_i} - 1)]$$

$$-[(G° - H_0°)/T]_{\text{vib.}} = -1.9872 \sum_i [(\ln (1 - e^{-x_i})]$$

$$(C_p°)_{\text{vib.}} = 1.9872 \sum_i [x_i^2 e^{x_i}/(e^{x_i} - 1)^2]$$

where $x_i = (\omega_i hc/kT)$, ω_i being one of the fundamental frequencies (in wave numbers).

[a] The value of the entropy is obtained from the basic relation:

$$S° = \left[\frac{H° - H_0°}{T} - \frac{G° - H_0°}{T}\right]$$

and the value of the heat capacity at constant volume, by $C_v° = C_p° - R$.

tional contributions are known as *Einstein Functions*, and have been evaluated and tabulated[10] in terms of the variable x. Having the fundamental assignment of frequencies from spectroscopic data, the vibrational contribution is readily calculated by reference to the tables of the *Einstein Functions* as given in Table 3 (Part II).

For calculation of the rotational contribution, the moment of inertia and the symmetry number must be known. The latter is readily evaluated by inspection of a sketch of the polyatomic molecule, being the number of ways the molecule may be superimposed upon itself by rotation of the entire molecule. Construction of the molecule from the conventional atom model kits is most helpful for more difficult systems. For calculation of the moment of inertia, a knowledge of the molecular parameters of bond lengths and bond angles in the polyatomic system is essential. The latter are experimental data from electron and X-ray diffraction, and related investigations. A best assignment of these parameters is made, using generalized bond lengths and angles from related molecules as compiled in Table 2 (Part II) when the data for a specific polyatomic molecule are not available. The data permit a diagram to be drawn to scale using projected angles and distances, thus obtaining the coordinates of each atom by direct measurement. The moment of inertia for a diatomic molecule is calculated from the following equations:

$$I = \mu r_0^2 \tag{2.18}$$

$$\mu = \frac{m_1 m_2}{m_1 + m_2} \tag{2.19}$$

where m_1, m_2, μ, and r_0 are the masses of the two atoms, the "reduced mass" of the molecule, and the internuclear distance. For a linear molecule, in general, the relations are:

$$I = \sum_{i=1}^{i=n} m_i x_i^2 \tag{2.20}$$

where the sum is over all the atoms, and x is the vector distance of each from the center of mass of the molecule, i.e.,

$$\sum_{i=1}^{i=n} m_i x_i = 0. \tag{2.20a}$$

For nonlinear polyatomic molecules, a number of spatial arrangements can be conceived. The configuration having as closely as possible a linear or planar structure, i.e., with all the carbon atoms in the xy plane, and the methyl and methylene hydrogens symmetrical to this plane, should be selected unless a specific stereochemical orientation is known as the arrangement for the system. Changes in configuration from the orientation will have only small effect on the products of the moments of inertia of the molecule as a whole.

[10] J. Sherman and R. B. Ewell, *J. Phys. Chem.*, **46**, 641 (1942).

When the positions of the principal axes are not obvious, the product of the three principal moments of inertia is readily evaluated by means of the secular equation:

$$I_A I_B I_C = \begin{vmatrix} +I_{xx} - I_{xy} - I_{xz} \\ -I_{xy} + I_{yy} - I_{yz} \\ -I_{xz} - I_{yz} + I_{zz} \end{vmatrix} \quad (2.21)$$

If the coordinate system is chosen so that the origin corresponds with the center of mass of the molecule, the moments and products of inertia, i.e., I_{xx}, I_{xy} ..., in this equation are given by the terms:

$$I_{xx} = \sum m_i (y_i^2 + z_i^2) \quad (2.22)$$

and

$$I_{xy} = \sum m_i x_i y_i \quad (2.23)$$

where m_i is the mass of atom i with coordinates, x_i, y_i, z_i. In the more general case, where the coordinate frame of reference does not coincide with center of mass, these relations involve cross product terms, i.e.,

$$I_{xx} = \sum m_i (y_i^2 + z_i^2) - \frac{1}{M} (\sum m_i y_i)^2 - \frac{1}{M} (\sum m_i z_i)^2 \quad (2.24)$$

and

$$I_{xy} = \sum m_i x_i y_i - \frac{1}{M} (\sum m_i x_i)(\sum m_i y_i) \quad (2.25)$$

where $M = \sum m_i$. The units of each of the principal moments of inertia in the expression of Table 2.1 are grams per cm².

Example 2.1. Compute the statistical thermodynamic properties,

$$-(G° - H_0°)/T, (H° - H_0°)/T, S° \text{ and } C_p°$$

at 298.1° K for trifluoroacetonitrile, CF_3CN.

The fundamental vibrational frequencies, from infrared and Raman studies[10a], for this system are: 2272, 1228, 1216 (2), 802, 623 (2), 521, 464 (2), and 192 (2) cm⁻¹. The bond lengths, from electron diffraction data, are: (C—F) 1.328 A°, (C—C) 1.472 A°, and (C≡N) 1.157 A°. The FCF and FCC bond angles, similarily, are 108°28 and 112°34, respectively.

SOLUTION:

(a) Translational Contributions (CF_3CN) (see Table 2.1).

At 298.1° K, the contributions are:

$$-\left(\frac{G° - H_0°}{T}\right)_{\text{trans.}} = 6.8635 \log_{10} 95.028 + 11.4392 \log_{10} 298.16 - 7.2820$$
$$= 13.5750 + 28.3056 - 7.2820 = 34.5986 \text{ cal./deg.mole}$$

$[(H° - H_0°)/T]_{\text{trans.}} = C_p°_{\text{trans.}} = 4.9680$ cal/deg.mole

[10a] S. C. Wait and G. J. Janz, *J. Chem. Phys.* **26**, 1554 (1957).

$$(S°_{298.1})_{\text{trans.}} = \left(\frac{H° - H_0°}{T}\right) - \left(\frac{G° - H_0°}{T}\right) = 39.5666 \text{ e.u.}$$

(b) Rotational Contributions (CF_3CN) (see Table 2.1).

(i) Symmetry. The symmetry number, σ, equals 3 for this molecule, there being a threefold axis of symmetry passing through the $[C-C\equiv N]$ atoms in this system.

(ii) Moments of Inertia (see Eqs. 2.21–2.25 inclusive). A sketch showing the bond angles and bond lengths in two projections as below is used as reference for calculating the coordinates of the atoms. The carbon and nitrogen atoms are distinguished by the subscripts, and the carbon atom of the CF_3 group, i.e., C_2, is arbitrarily selected as the origin for the x–y–z coordinate reference frame.

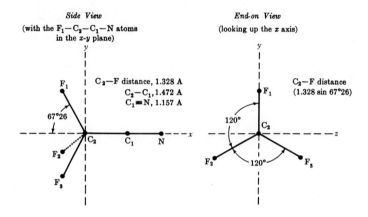

From the geometry of the system, accordingly, the coordinates of the atoms are as follows:

Atom	$m(\text{g/atom} \times 10^{23})$	$x(\text{A}°)$	$y(\text{A}°)$	$z(\text{A}°)$
C_1	1.994	1.472	0	0
C_2	1.994	0	0	0
N	2.326	2.629	0	0
F_1	3.155	$-1.328 \cos 67°26$ $= -0.5096$	$1.328 \sin 67°26$ $= 1.2263$	0
F_2	3.155	-0.5096	$-1.2263 \cos 60°$ $= -0.6132$	$-1.2263 \sin 60°$ $= -1.0620$
F_3	3.155	-0.5096	-0.6132	$+1.0620$

The terms of the secular equation 2.21 are computed from the above coordinates and atomic masses. These calculations are conveniently summarized as follows:

	C_1	C_2	N	F_1	F_2	F_3	Σ
x^2	2.1668	0	6.9116	0.2597	0.2597	0.2597	—
y^2	0	0	0	1.5038	0.3760	0.3760	—
z^2	0	0	0	0	1.1278	1.1278	—
xy	0	0	0	−0.6249	0.3125	0.3125	—
xz	0	0	0	0	0.5412	−0.5412	—
yz	0	0	0	0	0.6512	−0.6512	—
mx	2.9352	0	6.1150	−1.6078	−1.6078	−1.6078	4.2268
my	0	0	0	3.8689	−1.9346	−1.9346	0
mz	0	0	0	0	−3.3506	3.3506	0
mxy	0	0	0	−1.9716	0.9856	0.9858	0
mxz	0	0	0	0	1.7075	−1.7075	0
myz	0	0	0	0	2.0545	−2.0545	0
$m(x^2+y^2)$	4.3206	0	16.0764	5.5638	2.0056	2.0056	29.9320
$m(x^2+z^2)$	4.3206	0	16.0764	0.8194	4.3776	4.3776	29.9716
$m(y^2+z^2)$	0	0	0	4.7445	4.7445	4.7445	14.2335

From the above and Eq. 2.24 and 2.25, the terms for the secular equation 2.21 are readily obtained, i.e.,

$$I_{xx} = 14.2335 - 0 - 0 = 14.2335$$

$$I_{yy} = 29.9716 - 1.1323 - 0 = 28.8393$$

$$I_{zz} = 29.9720 - 1.1323 - 0 = 28.8397$$

and $I_{xz} = I_{xy} = I_{yz} = 0$. Using these in Eq. 2.21, the value of the products of the principal moments of inertia is:

$$I_A I_B I_C = 11838 \times 10^{-117} \text{ g. cm}^2.$$

(iii) The rotational contributions to the statistical thermodynamic functions at 298.1° K are accordingly:

$$-[(G° - H_0°)/T]_{\text{rot.}} = 2.2878 \log (11838) + 6.8635 \log 298.16 - 4.5757 \log 3 -$$

$$3.0140 = 9.3188 + 16.9833 - 2.1832 - 3.0140 = 21.1049 \text{ cal./deg.mole}$$

$$[(H° - H_0°)/T]_{\text{rot.}} = C_p° = 2.9808 \text{ cal./deg.mole}$$

and $S°_{\text{rot.}} = 2.9088 + 21.1049 = 24.0857$ cal./deg.mole

(c) Vibrational Contribution (CF_3CN)

The number of fundamental vibrational modes for trifluoroacetonitrile is $(3n - 6) = 12$ since the system is a nonlinear vibrator. Using the frequency assignment given, and the Table of Einstein Functions (Table 3, Part II) the contributions for each fundamental mode is obtained by interpolation and summation. These computations are summarized as follows:

ω (cm^{-1})	x	$\dfrac{x}{e^x - 1}$	$\left(\dfrac{x^2 e^x}{e^x - 1}\right)^2$	$-\ln(1 - e^{-x})$
2272	10.962	0.000192	0.002105	0.000018
1228	5.9251	0.015903	0.093681	0.002648
1216 (2)	5.880	0.016479	0.097170	0.002799
802	3.869	0.082441	0.325845	0.021079
623 (2)	3.006	0.156711	0.495098	0.050848
521	2.514	0.221262	0.605431	0.084320
464 (2)	2.238	0.249434	0.645893	0.101290
192 (2)	0.926	0.607806	0.931648	0.505061
$\sum_1^{12}$	—	2.380658	5.36680	1.428061
1.9872 ($\sum_1^{12}$)	—	4.7308	10.6647	2.8378

From the above summations, and the equations in Table 2.1, it is seen that the vibrational contributions at 298.1° K are:

$$[-(G° - H_0°)/T]_{\text{vib.}} = 2.8378 \text{ cal./deg.mole}$$

$$[(H° - H_0°)/T]_{\text{vib.}} = 4.7308 \text{ cal./deg.mole}; \quad C_p°_{\text{vib.}} = 10.6647 \text{ cal./deg.mole}$$

$$\text{and } S°_{\text{vib.}} = 4.7308 + 2.8378 = 7.5686 \text{ e.u.}$$

It is interesting to note the contributions of the individual frequencies; the largest part of the contribution is accounted for by the lower vibrational frequencies.

(d) Total Statistical Thermodynamic Properties (CF$_3$CN)

Summation of the translational, rotational, and vibrational contributions gives the following values for the statistical thermodynamic properties at 298.1° K for trifluoroacetonitrile:

$$-[(G° - H_0°)/T] = 58.5413 \text{ cal./deg.mole}; \quad C_p° = 18.6135 \text{ cal./deg.mole};$$

$$S° = 71.2209 \text{ e.u. and } (H° - H_0°)/T = 12.6796 \text{ cal./deg.mole},$$

as an ideal gas at one atmosphere pressure, assuming a rigid rotator-simple vibrator as model for this polyatomic system.[10b]

[10b] G. J. Janz and S. C. Wait, *J. Chem. Phys.* **26,** 1766 (1957).

Example 2.2. Compute the total entropy for ethylene at 298.1° K and one atmosphere pressure in the ideal gas state. The fundamental vibrational frequencies, taken from the literature, are:

3105, 3069, 3019, 2,989, 1623, 1444, 1342, 1055, 995, 950, 943, 825 cm^{-1}
(ANSWER: $I_A I_B I_C$, 5.47 × 10^{-117}; $S°_{\text{trans.}}$, 35.94; $S°_{\text{rot.}}$, 15.87; $S°_{\text{vib.}}$, 0.646; $S°_{\text{total.}}$, 52.46 e.u.)

For high temperatures, i.e., where $h^2/8\pi^2 IkT$ is much less than unity, evaluation of the partition function from a knowledge of the actual levels of rotational energies is necessary where exact calculations are required. Corrections to the equations for the rigid rotator for "stretching" of the molecule in these higher rotational levels may also be applied[11,12] to improve the calculations.

6. Internal Rotation

As is well known, the ordinary stereochemical formula for a polyatomic molecule does not specify molecular configuration with reference to internal rotation about a single C—C bond as axis. For some time it was supposed that such rotations were entirely free, and the energy and entropy contributions were calculated on this basis. The form of the partition function for free rotation of a group with respect to the molecule as a whole is:

$$Q_f = (8\pi^3 I_{\text{red.}} kT)^{\frac{1}{2}} (hn)^{-1} \tag{2.26}$$

or, expressing the "reduced" moment of inertia, $I_{red.}$, for the internal rotation in units of g.cm^2.:

$$Q_f = 2.7935 \, (10^{38} I_{\text{red.}} T)^{\frac{1}{2}} n^{-1} \tag{2.27}$$

where the temperature is in degrees Kelvin, and n is the symmetry number of the internal rotation. The "reduced" moment of inertia is calculated, in general, by the expression:

$$I_{\text{red.}} = A_m \left(1 - \sum_{i=1}^{3} A_m a_{mi}^2 / I_i\right) \tag{2.28}$$

in which A_m is the moment of inertia of this rotating group, a_{mi} is the direction cosine between the axis of this group and the ith principal axis, and I_i is the moment of inertia of the whole molecule about the ith principal axis. The set of equations:

$$\alpha(I_{xx} - I_i) - \beta I_{xy} - \gamma I_{xz} = 0 \tag{2.29}$$

$$-\alpha I_{xy} + \beta(I_{yy} - I_i) - \gamma I_{yz} = 0 \tag{2.30}$$

$$-\alpha I_{xy} - \beta I_{yz} + \gamma(I_{zz} - I_i) = 0 \tag{2.31}$$

where

$$\alpha^2 + \beta^2 + \gamma^2 = 1 \tag{2.32}$$

---
[11] E. B. Wilson, Jr., *J. Chem. Phys.* **4**, 526 (1936).
[12] J. E. Mayer and M. G. Mayer, "Statistical Mechanics," Wiley, New York, 1940.

can be solved for three different values of I_i, the three principal moments of inertia. The corresponding values of α, β, and γ are the direction cosines of the principal axes. The terms I_{xx}, I_{xy}, ... are the moments and products of inertia as defined earlier (Eqs. 2.22, 2.23, respectively).

For a group or molecule having two moments equal and the third a different value, all the axes in the plane of the two axes with equal moments have the same moments of inertia. Such systems are termed "symmetrical tops." For the case of a pair of symmetrical tops, coaxial as in ethane, the generalized equation (i.e. Eq. 2.28) for $I_{\text{red.}}$ simplifies to the expression:

$$\frac{1}{I_{\text{red.}}} = \frac{1}{I_A} + \frac{1}{I_B} \tag{2.28'}$$

where I_A, I_B are the moments of inertia of the A and B groups, respectively. In molecules with two rotating subgroups the coefficient of interaction of the two rotations is generally much less than either of the two reduced moments. Accordingly, in the first approximation, the two internal rotations are treated separately assuming the interaction effect to be negligible. If a second approximation is required, an interaction term can be introduced.

The contributions of free internal rotation to the thermodynamic functions are given by the equations:

$$\left(\frac{H° - H_0°}{T}\right)_f = (C_p°)_f = (C_v°)_f = (1/2)R \tag{2.33}$$

$$-\left(\frac{G° - H_0°}{T}\right)_f = R \ln Q_f \tag{2.34}$$

and

$$S_f° = (1/2)R + R \ln Q_f \tag{2.35}$$

where Q_f is the partition function for free internal rotation.

With the development of precision low temperature calorimetric techniques, the direct measurement of the entropies of polyatomic molecules with a high degree of accuracy was possible. Thus it was observed by 1936 that the third law (experimental) entropy of ethane did not agree with that for the model of free internal rotation calculated from statistical mechanics. The experimental entropy value was greater than that of the rigid rotator-simple vibrator approximation but less than the model permitting free internal rotation. The discrepancy between these values was resolved by Kemp and Pitzer[13] on the assumption that the theoretically calculated values were based on an incorrect molecular model, and that the difficulty must lie in the assumption of completely free internal rotation. Accordingly the statistical thermodynamic entropy was calculated assuming various potential barriers hindering the internal rotation. The results for ethane were as follows:

[13] J. D. Kemp and K. S. Pitzer, *J. Am. Chem. Soc.* **59**, 276 (1937).

Potential energy barrier hindering rotation (cal./mole)	Entropy of ethane $S°_{184}$ (cal./deg.mole)
0.0	51.21 ± 0.2
315	51.10 ± 0.2
3150	49.44 ± 0.2
Experimental value	49.64 ± 0.15

With the assumption that the internal rotation was hindered by a barrier of 3150 cal./mole, agreement between the data was realized. Similarly a discrepancy in the heats of hydrogenation was resolved. The development of the concept of hindered internal rotation followed rapidly after this. The general treatment and calculation of the hindered internal rotation contribution used at present is generally that of Pitzer[14] and Pitzer and Gwinn.[15]

The quantum mechanics of the torsion-oscillator-rotator had been solved by Nielsen[16] as a problem of theoretical interest, and the solutions to the differential equation based on a barrier of the simple cosine type:

$$V = (1/2)V_0 (1 - \cos n\theta) \qquad (2.36)$$

had been tabulated, in part, by Ince[17] and Goldstein.[18] It remained for Pitzer to interpret these, and to extend the treatment for the problem in hand. In the above equations, V_0 is the barrier height and n the number of equivalent minima. The ratio of the partition function for the hindered rotator to that for the free rotator was found to depend only on the two variables V/RT and n^2/IV. The numerical data for contributions of hindered internal rotation were accordingly calculated using this ratio and a range of values for each of the two variables above. Using the dimensionless parameters, (V_0/RT) and $(1/Q_f)$, tables were compiled for the various thermodynamic functions for molecules with one internal rotation. Because the free energy and entropy approach infinity as $(1/Q_f)$ approaches zero, the difference from the value for free rotation was tabulated. Thus for evaluating the contributions to the thermodynamic properties, it is necessary to know the value of $I_{red.}$, and V_0. The actual contributions of hindered internal rotation are calculated from the relations:

$$-(G°/T)_{r'} = R \ln Q_f - [(G_{rr}° - G_f°)/T] \qquad (2.37)$$

and

$$(S°)_{r'} = R(1/2 + \ln Q_f) - (S_f° - S_{rr}°) \qquad (2.38)$$

[14] K. S. Pitzer, *J. Chem. Phys.* **5**, 469, 473, 752 (1937).
[15] K. S. Pitzer and W. D. Gwinn, *J. Chem. Phys.* **10**, 428 (1942).
[16] H. H. Nielsen, *Phys. Rev.* **40**, 445 (1932).
[17] E. L. Ince, *Proc. Roy. Soc. Edinburgh* **46**, 316 (1925–26).
[18] J. Goldstein, *Trans. Cambridge Phil. Soc.* **23**, 303 (1927).

The tables for the free energy and entropy corrections for hindered internal rotation are found in Tables 4 and 5, and 6 and 7, respectively, in Part II. The contributions of hindered internal rotation to the heat content $(H_T-H_0)/T$ and heat capacity $(C_f°)$ are given in Tables 8 and 9, also in Part II.

The tables are calculated for a single internal rotation, where the potential barrier is the symmetrical threefold cosine type function (i.e., Eq. 2.36). This is illustrated in Fig. 2.1 for the case of an ethane-like molecule where the

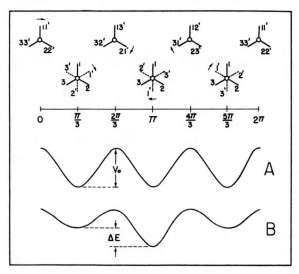

FIG. 2.1. The cosine type barrier for hindered internal rotation in an ethane-like molecule (symmetrical top). The rotating groups are viewed in projection down the C—C axis, the primed numbers referring to the more remote bonds. The shape of the barrier A corresponds to the case where all the atoms are equivalent or approximately so. A barrier of type B corresponds to the case where two of the atoms, e.g., 1 and 1', are markedly different from the remainder, as in 1,2-dichloroethane.

group is a symmetrical top about the axis of internal rotation. For $V_0 = 0$, the system would correspond to a state of free internal rotation, and for V_0 very large, the torsion would become equivalent to a vibrational mode of the harmonic oscillator type. For intermediate values of V_0, e.g., 3 kcal/mole, the system is similar to that of ethane with hindered internal rotation. For all the atoms 1, 1', 2, 2', 3, 3', which are equivalent or approximately so, the cosine type barrier has three equal or very nearly equal minima for a rotation of 360°.

The potential barriers which hinder internal rotation, as already indicated, are evaluated by comparison of an experimentally determined property, such as entropy, heat of hydrogenation, or heat capacity, with the calculated values. The chief objection to this method is that it assumes that there is no random orientation at the absolute zero. It is rare that the energy levels of the hindered rotor give rise to lines in the Raman or infrared. When they do, these lines may be used to calculate the barrier hindering internal rotation by the relation:

$$\frac{h\nu}{kT} = \left(\frac{nh}{2\pi kT}\right)\left(\frac{v}{2 I_{\text{red.}}}\right)^{\frac{1}{2}} = 4.463 \left(\frac{n}{T}\right)\left(\frac{V}{10^{40} I_{\text{red.}}}\right)^{\frac{1}{2}} \tag{2.39}$$

where ν is the torsional frequency, v and V the height of the barrier in ergs per molecule and cal. per mole respectively, and the remaining symbols have their conventional significance.

No satisfactory generalization for potential barriers has been developed, but intelligent approximations can be achieved by reference to known examples. The summary of potential barriers for polyatomic systems given in Table 10, Part II, illustrates the various types of systems studied, and the values observed. At the present there is no reason to believe that the hydrocarbons used for key barrier determinations have zero point entropies. The cause of the potential barriers hindering internal rotation awaits complete explanation. It seems that interaction through space of the symmetrical top and the rest of the molecule, the interaction being electrostatic (dipole, quadrupole) or exchange interaction, contributes in part at least to a force of repulsion that would be a periodic function of the internal rotation. The theoretical treatment of Lassettre and Dean[19] shows that quadrupole forces contribute an important part of this interaction.

If independent rotation of the methyl groups about the C—C axis in complex molecules is assumed, the symmetry number is threefold greater for each symmetrical top than that of the rigid rotator model. Thus the *total symmetry* number of ethane based on the above assumption is 18, in contrast to the value of 6 for the molecule as a rigid rotator. In any considerations of the additive properties with variations of molecular structure for free energy or entropy functions due note must be made of changes of the total symmetry number (i.e., external and internal rotation) for precise work.

Example 2.3. Compute the hindered internal rotational contribution to $S°$ for ethyl cyanide, C_2H_5CN, in the ideal gaseous state at 298.1°, 500°, and 1000° K and one atmosphere. Assume a symmetrical threefold cosine type barrier with $V_0 = 5200$ cal. The three principal moments of inertia I_A, I_B, I_C, and the corresponding direction cosines have been found to be, respectively:

31.2×10^{-40} g. cm^2, 173.9×10^{-40} g. cm^2, 194.5×10^{-40} g. cm^2

0.653   0.757   0.0

The moment of inertia of the CH_3 group is 5.303×10^{-40} g. cm^2.

[19] E. L. Lassettre and L. B. Dean, Jr., *J. Chem. Phys.* **16**, 157, 553 (1948).

SOLUTION:

(a) $I_{red.}$ (Eq. 2.28)

$$I_{red.} = 5.303 \times 10^{-40} \left[1 - \frac{5.303 \times 10^{-40}}{31.2 \times 10^{-40}}(.653)^2 - \frac{5.303 \times 10^{-40}}{173.9 \times 10^{-40}}(.757)^2\right]$$

$$= 4.826 \times 10^{-40} \text{ g. cm}^2$$

(b) Q_f (Eq. 2.27)

$$Q_f = 2.7935 \, (10^{38} \, I_{red.} \, T)^{\frac{1}{2}}/n$$

Substitute for $I_{red.}$ above and $n = 3$. Therefore $Q_f = 0.20778 \, T^{\frac{1}{2}}$

(c) Computation of hindered internal rotational increment (Refer to Tables 6 and 7, Part II).

(i) The parameters, $1/Q_f = 4.813T^{-\frac{1}{2}}$ and $V/RT = 2617T^{-1}$, are evaluated at the temperatures in question and used with Tables 6 and 7.

T (°K)	$T^{\frac{1}{2}}$	Q_f	$1/Q_f$ [a]	V/RT	$(S_f° - S_{r'}°)$ (e.u.)
298.1	17.267	3.584	0.279	8.77	2.027
500	22.361	4.651	0.215	5.23	1.429
1000	31.623	6.579	0.152	2.62	0.627

[a] The values of $1/Q_f$ are such that the increment $(S_f° - S_{r'}°)$ rather than $(S_{r'})$ must be computed (see Table 6). To gain the contribution $S_{r'}°$, it is necessary to calculate the entropy increment accompanying free internal rotation, i.e. $S_f°$.

(ii) $S_f° = 0.993 + 4.576 \log Q_f$ (Eq. 2.35). Having $S_f°$ at the desired temperature, the contribution to hindered internal rotation is readily gained from $S_{r'}° = S_f° - (S_f° - S_{r'}°)$. The last column in the table thus gives the hindered internal rotational contribution to $S°$ as 1.503, 2.619, and 4.110 at 298°, 500°, and 1000° K, respectively[20].

Example 2.4. Calculate the hindered internal rotational contribution to the free energy of methyl alcohol at 1000° K and one atmosphere pressure in the ideal gaseous state. Assume the barrier hindering internal rotation is 2700 cal./mole, with $n = 3$. (ANSWER: $I_{red.}$, 1.068×10^{-40}; Q_f, $0.09625T^{\frac{1}{2}}$; $-[(G° - H_0°)/T]_{1000°}$, 1.71 cal./deg.mole.

In polyatomic systems, where the minima of the potential energy function are not equivalent, e.g., 1,2-dichloroethane, the simplest procedure is to consider that rotational isomers exist, differentiated by the fact that their lower energy states are oscillations about each of the positions of minimum energy. Of the three minima in the potential barrier for a rotation of 360°, one is markedly different from the other two. This is illustrated in the shape of the barrier B, Fig. 2.1. The energy difference ΔE or ΔH, may be regarded as a heat of reaction for isomerization. Adopting the staggered model of

[20] N. E. Duncan and G. J. Janz, *J. Chem. Phys.* **23**, 434 (1955).

ethane, the rotational isomers (conformations) seen in projection down the C—C axis, would be:

```
        Cl                              H
        ⋮                               ⋮
   H╲   ⋮   ╱H                     H╲   ⋮   ╱H
       C                               C
   H╱ ⋰ ╲ ╲H                      H╱ ⋰ ╲ ╲Cl
        |                               |
        Cl                              Cl
        (T)                             (G)
```

in which the more remote bonds are designated by the broken lines. Conformations T and G have been termed *trans* or *anti* and the *gauche*, *skew*, or *syn*, respectively, relative to the positions of the substituent groups. The existence of such conformations has been confirmed by Raman and infrared spectroscopy. For such systems the number of lines in the spectra are much larger than would be predicted for only one configuration of the system. At sufficiently low temperatures the spectrum shows that the molecule exists predominantly in the form of lowest energy, e.g., dichloroethane, *n*-butane, both *trans*; succinonitrile, *gauche*[20a]. From the temperature dependence of these spectra the barrier heights between these conformations has been determined. Thus for *n*-butane, the difference in energy (ΔH) between the *trans* and *gauche* forms was found to be 800 cal. Accordingly, the entropy for *n*-butane was calculated by the relation:

$$S° = N_T S_T° + N_G S_G° - R(N_T \ln N_T + N_G \ln N_G) \tag{2.40}$$

where the subscripts T, G, refer to the *trans* and *gauche* conformations, and N and $S°$ are the mole fractions and entropy. The last term in the above expression is the entropy of mixing. The mole fractions are determined experimentally from the temperature-dependent studies of the Raman or infrared absorption spectra for the systems. More complex hydrocarbons have additional possibilities of internal rotational minima corresponding to nonequivalent configurations with different energies. The treatment for long chain hydrocarbons is the subject of the next chapter.

Example 2.5. From a comparison of the experimental and theoretical entropies of methylhydrazine, $H_2N \cdot NH(CH_3)$ at 298.1° K and one atmosphere as a gas, investigate the number of forms present and the relative amounts of each. It is assumed that the preferred conformation is a *skew* (*gauche*) form, and that the system has no zero point entropy.

The experimental entropy value ($S°_{298.1°}$) is 66.61 ± 0.20 e.u. The value calculated from statistical thermodynamics for the system, including the contributions of hindered internal rotation, is found to be 64.62 e.u.

SOLUTION:
(a) Optical Isomers: Entropy of Mixing

Methylhydrazine is theoretically capable of existing in a *trans* form, and two *skew* configurations. These may be readily distinguished by inspection

[20a] W. E. Fitzgerald and G. J. Janz, *J. Molec. Spectroscopy* **I**, 49 (1957).

of an *end-on view projection* (cf. Example 2.1), i.e., looking up the N—N bond as axis:

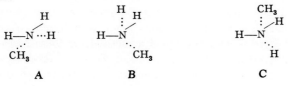

      A   B   C

where the dotted bonds refer to the groups attached to the nitrogen atom farthest from the observer. Of these three configurations, A is the *trans* form, and B and C are the *outer-* and *inner-skew* forms respectively, the reference being made to the position of the CH_3 relative to the two hydrogen atoms on the nitrogen atom nearest to the observer. From theoretical and experimental evidence it is predicted that the stable conformation is a *skew* form rather than the *trans* form. Accordingly, the contribution of the latter is assumed to be negligible to the entropy of the system as a whole.

Both the *skew* forms, B and C, may exist, each in two *dl*-isomeric forms. From Eq. 2.40, assuming an equimolal mixture of the two optical isomers (of the outer or inner forms), the entropy contribution is found to be:

$$S°(\text{mixing, 2 o.isomers}) = 1.379 \text{ e.u.}$$

Therefore, an entropy difference of 0.57 e.u. between the experimental and theoretically calculated values, i.e.,

$$\Delta S = 66.61 - 64.62 - 1.38 = 0.57 \text{ e.u.}$$

remains to be accounted for. Since there is no zero point entropy, the possibility of a mixture of the two *skew* forms is considered.

(b) Rotational Isomers

While the *outer-skew* form (B) would be expected to be more stable than the *inner* form, one might expect a certain fraction of the system at equilibrium to exist in the "inner" configuration. By a series of approximations starting with a large preponderance of the outer isomer in the mixture and with Eq. 2.40, it is found that:

$$S°_{\text{mixing}} = -4.576 \, (0.92 \log .92 + 0.08 \log .08)$$
$$= 4.210 \, (+ .0362) + .366 \, (+ 1.0969)$$
$$= 0.152 + 0.401 = 0.55 \text{ e.u.}$$

Thus, the experimental and calculated entropies can be brought into agreement if methylhydrazine is assumed to exist as an equilibrium mixture having 92 mole per cent and 8 mole per cent of the *outer* and *inner* forms, respectively.[21]

[21] G. J. Janz and K. E. Russell, *J. Chem. Phys.* **17**, 1352 (1949).

Most of the treatments of hindered internal rotation are based on the assumption of the simple cosine type potential energy barrier. The tables prepared by Pitzer and Gwinn for the thermodynamic properties hold for almost all molecules. A few exceptions, such as methanol, have been considered by Halford,[22] who developed a method for these cases. Halford,[22] and Blade and Kimball[23] have also discussed general methods by which these contributions were calculated conveniently for shapes of the potential curves other than the cosine type. The status of the problem of hindered rotation about single bonds in hydrocarbons and related molecules has been the subject of recent reviews by Pitzer[24] and Aston.[25]

[22] J. O. Halford, *J. Chem. Phys.* **15,** 645 (1947); **16,** 410, 560 (1948).
[23] E. Blade and G. E. Kimball, *J. Chem. Phys.* **18,** 630 (1950).
[24] K. S. Pitzer, *Discussions Faraday Soc.* **10,** 66 (1951).
[25] J. G. Aston, *Discussions Faraday Soc.* **10,** 73 (1951).

CHAPTER 3

Thermodynamic Properties of Long Chain Hydrocarbons

1. Introduction

It follows from the preceding discussion that thermodynamic properties for the simpler molecules frequently more accurate than direct experimental values can be calculated by statistical thermodynamics, providing the molecular structure data are complete. Atomic masses are too well known to require comment. Bond distances and bond angles, from electron and X-ray diffraction investigations, are well established for many of the simpler polyatomic systems and are directly transferable in closely related structures. One of the basic problems is the fundamental vibrational frequency assignment, based on the experimental infrared and Raman spectra and a theoretical normal coordinate analysis. The calculations rapidly become exceedingly involved and laborious with increasing complexity in molecular structure. For a five atom system the treatment already leads to a ninth order determinant unless the system has a high degree of symmetry, so that, even with molecular structure data available, solutions are not readily achieved without electronic or automatic computers. In addition, for the more complex molecules, the molecular structure data are frequently too incomplete for application of the rigorous methods of the preceding chapter.

The problem of calculating the thermodynamic functions for long chain molecules, with particular attention to the normal and branch chain paraffins, has been most successfully treated by Pitzer.[5] In this section the procedure developed by Pitzer, based on somewhat different statistical methods, is described. It is especially suited to the complex hydrocarbons and the present incomplete knowledge of their molecular constants. The scheme, in principle, is based on the integral or classical form of the partition function which is approached at high temperatures. The basic equation does not require a knowledge of the normal coordinates of vibration or their frequencies, but only the masses of the particles and the potential energy as a function of their positions. Tables of numerical values, developed by Pitzer[5] based on this method, and revised in the light of more modern data by Person and Pimentel,[26] are given, to make possible calculation of the thermodynamic properties, $-(G°-H_0°)/T$, $(H°-H_0°)/T$, $S°$, and $C_p°$ for normal paraffins.

[26] W. B. Person and G. C. Pimentel, *J. Am. Chem. Soc.* **75**, 532 (1953).

2. Construction of the Correlations

The model assumed for the structure of the long chain hydrocarbons is one in which the carbon atoms lie at equilibrium in a planar zig-zag chain, with the hydrogen atoms grouped in pairs in perpendicular planes through the carbon atoms. Theoretical treatments of the vibrations of chains of similar dynamical systems have been described by Kirkwood,[27] Whitcomb and his associates,[28] and Barriol.[29] Theory indicates that the normal modes of vibration of chain molecules fall into two classes, end vibrations in which the amplitude falls off exponentially from the ends towards the middle, and chain vibrations in which it varies harmonically as a standing wave. For a long chain only those chain vibrations will be infrared active for which the place of the electric moment is constant along the chain. Pitzer, after Kirkwood, treats the carbon atom skeleton only in the above model as the problem of an infinitely long chain to obtain the fundamental vibrational modes. The vibrational analysis shows that all skeletal frequencies for molecules of the normal hydrocarbons fall into two groups, a fairly narrow band near 1000 cm.$^{-1}$ and a broader band extending from 0 to 460 cm.$^{-1}$. Contributions from hydrogen atom vibrations are considered separately. It is found that the higher or stretching frequencies are surprisingly well approximated by the infinite chain method, but that the lower or in-plane bending modes are less satisfactorily approximated. Calculation of the frequencies for the out-of-plane carbon atom vibrations shows that the infinite chain method is even less satisfactory for such modes. It is concluded that all these modes fall within the limits of the above two bands.

A partition function is set up for the long chain system:

$$Q = Q^{\text{classical}}_{\text{(trans., rot., tors., bend.)}} Q^{n-1}_{(1000 \text{ cm.}^{-1} \text{ vib.})} \tag{3.1}$$

assuming that the motions in the low frequency group can be treated classically and that the high frequency band can be replaced by a suitable number $(n-1)$ of 1000 cm.$^{-1}$ frequencies. For convenience a transformation is made to permit calculations with a complete classical partition function, i.e.,

$$Q = \left[Q^{\text{classical}}_{\text{(trans., rot., tors., bend.)}} \left(Q^{\text{classical}}_{(1000 \text{ cm.}^{-1} \text{ vib.})} \right)^{n-1} \right] \times \left[\frac{Q_{(1000 \text{ cm.}^{-1} \text{ vib.})}}{Q^{\text{classical}}_{(1000 \text{ cm.}^{-1} \text{ vib.})}} \right] \tag{3.2}$$

Furthermore, since the ratio of the vibrational partition function of quantum theory to that of classical theory is given by:

$$\left(\frac{Q_v}{Q_v^{\text{classical}}} \right) = h\nu/kT(1 - e^{-h\nu/kT}) \tag{3.3}$$

$$= 1 + \frac{1}{2}(h\nu/kT) + \frac{1}{12}(h\nu/kT)^2 + \cdots \tag{3.4}$$

[27] J. G. Kirkwood, *J. Chem. Phys.* **7**, 506 (1939).
[28] S. E. Whitcomb, H. H. Nielsen, and L. H. Thomas, *J. Chem. Phys.* **8**, 143 (1940).
[29] J. Barriol, *J. phys. radium* **10**, 215 (1939).

3. LONG CHAIN HYDROCARBONS

it follows that where $h\nu/kT$ is less than about 1.7, mean frequencies may be used to calculate the correct contributions for the bending and torsion modes. Evaluation of the partition function itself in the classical sense requires a geometric mean frequency. For the lower band this would be grossly affected by the continuance of the band to zero. Accordingly the partition function may be calculated as a product of the complete classical functions and correcting factors for the higher vibrational frequencies, provided that the exact frequencies are used, or that mean frequencies are used for narrow bands and when $h\nu/kT$ is less than about 1.7. The latter corrections are only of the order of 0.1 cal. per degree or less. The force constants used in the classical function must be consistent with the vibration frequencies.

With this method, the partition function for a generalized normal paraffin was realized in the following way. The complete classical partition function is given by:

$$Q = \frac{1}{h^{3n}} \int_{-\infty}^{\infty} \cdots \int_{-\infty}^{\infty} \int_{0}^{x_0} \cdots \int_{0}^{z_0} x e^{-(\text{P.E.} + \text{K.E.})/kT} dp_{x_1} \cdots dp_{z_n} dx_1 \cdots dz_n \quad (3.5)$$

where P.E. and K.E. are the potential and kinetic energies respectively, $p_{x_1}\ldots$, momenta, and $x_1 \ldots$, the corresponding coordinates. Only an expression of P.E. and K.E. in terms of coordinates and momenta is required, moments of inertia and the normal coordinate treatment are not needed for this form of Q. In the model assumed, the CH_2 (or CH, and CH_3) groups are considered as single units, the hydrogen motions being considered separately. The kinetic energy may be expressed in terms of the momenta as:

$$\text{K.E.} = \sum_{i=1}^{n} \tfrac{1}{2} m \, (p_{x_i}^2 + p_{y_i}^2 + p_{z_i}^2) \quad (3.6)$$

if all n-particles have the same mass, m. The partition function now is:

$$Q = \left(\frac{2\pi m kT}{h^2}\right)^{\frac{3n}{2}} \int_{0}^{x_0} \cdots \int_{0}^{z_0} x e^{-(\text{P.E.})/kT} dx_1 \cdots dz_n \quad (3.7)$$

The potential energy of the system must embrace the bond stretching and bending motions, internal rotation, for which a threefold symmetrical cosine type barrier $[\tfrac{1}{2}V_0 (1 - \cos 3\theta)]$ is used, and those configurations of high energy owing to serious steric repulsions.

Selecting any atom at random, assuming a perfect gas state, the second atom may be oriented in any direction relative to the first, but a potential energy term $[\tfrac{1}{2}k_1 (r - r_0)^2]$ must always be included for the distance between these two. Integration for these atoms leads to the expression for the partition function:

$$Q = \left(\frac{2\pi m kT}{h^2}\right)^{\frac{3n}{2}} V \, (4\pi r_0^2) \left(\frac{2\pi kT}{k_1}\right)^{\frac{1}{2}} \int_{0}^{x_0} \cdots \int_{0}^{z_0} x e^{-\text{P.E.}/kT} dx_3 \cdots dz_n \quad (3.8)$$

where r_0, V, and k_1 are the C—C distance, volume, and force constant, respectively. The motion of the next atom involves another bond stretching term and, in addition, a bond bending term, $[\tfrac{1}{2}k_2\delta^2]$ where δ is the displacement from the equilibrium position, and a third rotation. When γ is the [C—C—C] angle, the circumference of the allowed circle is $2\pi r_0 \sin \gamma$. Accordingly the expression of Q for this system is given by:

$$Q = \left(\frac{2\pi mkT}{h^2}\right)^{\frac{3n}{2}} V \, (8\pi^2 r_0^3 \sin \gamma) \, \frac{(2\pi kT)^{\frac{3}{2}}}{(k_1^2 k_2)^{\frac{1}{2}}} \int_0^{x_0} \cdots \int_0^{z_0} e^{-\text{P.E.}/kT} dx_4 \cdots dz_n \qquad (3.9)$$

After the third atom, all atoms along a normal chain have bond stretching and bending motions and a restricted internal rotation. Therefore a factor given by:

$$r_0 \sin \gamma \int_0^{2\pi} e^{-(V_0/2kT)(1-\cos 3\theta)} d\theta$$

enters the partition function for each internal rotation, assuming the usual sinusoidal type barrier of three equal maxima and minima. Since it is certain that some configurations of the long chain hydrocarbon will lead to considerable steric repulsion, a correction to the internal rotational contribution for the energies of these configuration is necessary. The following procedure was adopted. The upper limit of integration in each [I.Rot.] factor was changed from 2π to $2\pi/3$, and a term:

$$G_{\text{steric}} = \sum e^{-E_i/kT} \qquad (3.10)$$

was introduced, in which E_i is the energy of the i th configuration. It is assumed that the configuration of lowest energy is the planar, zig-zag model. The same interaction, leading probably to an increase in energy, is associated with each rotation of $2\pi/3$ about any bond. This energy increment is called a. It was found by inspection of molecular models that rotation about several bonds led to either that number of same interactions, or sterically impossible configurations. By counting the number of each kind of positions and assuming a value of a, the G_{steric} summation was readily calculated. The molecular symmetry number is taken as 2 since both ends of the long chain paraffin are the same.

The complete classical partition function thus may be expressed by the relation:

$$Q^{\text{classical}} = \tfrac{1}{2} G_0 (T) \, [\text{C—C}_{\text{str.}}]^{n-1} \, [\text{C—C}_{\text{bend.}}]^{n-2} \times [\text{I.Rot.}]^{n-3} \, (G_{\text{steric}}) \qquad (3.11)$$

in which
$$G_0 (T) = (2\pi mkT/h^2)^3 V \, (8\pi^2 r_0^3 \sin \gamma)$$

$$[\text{C—C}_{\text{str.}}] = (2\pi mk/h)(m/k_1)^{\frac{1}{2}}$$

$$[\text{C—C}_{\text{bend.}}] = (2\pi kT/h)(m/k_2)^{\frac{1}{2}}$$

and
$$[\text{I.Rot.}] = (2\pi mkTr_0^2 \sin^2 \gamma/h^2)^{\frac{1}{2}} \int_0^{\frac{2\pi}{3}} e^{-(V_0/2kT)(1-\cos 3\theta)} d\theta$$

$$G_{\text{steric}} = \sum e^{-E_i/kT}$$

For N_0 identical molecules, Q is raised to that power and divided by $(N_0/e)^{N_0}$ which approximates $(N_0!)$. In addition, factors of the type $[Q_v/Q_v^{\text{class.}}]$ for $(n-1)$ 1000 cm^{-1} frequencies, $(n-2)$ 290 cm.$^{-1}$ frequencies, and $(n-3)$ 120 cm^{-1} frequencies are to be added to correct for deviations from the classical case. It should be noted that the [C—C$_{\text{str.}}$] term of the partition function in (3.11) is somewhat larger than the classical factor for a 1000 cm^{-1} vibration. The difference may be thought of as the contribution of the additional atom in the molecule to the mass and moment of inertia terms in the usual formulas. The same applies to the [C—C$_{\text{bend.}}$] and [I.Rot.] terms. The partition function is completed by addition of the hydrogen vibrations. This was achieved by using the characteristic frequencies of the CH$_3$, CH$_2$, and CH groups.

Based on these considerations, the thermodynamic function of a long chain hydrocarbon of N atoms may be expressed by the relatively simple equations:

$$G(X) = G_0(T) + (N-1)\,[\text{C—C}_{\text{str.}}] + (N-2)\,[\text{C—C}_{\text{bend.}}] + (N-3)\,[\text{I.Rot}] +$$
$$G_{\text{steric}} + G_\sigma + 2\,[\text{CH}_3] + (N-2)\,[\text{CH}_2] \quad (3.12)$$

in which $G(X)$ is the desired thermodynamic function, [CH$_3$] and [CH$_2$] represent the contributions of these groups to the thermodynamic functions, G_σ adjusts for the symmetry number in the usual way, and the remaining terms have the significance already discussed.

3. Calculation of Thermodynamic Properties

The expressions used in calculating the thermodynamic properties of normal paraffins, with the revised parameters of Person and Pimentel,[26] are given in Table 3.1. Some comments on the terms are in order.

The term $G_0(T)$ is perfectly general, being a function of T and the effective CH$_2$ mass, and not dependent on the molecule. A value of $m = 18.6$ has been used both by Pitzer, and Person and Pimentel. The former based his value on calculations for propane in which only translational and rotational degrees of freedom were considered. The latter calculated the effective mass assuming as a model a normal paraffin with n CH$_2$ groups, and taking into account vibrational and internal rotational degrees of freedom in addition to the translational-rotational contributions. The agreement in values (18.6) is considered fortuitous. Person and Pimentel indicate that for short chains, the effective CH$_2$ mass may be a function of the chain length. Tables of thermodynamic properties for the normal paraffins shorter than n-heptane are given by these investigators, in which the value of m was adjusted in the calculations in the ratio of the effective rotational masses.

In the [C—C$_{\text{str.}}$] term the average frequency was changed[26] from 1000

TABLE 3.1 FORMULAS FOR CALCULATING THERMODYNAMIC FUNCTIONS OF NORMAL PARAFFINS[5]

	$-(G°-H_0°)/T$	$(H°-H_0°)/T$	$C_p°$
$G_0(T)$	$R\ln\dfrac{64\pi^5 m^3 v_0{}^3 k^3 \sin\gamma R}{h^6 N_0} + 4R\ln T - R\ln P$ $= 4.235 + 4R\ln T - R\ln P$	$4R$	$4R$
C—C$_{\text{Str.}}$	$R\ln(2\pi v_1(m/k_1)^{1/2}) + \text{Ein}(hv_1/kT)$ $= 0.689 + \text{Ein}(hc\,975/kT)$ $v_1/c = 975\text{ cm.}^{-1}$	$\text{Ein}(hv_1/kT)$	$\text{Ein}(hv_1/kT)$
C—C$_{\text{Bend}}$	$R\ln(2\pi k m^{1/2}/hk_2{}^{1/2}) + R\ln T + \dfrac{R}{2}(hv_2/kT)^2 - \dfrac{R}{24}(hv_2/kT)^2$ $= -11.296 + R\ln T + \dfrac{413}{T} - \dfrac{14{,}200}{T^2}$	$R - \dfrac{R}{2}(hv_2/kT) + \dfrac{R}{12}(hv_2/kT)^2$ $= R - \dfrac{413}{T} + \dfrac{28{,}400}{T^2}$	$R - \dfrac{R}{12}(hv_2/kT)^2$ $= R - 28{,}400/T^2$ $(T > 250)$
I.Rot.	$R[-1.275 + \tfrac{1}{2}\ln T + \tfrac{1}{2}\ln(mv_0{}^2\sin^2\gamma \times 10^{40}) - \ln 3]$ $-\dfrac{(G-G_f)}{T}\left(V_0'/RT,\dfrac{1}{Q_f}=0\right) + \dfrac{hv_3}{2T} =$ $= \tfrac{1}{2}R\ln T - 0.568 + \dfrac{170}{T} - \dfrac{(G-G_f)}{T}\left(\dfrac{3260}{RT}\right)$	$\dfrac{H}{T}\left(\dfrac{V_0'}{RT},\dfrac{1}{Q_f}=0\right) - \dfrac{hv_3}{2T}$ $\dfrac{H}{T}\left(\dfrac{3260}{RT}\right) - \dfrac{170}{T}$	$C\left(\dfrac{V_0'}{RT},\dfrac{1}{Q_f}=0\right)$ $= C\left(\dfrac{3260}{RT}\right)$
G_{steric}	$R\ln Q$	$R(Q'/Q)$	$R\left[\dfrac{Q''}{Q} - \left(\dfrac{Q'}{Q}\right)^2\right]$
	$Q = \sum_i e^{-E_i/kT}$ $(a = 500\text{ cal./mole})$	$Q' = \sum_i \left(\dfrac{E_i}{kT}\right) e^{-E_i/kT}$	$Q'' = \sum_i \left(\dfrac{E_i}{kT}\right) e^{-E_i/kT}$
$G\sigma$	$-R\ln\sigma = -R\ln 2$	0	0
CH_3	$\sum_i \text{Ein}(hv_i/kT) + R(-1.275 + \tfrac{1}{2}\ln T + \tfrac{1}{2}\ln T +$ $\tfrac{1}{2}\ln I_R \times 10^{40} - \ln 3) - \dfrac{(G-G_f)}{T} - \dfrac{H}{T}\left(\dfrac{V_0}{RT}\right)$ $= -3.077 + \tfrac{1}{2}R\ln T + \sum_i \text{Ein}\left(\dfrac{hv_i}{kT}\right) - \dfrac{(G-G_f)}{T} - \left(\dfrac{V_0}{RT}\right)$ $v_i/c = (3)\ 2950,\ (2)\ 1460,\ 1375,\ 1170,\ 827;\ V_0 = 3480;\ I_R = 5.211 \times 10^{-40}$	$\sum_i \text{Ein}\left(\dfrac{hv_i}{kT}\right) + \dfrac{H}{T}\left(\dfrac{V_0}{RT}\right)$	$\sum_i \text{Ein}\left(\dfrac{hv_i}{kT}\right) + C\left(\dfrac{V_0}{RT}\right)$
CH_2	$\sum_i \text{Ein}(hv_i/kT)$ $v_i/c = 2936,\ 2865,\ 1469,\ 1407,\ 1297,\ 912$	$\sum_i \text{Ein}(hv_i/kT)$	$\sum_i \text{Ein}(hv_i/kT)$

cm $^{-1}$ to 975 cm $^{-1}$ in the light of more recent data. The carbon-carbon stretching and bending force constants were taken as 4.1×10^5 and 3.6×10^4 dynes per centimeter, respectively. Values for V_0' and a of 3260 cal. per mole and 500 cal. per mole have been used in the terms [I.Rot.] and G_{steric}. The frequencies of the normal vibrations for the [CH$_2$] term are those calculated by Person and Pimentel based on a normal coordinate analysis of an infinite CH$_2$ chain, and extended to values for phase angles other than 0 and π. The term [CH$_3$] does not directly depend on chain length, but rather is a function of I_R, the effective moment of inertia of the CH$_3$ group, and V_0, the barrier restricting rotation, and the fundamental vibrational frequencies. The latter are well established. According to the generalized definition for I_R given earlier (Eq. 2.28) the value of the reduced moment should be a function of chain length. Based on this equation the expression:

$$I_R = 5.40 \, (1 - 5.40/30n) \, 10^{40} \tag{3.13}$$

may be derived for molecules when n is greater than 6. Using this expression for I_R, the best fit between experimental and calculated data was gained with $V_0 = 3480$ cal. per mole for this group. A value of $I_R = 5.319 \times 10^{-40}$ was used for simplicity in the calculation for normal paraffins from n-hexane to n-eicosane (C$_{20}$H$_{42}$), i.e., a weighted average value corresponding approximately to $n = 12$. The terms in Table 3.1 have been calculated accordingly by Person and Pimentel and compiled as tables of numerical increments over the temperature range from 298.1° to 1500° K. The data are given in Tables 11, 12, and 13 (Part II) in this book.

The basic equations for the thermodynamic properties of normal paraffins of N carbon atoms are:

Free Energy Functions:

$$-(G° - H_0°)/T = [G_0 \, (T)] + (N-1) \, [\text{C—C}_{\text{str.}}] + (N-2) \, [\text{C—C}_{\text{bend.}}] +$$
$$(N-3) \, [\text{I.Rot.}] + G_{\text{steric}} + G_\sigma + 2 \, [\text{CH}_3] + (N-2) \, [\text{CH}_2] \tag{3.14}$$

where $G_{\text{steric}} = A + (N-7) \, B$ for $N \geqslant 7$, but must be calculated individually for $N < 7$. For hexane, pentane, and butane, $G_0(T)$ and [I.Rot.] are decreased by 0.064, 0.162, 0.195, and 0.033, 0.054, and 0.049, respectively. Since σ is 2 for long chain paraffins, $G_\sigma = -R \ln 2 = -1.379$.

Heat Content Function:

$$(H° - H_0°)/T = [G_0 \, (T)] + (N-1) \, [\text{C—C}_{\text{str.}}] + (N-2) \, [\text{C—C}_{\text{bend.}}] +$$
$$(N-3) \, [\text{I.Rot.}] + G_{\text{steric}} + 2 \, [\text{CH}_3] + (N-2) \, [\text{CH}_2] \tag{3.15}$$

where G_{steric} is calculated as above.

Heat Capacity

$$C_p° = [G_0 \, (T)] + (N-1) \, [\text{C—C}_{\text{str.}}] + (N-2) \, [\text{C—C}_{\text{bend.}}] +$$
$$(N-3) \, [\text{I.Rot}] + G_{\text{steric}} + 2 \, [\text{CH}_3] + (N-2) \, [\text{CH}_2] \tag{3.16}$$

in which G_steric is expressed again as a function of the parameters A and B, as for the free energy function.

Using these equations and the tables of numerical increments, the thermodynamic properties for any n-paraffin from C_4 to C_{20} can readily be calculated. Values calculated for longer chains based on these tables and equations are predicted to be fairly good approximations.

Comparisons of calculated and experimental entropies are shown in Table 3.2. The calculated entropy increment and experimental value for the CH_2 group are in very close agreement. With reference to the heat capacities (Table 3.2), it is felt that there is enough uncertainty in the experimental values so that an adjustment of the calculated values to remove the dis-

TABLE 3.2

COMPARISON OF CALCULATED AND OBSERVED ENTROPIES AND HEAT CAPACITIES OF IDEAL GASEOUS PARAFFINS[26]

		n-Hexane	n-Heptane	n-Octane
$S^\circ_{298.1}$ (cal./deg.mole)				
	Calculated	92.83	102.24	111.55
	Experimental	92.90	102.26	111.65
$\Delta S/[CH_2]$				
	Calculated	—	9.309	9.309
	Experimental	—	9.31	9.31
C_p° (cal./deg.mole)				
400° K	Calculated	43.47	50.42	57.36
	Experimental	43.38	50.29	57.1
500° K	Calculated	51.83	60.07	68.32
	Experimental	52.15	60.13	68.3

crepancies is not justified. For example, the experimental heat capacity value for n-octane at 400° K may be low by 0.1 – 0.2 cal. per deg. per mole just due to the gas imperfection correction. For the normal hydrocarbons below $C_{20}H_{42}$ the calculated and experimental entropies thus agree well within experimental error.

The thermodynamic properties for n-heptane and the numerical values for the $[CH_2]$ increment for normal paraffins are given in Tables 14, and 15, Part II, respectively. The latter were calculated by the relation:

$$\Delta G\ (X)_{CH_2} = [\text{C---C}_\text{str.}] + [\text{C---C}_\text{bend.}] + [\text{I.Rot.}] + [B] + [CH_2] \quad (3.17)$$

where $\Delta G(X)_{CH_2}$ is the increment in the free energy function, heat content function or heat capacity. The numerical values in Tables 11, 12, and 13, Part II were used to compile the values of the $[CH_2]$ increment. The thermodynamic properties for normal paraffins for $N > 7$ can be readily estimated by adding the required number of $[CH_2]$ increments to the values of the parent molecule, n-heptane.

3. LONG CHAIN HYDROCARBONS

Example 3.1. Estimate the statistical thermodynamic properties for n-decane at $1000°$ K and one atmosphere pressure in the ideal gas state.

SOLUTION:

(a) In accord with the preceding discussion, the thermodynamic properties of n-decane may be estimated by the simple calculation:

$$G\ (n\text{-decane}) = G\ (n\text{-heptane}) + 3\Delta G\ (CH_2)$$

where G is the desired statistical thermodynamic property, and the contributions of n-heptane and the $[CH_2]$ increment are taken from Tables 14 and 15, Part II. Thus at $1000°$ K:

	$-(G° - H_0°)/T$ cal./deg.mole	$(H° - H_0°)/T$ cal./deg.mole	$S°$ cal./deg.mole	$C_p°$ cal./deg.mole
n-heptane	123.70	56.52	180.22	91.22
3 $[CH_2]$	36.90	22.99	59.89	37.00
n-decane	160.60	79.51	240.11	128.22
(cf. lit.[1])	160.20	80.21	240.41	128.42

(b) An alternate procedure would be to use Eq. 3.14 – 3.16, and Tables 11–13, Part II, i.e., for n-decane at $1000°$ K:

	$-(G° - H_0°)/T$ cal./deg.mole	$(H° - H_0°)/T$ cal./deg.mole	$C_p°$ cal./deg.mole
$[G°(T)]$	59.143	7.949	7.949
9 $[C-C_{str.}]$	11.250	8.181	15.219
8 $[C-C_{bend.}]$	22.640	12.824	15.672
7 $[I.Rot.]$	36.057	12.845	10.647
$G_\sigma(\sigma=2)$	– 1.379	—	—
2 $[CH_3]$	9.546	11.338	21.452
8 $[CH_2]$	12.456	24.712	57.120
A	6.354	0.977	0.096
3 B	4.536	0.684	0.066
	160.60	79.51	128.22

and $S°_{1000} = 240.11$ e.u.

4. Branch Chain Hydrocarbons

In adapting the formulas of Table 3.1 to branch chain paraffins the basic equation for the estimation of thermodynamic properties for such molecules may be expressed as:

$$G(X) = [G_0(T)] + n_1 [\text{C—C}_{\text{str.}}] + n_2 [\text{C—C}_{\text{bend.}}] + n_3 [\text{I.Rot.}] +$$
$$G_{\text{steric}} + G_\sigma + n_4 [\text{CH}_3] + n_5 [\text{CH}_2] + n_6 [\text{CH}] +$$
$$n_7 [\text{Isobutane Correc.}] + n_8 [\text{Neopentane Correc.}] \quad (3.18)$$

where n_1 is the number of C—C bonds, n_2 the number of C—C bond bendings, n_3 the number of skeletal rotations, n_4, n_5, n_6, n_7, and n_8, the number of such groups or bondings in the molecule. The numerical values listed in Table 16, Part II, combined with the values of Tables 11–13, Part II, may be used to estimate the desired data for branch chain paraffins.

It should be noted that the steric factor, G_{steric} must be considered, and is not necessarily as simple as for normal paraffins. The symmetry number for skeletal internal rotations, as of the tertiary butyl groups, must also be taken into account.

The correction factors (isobutane and neopentane) were obtained as the difference between the thermodynamic values calculated by the more exact statistical methods of the preceding chapter and the values estimated by the approximate methods of this chapter, using the same force constants and potential barriers as for the normal paraffins. These differences should be applied as "corrections" whenever a carbon atom in the branch chain molecule is bonded to three or four other carbon atoms, respectively. Owing to the lack of knowledge of force constants, vibration frequencies, and experimental values for the thermodynamic properties in the branch chain paraffins, refinements in the method of estimation have not been attempted. From a comparison with the limited experimental data it is estimated that confidence can be placed in the values of entropies estimated for branch chain paraffins only to an accuracy of 2 or 3 cal. per deg. per mole.

5. Unsaturated Hydrocarbons

No detailed method, as developed for n-paraffins, has been reported for the olefins or acetylenes. Estimates can be made in particular cases using the difference between the most analogous butene and butane together with the value for the corresponding paraffin calculated by the methods just discussed. Changes in symmetry numbers and possibly in heats of hydrogenation should be corrected for in such estimates. This procedure is basic to the method of group equations discussed in detail in Chapter 6.

6. The Steric Factor

As already indicated, the steric factor is a function of energy assigned to various configurations of the molecule on the basis of steric repulsions. Each configuration was assigned an energy in terms of a. The number of these as multiples of a was obtained by inspection of "Fisher-Hirschfelder" models, which approximate proportional atomic sizes. In Table 3.3 a summary is

TABLE 3.3

POSITIONS CONTRIBUTING TO STERIC FACTOR FOR DIFFERENT VALUES OF E_i.[5]

Hydrocarbon	Number of positions with E_i (Steric) =						
	0	a	$2a$	$3a$	$4a$	$5a$	∞
Butane	1	2	0	0	0	0	0
Pentane	1	4	2	0	0	0	2
Hexane	1	6	8	2	0	0	10
Heptane	1	8	18	12	2	0	40
Octane	1	10	32	38	16	2	144
2-Methylbutane	2	0	1	0	0	—	0
2,2-Dimethylbutane	1	0	0	0	0	—	0
2,3-Dimethylbutane	1	0	2	0	0	—	0
2-Methylpentane	2	0	3	0	0	—	4
2,2,3-Trimethylbutane	1	0	0	0	0	—	0
2,2-Dimethylpentane	1	0	0	0	0	—	2
3-Ethylpentane	11	0	0	0	0	—	16
3-Methylhexane	4	4	4	0	0	—	15
2,2,3,3-Tetramethylbutane	1	0	0	0	0	—	0
2,2,4-Trimethylpentane	2	0	0	0	0	—	1

given of the number of positions contributing to the steric factor for normal and branch chain paraffins. Person and Pimentel[26] found that with V_0 equal to 3260 cal. a value of 500 cal. per mole for the steric parameter a gave a best agreement between the calculated and experimental entropies of n-hexane and n-octane. In branch chain compounds, fortunately, the number of different positions contributing to G_{steric} is less because of the increased number of symmetrical groups.

7. Random Kinking and Ball-Like Molecules

The preceding method was developed for a planar zig-zag carbon skeletal model for the long chain paraffins, and is the most complete procedure for accurate estimates of thermodynamic properties. Calculations, restricted to entropies, for long chain molecules have been discussed by Huggins[30]

[30] M. L. Huggins, *J. Chem. Phys.* **8**, 181 (1940).

using quite different methods. It is assumed that the long chain molecules are kinked in random manner. The total entropy is considered to be equal to:

$$S° = S°_{trans.} + S°_{vib.} + S°_{rot.} + S°_{i.r.} \qquad (3.19)$$

where the last term is the contribution of internal randomness due to the flexibility of the chain.

The molal translational entropy (Eqs. 2.11 and 2.13, Chapter II) may be calculated classically from:

$$S°_{trans} = R \ln \frac{V}{N} \left(\frac{2\pi mkT}{h^2}\right)^{\frac{3}{2}} e^{\frac{5}{2}} \qquad (3.20)$$

in which R is the gas constant per mole, V the molal volume, m the mass per molecule, and e the base of natural logarithms. In the long chain compound, where m_0 is the mass of one of the n "sub-molecules" in each polyatomic system, the mass per molecule, m is given by:

$$m = nm_0 = \frac{M_0 n}{N} \qquad (3.21)$$

M_0 being the formula weight of the atoms in the sub-molecule. Substituting for V from the ideal gas law with R in units per mole in cc. atmos. per degree and P in atmospheres, it follows that:

$$S_t° = R \ln \left[\frac{R(2\pi k)^{\frac{3}{2}} e^{\frac{5}{2}}}{h^3 N^{\frac{5}{2}}} \cdot \frac{T^{\frac{5}{2}}}{P} \cdot M_0^{\frac{3}{2}} n^{\frac{3}{2}}\right] \qquad (3.22)$$

At one atmosphere pressure and 298.1° K, the molal translational entropy of a long chain molecule, as a dilute gas, may thus be calculated by:

$$S_t° = 26.0 + \tfrac{3}{2} R \ln M_0 + \tfrac{3}{2} R \ln n \qquad (3.23)$$

or

$$S_t° = 33.8 + \tfrac{3}{2} R \ln n \qquad (3.24)$$

where the latter is for a long hydrocarbon chain, $(CH_2)_n$, taking M_0 as 14.03.

The calculation of the vibrational entropy for a long chain normal paraffin was based on the frequency assignment by Pitzer[14] using generalized characteristic frequencies for the C—H, C—C bond stretching modes and the H—C—H, H—C—C, and C—C—C bond deformations. These data may be summarized as follows:

| Type | Vibrational modes | | Entropy |
	Number	Frequency	$S_v°$ (298° K)
C—H	$2n + 2$	3000 cm^{-1}	0.000
H—C—H	$n + 4$	1440	$0.064 + 0.016n$
C—C	$n - 1$	1000	$-0.092 + 0.092n$
H—C—C	$3n - 2$	950	$-0.226 + 0.339n$
C—C—C	$n - 2$	320	$-2.628 + 1.314n$

Thus for normal paraffin hydrocarbons, the total vibrational entropy is readily calculated from:

$$S^\circ_{\text{vib.}} = -2.88 + 1.76\,n \tag{3.25}$$

where, from the above data, the vibrational contributions have been evaluated in the usual way as the sum of Einstein functions. A more general expression for the vibrational entropy contribution for a long chain molecule, by analogy, is:

$$S^{\text{vib.}}_\circ = k'_v + k_v n \tag{3.26}$$

where the constants k'_v and k_v are computed from the generalized characteristic frequencies and vibrational assignment as already illustrated.

The entropy of rotation of the molecule as a whole (Eqs. 2.11, 2.17) may be expressed as:

$$S^\circ_{\text{rot.}} = R \ln \left[\frac{1}{\pi\sigma} \left(\frac{8\pi^3 k e T}{h^2} \right)^{\frac{3}{2}} (I_x I_y I_z)^{\frac{1}{2}} \right] \tag{3.27}$$

For randomly kinked long chain molecules an average value of the products of the moments of inertia must be used, since all molecules will not have the same moments of inertia. Although the individual moments (I_x, etc.) will vary considerably, the moment products and the logarithms of the products may be expected to be closely bunched about the average values. Assuming therefore that:

$$[\ln (I_x I_y I_z)]_{\text{av.}} = \ln [(I_x I_y I_z)]_{\text{av.}} \tag{3.28}$$

the rotational entropy for the randomly kinked polyatomic molecule at 298.1° K is given by:

$$S^\circ_{\text{rot.}} = 284.5 + \tfrac{1}{2} R \ln [(I_x I_y I_z)]_{\text{av.}} \tag{3.29}$$

Two models are considered for calculation of $S^\circ_{\text{rot.}}$. Assuming that the van der Waals attractions between various parts of the molecular chain cause coiling, one model is for spherical molecules having uniform density, ρ, equal to the density the substance would have in the liquid state at that temperature. The other model is for symmetrical "average" molecules, which on the assumption of random kinking would have dimensions equal to the root mean square dimensions of such systems. The expressions resulting from these considerations are:

$$S^\circ_{\text{rot.}} = 7 + \tfrac{5}{2} R \ln M_0 - R \ln \rho + \tfrac{5}{2} R \ln n \tag{3.30}$$

$$S^\circ_{\text{rot.}} = 116 + \tfrac{3}{2} R \ln (B_\infty M_0 l^2) + 3 R \ln n \tag{3.31}$$

for spherical and randomly kinked polyatomic molecules, respectively. In the above expressions, B_∞ is defined by:

$$B_\infty = \frac{1 + \cos \alpha}{1 - \cos \alpha} \tag{3.32}$$

where a is the angle between each bond in the chain, and l is the distance between neighboring atoms in the chain skeleton. For a paraffin chain where a is the tetrahedral angle, and l is 1.54×10^{-8} cm., at $298.1°$ K, the equations above reduce to:

$$S°_{\text{rot.}} = 21 + \tfrac{5}{2} R \ln n \qquad (3.33)$$

and

$$S°_{\text{rot.}} = 23 + 3 R \ln n \qquad (3.34)$$

for large paraffin chain molecules treated as spherical and randomly kinked systems, respectively.

The entropy of internal randomness, which takes into account the contributions due to alternative orientations and restricted rotations about the bonds, would be expected to be a linear function of n. For randomly kinked molecules with no internal rotation and with free internal rotation it is found that the entropy contributions are $(-6 + 2.18 n)$ and $(-26 + 9 n)$ respectively, or more generally in the first approximation by the linear equation:

$$S_{\text{i.r.}} = -3 k_{\text{i.r.}} + k_{\text{i.r.}} n \qquad (3.35)$$

in which the parameter $k_{\text{i.r.}}$ is a constant that depends on the flexibility of the chain. It is assumed that this relationship also holds for the long chain paraffins in which the rotation is hindered. The value of $k_{\text{i.r.}}$ is adjustable, and is selected to give best agreement with experimental entropies for known systems.

The total entropy of long chain compounds at $298°$ K and one atmosphere pressure may thus be readily estimated from the general equations:

$$S° = 33 + k_v' + k_{\text{i.r.}}' + R \ln \left(\frac{M_0^4}{\rho}\right) + 4 R \ln n + (k_v + k_{\text{i.r.}}) n \qquad (3.36)$$

and

$$S° = 142 + k_v' + k_{\text{i.r.}}' + 3 R \ln (M_0 B_\infty^{\tfrac{1}{2}} 1) + \frac{9}{2} R \ln n + (k_v + k_{\text{i.r.}}) n \qquad (3.37)$$

for ball-like and randomly kinked molecules. For long chain paraffins these reduce to:

$$S° = 52 - 3 k_{\text{i.r.}} + 8 \ln n + (1.8 + k_{\text{i.r.}}) n \qquad (3.38)$$

and

$$S° = 54 - 3 k_{\text{i.r.}} + 9 \ln n + (1.8 + k_{\text{i.r.}}) n \qquad (3.39)$$

respectively, where $k_{\text{i.r.}}$ is the only adjustable parameter.

The agreement found between the predicted values, based on the assumption of spherical molecules (Eq. 3.38) and completely random kinking (Eq. 3.39) and the literature data[1] is shown in Table 3.4. If a value of 6.2 is selected for $k_{\text{i.r.}}$, the predicted value of the entropy for ethane is in exact

TABLE 3.4

COMPARISON OF ESTIMATED ENTROPIES ASSUMING BALL-LIKE AND RANDOMLY KINKED MOLECULES FOR n-PARAFFINS

Value of $k_{i.r.}$	Molecular type	Entropy, $S°_{298}$ (g) cal./deg./mole			
		C_2H_6	C_6H_{14}	$C_{12}H_{26}$	$C_{20}H_{42}$
6.2	ball-like	54.85	95.7	149.3	217.4
	randomly kinked	57.63	99.5	153.8	222.4
6.0	ball-like	55.0	95.1	147.5	214.0
	randomly kinked	57.8	98.9	152.0	219.0
5.5	ball-like	55.6	93.6	143.0	204.5
	randomly kinked	58.3	97.4	147.5	209.5
Lit.[1]		54.85	92.45	147.55	221.02

agreement with the experimental value, the molecules being assumed ball-like. With this value of $k_{i.r.}$ the use of Eq. 3.38 to predict the entropies seems to be justified for hydrocarbons up to dodecane, but the entropy values for very long chain compounds would appear to be better approximated by the use of Eq. 3.39 (cf. $C_{20}H_{42}$). Adjusting the value of $k_{i.r.}$ to 6.0 gives very close agreement with the literature value for dodecane, assuming the latter may be treated as ball-like rather than randomly kinked molecules. Inspection of the estimated values relative to this molecule as standard shows that, as in the preceding case, the use of Eq. 3.38 seems well justified for moderately long chains, but that Eq. 3.39 may be expected to give better estimates for very long chains. With a value of 5.5 for $k_{i.r.}$, the entropy of dodecane can be brought into accord with the literature value, assuming the molecules are randomly kinked rather than ball-like, i.e., by Eq. 3.39. Based on this assumption, however, the values estimated for the very long chain compounds would appear to be less reliable than in the preceding two examples. Using "best values" for $k_{i.r.}$, selected by comparison with experimental entropies, approximate entropy values for very long chain compounds in the gaseous state can thus be readily estimated.

CHAPTER 4

The Method of Structural Similarity

1. Introduction

The method of structural similarity is purely empirical in its approach. The regularity and systemization found in organic compounds applies to both chemical and physical properties to a very large extent. With reference to thermodynamic data, the interest has been to apply this systemization empirically, seeking correlations that enable the prediction of free energies, heats of formation, entropies, and heat capacities for compounds for which the experimental or statistical thermodynamic data are not available. The work of Parks and Huffman[4] is classic in this field and has proved a guide to many on first undertaking the application of thermodynamics in the field of organic chemistry, and in the development of the more elaborate semi-empirical methods that have followed this work. While many of the numerical values for the increments in the thermodynamic properties accompanying various molecular structural modifications can be revised in the light of more recent data, the principles of the method as developed by Parks and Huffman remain essentially unchanged. This chapter is devoted to a discussion of the work of Parks and Huffman in developing the method of structural similarity for estimation of thermodynamic properties.

2. Construction of the Correlation

Stimulated by the publication of several earlier theoretical papers and the increasing interest in the thermodynamics of hydrocarbons, Parks and Huffman undertook the task of extending experimental data and seeking correlations systematically of thermodynamic free energy and entropy with the molecular structure. A comprehensive investigation of the data for the paraffins, olefins, cyclic and aromatic hydrocarbons, and organic compounds containing oxygen, halogens, nitrogen, and sulfur led to the compilation of tables of structural modifications and thermodynamic increments. A brief summary of some of the more basic results follows. It should be noted that owing to the limited data then available, the correlations are largely restricted to 25° C, and the liquid phase.

3. Paraffins

A systematic study of the thermodynamic properties and molecular structures for the normal and branch chain paraffins led to two generalized equations for calculating the entropy and free energy for any *liquid* paraffin:

$$S^\circ_{298} = 25.0 + 7.7n - 4.5r \qquad (4.1)$$

$$\Delta G^\circ_{298} = -11700 + 1080n + 800r \qquad (4.2)$$

where n is the total number of carbon atoms in the molecule and r the number of methyl (or aliphatic) branches on the main straight chain. On the assumption of a linear relationship between ΔG° and the number of carbon atoms n in normal paraffins, the equation:

$$\Delta G^\circ = -14700 - 6000n + 1.0T + 25.0nT \qquad (4.3)$$

was proposed for calculating the free energy between 400° and 1000° K for any hydrocarbon between ethane and tetradecane.

The increments in entropy and free energy for the first five structural modifications in Table 4.1 were developed by Parks and Huffman from comparisons within the paraffin hydrocarbons. It was qualitatively observed that within a series of paraffin isomers, the branched chain structures were thermodynamically unstable relative to the normal compound, especially at higher temperatures.

4. Unsaturated Hydrocarbons

Owing to the limited data available, the correlations were restricted to the olefin series only. From the free energies for gaseous normal olefins up to hexene, a general equation:

$$\Delta G^\circ = 5700 - 6500 (n-2) + 21.1T + 24.0 (n-2) T \qquad (4.4)$$

quite similar to that for the paraffins was derived. The influence of the position of the double bond in the molecule could only be investigated very qualitatively. It was concluded that in a group of olefin isomers the free energy decreases as the double bond becomes more centrally placed within the structure. An increment in the entropy and free energy accompanying the conversion of a single bond into an olefinic link (Table 4.1) was proposed.

5. Cyclic Hydrocarbons

It was found that the entropies for *liquid* cyclic hydrocarbons could be calculated by:

$$S^\circ_{298} = 25.0 + 7.7n - 4.5r + 19.5p_1 + 26.5p_2 \qquad (4.5)$$

where n, r, p_1 and p_2 are the total number of carbons outside the ring, the number of hydrocarbon groups in excess of two attached to any carbon atom in the aliphatic chain, the number of phenyl groups, and the number

TABLE 4.1

THE CHANGES IN MOLAL ENTROPY AND FREE ENERGY ACCOMPANYING VARIOUS STRUCTURAL MODIFICATIONS AT 298.1 °K[4]

Structural modification	Change in molal entropy			Change in $\Delta G_{298}°$
	Solid cal./deg.mole	Liquid cal./deg.mole	Gaseous cal./deg.mole	cal./mole
(1) Insertion of CH_2 into a hydrocarbon chain	5.8	7.7	10.0	1080
(2) Substitution of CH_3 for H attached to a main hydrocarbon chain	5.0(?)	3.2	5.0(?)	1900
(3) Substitution of C_2H_5 for H attached to a hydrocarbon chain	—	10.9	—	3000
(4) Substitution of CH_3 for H in a hydrocarbon ring	5.8	7.7	—	0(?)
(5) Substitution of C_2H_5 for H in a hydrocarbon ring	11.6(?)	15.4	—	1100
(6) Substitution of a phenyl group for H attached to carbon	17.0	19.5	—	36000
(7) Substitution of a cyclohexane or cyclopentane ring for H attached to carbon	—	26.5	—	13000
(8) Conversion of a single bond into an ethylenic double bond	−2.7	−2.7	−2.7	very irregular; mean about 20000
(9) Substitution of OH for H to form a monohydroxy, primary alcohol	0(?)	−1.5	13.0	−34000
(10) Substitution of OH for H to form a monohydroxy, secondary alcohol	0.5(?)	−4.0	9.0(?)	−37000
(11) Substitution of OH for H to form a monohydroxy, tertiary alcohol	0.5(?)	−6.0	7.0(?)	−41000
(12) Substitution of OH for H to form a phenol	0.0	0(?)	—	−41000
(13) Substitution of OH for H to form polyhydroxy compounds	0.5(?)	0.5(?)	—	primary OH −34000 secondary OH −37000
(14) Insertion of —O— linkage in a chain to form an ether	—	5.0	8.0	−20000(?)
(15) Substitution of O for 2H to form an aldehyde	1.0(?)	5.0(?)	7.3(?)	−23000(?)
(16) Substitution of O for 2H to form a ketone	1.0	0.5(?)	6.0	−30000(?)
(17) Substitution of CO_2H for H to form a carboxylic acid	5.8	7.7	—	−83200
(18) Insertion of $-\overset{O}{\underset{\|}{C}}-O-$ group into a hydrocarbon chain to form an ester	—	13.2	21.0	−70000
(19) Substitution of NH_2 for H to form an amine	0.0	0.0	—	6000
(20) Substitution of NO_2 for H to form a nitro compound	7.0	8.0	—	7000
(21) Substitution of Cl for H	6.0	7.0(?)	9.0(?)	−1600
(22) Substitution of Br for H	7.5	9.0(?)	11.5(?)	4500(?)
(23) Substitution of I for H	9.0(?)	11.0(?)	14.0(?)	10000(?)
(24) Substitution of bivalent S for O in a compound	2.0(?)	2.0(?)	2.0(?)	36000(?)

of saturated rings, respectively, in the molecule. From a compilation of the free energy data of the alicyclic and aromatic hydrocarbons at 298.1° K, the structural increment for the introduction of a phenyl group into a saturated hydrocarbon (Table 4.1) was evident. The difference in free energy between the liquid and solid states at 298.1° K was estimated for compounds by the approximate equations:

$$\Delta G^\circ_{298} = \Delta S^\circ_{fus.} (T_M - 298.1°) \tag{4.6}$$

and

$$\Delta S^\circ_{fus.} = \Delta H^\circ_{fus.}/T_M - \tfrac{1}{2} \Delta C_p \ln (T_M/298.1) \tag{4.7}$$

where the latter equation was used when $(T_M - 298.1°)$ was large. Free energies in the gaseous state for cyclohexane, benzene, toluene, xylene, naphthalene, and anthracene of the general type:

$$\Delta G^\circ = A + BT \ln T - CT^2 - DT \tag{4.8}$$

were calculated from a knowledge of the vapor pressures and heats of vaporization at 298.1° K, and the heat capacities of the vapors. The thermodynamic stability of various rings, C_2, C_3, C_4, C_5, and C_6 was considered. The free energies of formation per carbon atom in these rings were 7.3, 7.1, 6.5, 0.88, and 0.92 kcal./mole, respectively. Owing to the approximations involved, these results are orders of magnitude only.

6. Organic Compounds Containing Oxygen

The free energy data for ethers, aldehydes, alcohols, acids, acetals, polyhydroxy-alcohols, esters, ketones, and others were compiled to detect tendencies and relationships in the oxygen containing compounds.

For the series of normal aliphatic alcohols after n-propanol the free energy increment was small and regular, similar to that found among the normal paraffins. The increments of ΔG°_{298} per hydroxyl group for formation of a primary, secondary, and tertiary alcohol, and a phenol derived from comparison of the data for alcohols with hydrocarbons are found in Table 4.1. In the calculation of the data for *gaseous* alcohols, when no heat capacity data was available, it was found that a fairly reliable equation could be gained by adding the effect for a [CH_2] increment used earlier for normal paraffins. The following general equation for heat capacity:

$$C_p^\circ \text{ (per mole)} = (1.9 + 1.3n) + (0.014 + 0.012n) T \tag{4.9}$$

was thus gained. The application of these data to processes taking place in solution rather than the gaseous state was considered to stress the need for activity measurements. Since in general such solutions deviate markedly from ideal behavior, activity data are essential for such free energy calculations.

In the aliphatic *acid* series, the equation:

$$\Delta G°_{298} = -96000 + 1080n \tag{4.10}$$

was suggested for estimating the molal free energy for a normal, saturated aliphatic acid containing n carbon atoms. The [CH$_2$] increment is the same as that obtained in the normal paraffin series. The free energy relations among the other organic compounds containing oxygen were likewise developed, leading to the increments for the modifications of a paraffin structure as listed in Table 4.1. The relationships for each series were examined graphically, plotting $\Delta G°_{298}$ against n, the number of carbon atoms in the molecule. It was observed in this analysis that after the first few members were passed in any homologous series of oxygen containing compounds the change in $\Delta G°_{298}$ was very similar to that found in the normal paraffin series. Parks and Huffman accordingly note that the first one or two members of any series containing oxygen must be considered as unique or abnormal in their free energy values. It was suggested that this may be due, in part, to the influence of oxygen in increasing the polar nature of the molecule.

7. Nitrogen, Halogen, and Sulfur-Containing Organic Compounds

The entropy and free energy effects of the NH$_2$, NO$_2$, Cl, Br, I, and S structural modifications are given in Table 4.1. The values for these important increments were based on the data of twenty-four compounds containing nitrogen, eight or nine containing the halogens, and only five containing sulphur. Of the latter, three were estimated values, at best only first approximations. The entries in Table 4.1 corresponding to these data are marked accordingly by question marks in brackets. The work of Parks and Huffman called attention to the complete lack of data for many important classes of substances, as, for example, aliphatic amines, azo-compounds, and organic compounds containing phosphorous, arsenic, boron, and metallic elements, at the time of their monograph.

In summary, the method of structural similarity as developed by Parks and Huffman is purely empirical in its approach. It is based on the increments obtained by comparison of the data for related compounds, assuming that they may be applied quite simply by mathematical addition for a corresponding structural modification in another molecule. No cognizance is taken in this approach of the influence of the symmetry of the molecule, hindered internal rotation, and the products of the moments of inertia which enter into the additivity of the free energy function and entropy (cf. Table 2.1). The increments are thus likely to be rather specific, and should be applied only in estimating the data for quite comparable molecular species.

8. Entropy and Free Energy Regularities

The basic molal entropies for a normal paraffin in the crystalline, liquid, and gaseous states were found to be 18.0, 25.0, and 34.0, respectively. Accordingly the molal entropy for any normal paraffin may be calculated by the simple equations:

$$S^\circ_{298 \text{ (s.)}} = 18.0 + 5.6n \tag{4.11}$$

$$S^\circ_{298 \text{ (liq.)}} = 25.0 + 7.7n \tag{4.12}$$

$$S^\circ_{298 \text{ (g.)}} = 34.0 + 10.0n \tag{4.13}$$

where n is the number of carbon atoms, and the entropy increment is taken from the first line entry of Table 4.1. The various branch chain paraffins are considered to be derived from one of the normal paraffins by replacement of one or more hydrogen atoms by the appropriate number of methyl, ethyl, or longer aliphatic groups. Cyclic hydrocarbons may be considered as derived from the paraffins through modifications 6 and 7 or from the appropriate cyclic compounds and modifications 4 and 5 as listed in Table 4.1. The entropy increment accompanying the formation of a double bond is recognized to be rather more specific in character than that of most other increments. It is suggested that the use of the tabulated value (– 2.7 e.u.) in estimating the entropy for mono-olefins will probably not lead to very serious errors. In all instances where the increment was an approximation in itself, a question mark indicates doubt as to its magnitude. Where the data for the solid state are absent, Parks and Huffman recommend the use of the increment for the change in the liquid state for estimates involving crystalline compounds. The free energy increments were derived primarily for the liquid state as a matter of convenience since a majority of organic compounds at 25° C are liquids. The values should be regarded as average values, based in many instances on very limited data, and frequently rather approximate in character. The use of these increments for anything but estimating very tentative free energy and entropy values is not recommended in practice.

The rules for estimation summarized in Table 4.2 were prepared by Ewell[31] after a critical correlative study of the entropy and heat of formation data of hydrocarbons. The increment in entropy for the simple side chain branch was specified more accurately, and the more accurate extension of the method of structural similarity to olefins was possible in the light of additional data. Regularities in heats of formation rather than the free energies were considered since the free energy is a derived property dependent on a knowledge of the entropy and heat of reaction. The linear equation for calculating the molal entropy of a normal paraffin (4.13) was

[31] R. B. Ewell, *Ind. Eng. Chem.* **32**, 778 (1940).

TABLE 4.2

Increments for Estimation of Entropy and Heat of Formation of Gaseous
n-Paraffins or 1-Olefins with Same Number of Carbon Atoms[31]

Structural modification	ΔS°_{298} (g) e.u./mole	$-\Delta H^\circ_{298}$ (g) kcal./mole
(a) Simple chain branch	− 4.0	1.8
(b) Neopentyl grouping	− 10.5	4.7
(c) Chain branch adjacent to the double bond	− 4.0	3.5
(d) Double bond not in terminal position, *cis*	− 1.8	1.6
(e) Double bond not in terminal position, *trans*	− 2.4	2.6
(f) 3 or 4 Substituents on double bond	− 1.8	1.6
(g) 4 Substituents on double bond	1.6	− 1.6

considered in the light of more recent data. Two equations were recommended:

$$S^\circ_{298} = 35.3 + 9.8n \tag{4.14}$$

and

$$S^\circ_{298} = 35.6 + 9.7n \tag{4.15}$$

the former for paraffins from n-butane to n-heptane, and the latter for paraffins from n-octane to undecane. No attempt was made to estimate the probable errors of the generalized rules (Table 4.2). The estimations are largely intuitive so that the limits of error are best evaluated for each application in practice.

Example 4.1. Estimate the entropy of pentanethiol at 298° K and one atmosphere pressure in the ideal gas state.

SOLUTION:

Using the method of structural similarity, and the correlations of Parks and Huffman, and Ewell, to estimate $S^\circ_{298(g)}$ the steps in the procedure would be:

(a) For n-pentane, the entropy may be estimated by Eq. 4.13 or 4.14. Using the latter, being based on more recent work, it follows that:

$$S^\circ_{298} = 35.3 + 9.8(5) = 84.3 \text{ e.u.}$$

(b) The estimate for pentanethiol is next achieved by the following structural modifications, using the corresponding increments from Table 4.1:

S°(pentanethiol) = S° (pentane) + ΔS° (OH for H to form primary alcohol) +

ΔS° (bivalent S for O in a compound)

Accordingly,

$$S^\circ_{298} \text{ (pentanethiol)} = 84.3 + 13.0 + 2.0 = 99.3 \text{ e.u.}$$

This is to be compared with the experimental value of 99.18 e.u. reported[32] in 1952 for the ideal gas entropy of pentanethiol. The agreement leaves little to be desired.

Example 4.2. Estimate the value of $S°_{298(g)}$ for 1,2-dibromoethane. (ANSWER: 77.0 e.u., lit.[33] 78.7).

The recent literature[1,2] should be consulted to obtain best values for the increments in thermodynamic properties following the basic principles of Parks and Huffman in any application of the method of structural similarity for estimation purposes.

[32] H. L. Finke, D. W. Scott, M. E. Gross, G. Waddington, and H. M. Huffman, *J. Am. Chem. Soc.* **74**, 2804 (1952).
[33] J. W. Andersen, G. H. Beyer, and K. M. Watson, *Natl. Petroleum News* **36**, 476 (1944).

CHAPTER 5

The Methods of Group Contributions

1. Introduction

The various methods based on the properties of group contributions were developed after the work of Parks and Huffman to calculate the numerical values of thermodynamic properties in a simple manner with the help of as few data as possible. Parks and Huffman had established empirically a number of more or less rough correlations of thermodynamic properties with molecular structure. With the theoretical contributions of statistical thermodynamics and the greatly increased experimental activity on the investigation of thermodynamic properties of hydrocarbons and all classes of organic compounds, it proved possible to state more accurate relations between structure and thermodynamic properties. The scope of this chapter is limited to the contributions of four groups of investigators, Andersen, Beyer, and Watson,[33] Franklin,[34] Souders, Matthews, and Hurd,[35] and Van Krevelen and Chermin,[36] to the development of more modern estimation methods for calculating the thermodynamic data for the vast number or organic molecules for which these data are lacking. The principle that within certain recognized limits molecular structural groups have the same contributions to the thermodynamic property no matter what may be the total molecular structure is basic to all methods, but is applied somewhat differently in each case. Reference to these methods in practice is most readily made by the names of the investigators.

2. Methods

In the method of *Andersen, Beyer, and Watson*, each compound is considered as composed of a *parent* molecule which has been modified by substitution of the appropriate groups for atoms on it to achieve the final molecule. Thus all paraffins are derived from the parent molecule, methane, by substituting the hydrogen atoms by CH_3 groups. Similarly the values for all ethers are obtained, starting from dimethyl ether as parent, and summing the increments in thermodynamic data corresponding to the structural modifications by group substitutions. The thermodynamic increments are correlated with structure for three properties, $\Delta H^\circ_{f\,298.1(g)}$, $S^\circ_{298.1(g)}$, and the

[34] J. L. Franklin, *Ind. Eng. Chem.* **41**, 1070 (1949).
[35] M. Souders, C. S. Matthews, and C. O. Hurd, *Ind. Eng. Chem.* **41**, 1037, 1048 (1949).
[36] D. W. Van Krevelen and H. A. G. Chermin, *Chem. Eng. Sci.* **1**, 66 (1951).

heat capacity expressed as the three term power series equation, $C_p° = a + bT + cT^2$. While it is not directly stated, possible deviations from the simple principle of additivity (owing to symmetry effects and hindered internal rotation) are taken into account as much as possible by specifying the related molecular structural environment for each increment. The correlation of heat capacity, together with the other two properties, makes possible the calculation of the free energies of formation, heats of formation, and entropies at temperatures other than 25° C.

The method of *Franklin* is based on an extension of the relations and principles developed by Pitzer[5] for the long chain paraffins. According to the latter, it was shown from theoretical considerations that the heat content and free energy functions for gaseous normal paraffins can be expressed as additive functions of the number of carbon atoms, and constants characteristic of temperature, bond stretching, bending, internal rotation, and the symmetry number for the molecule. To extend this approach to branched chain paraffins, Franklin revised Pitzer's basic equation (3.12) in the form:

$$X = G° + (n_1 + n_2 + n_3 + n_4 - 1) S + (2n_1 + n_2 + n_3 + n_4 - 4) B$$
$$+ (n_1 + n_2 - n_4 - 3) I + n_1 V_1 + n_2 V_2 + n_3 V_3 + n_4 V_4 + G_{\text{steric}} + F\sigma \quad (5.1)$$

to account for the CH and C (i.e., n_3 and n_4) groups as well as the CH_3 and CH_2 groups (i.e., n_1 and n_2) which occur in such molecules; S, B, I, V_2, V_3, and V_4 are the empirical constants for C—C stretching, bending, internal rotation, CH_2, CH, and the quaternary C group within the molecule. On rearrangement, this gives:

$$X = 2(G°/2 + S/2 - I/2 + V_1) + (n_1 - 2)(S + 2B + I + V_1)$$
$$+ n_2(S + B + I + V_2) + n_3(S + B + V_3) + n_4(S + B - I + V_4) + G_{\text{steric}} + F\sigma \quad (5.2)$$

which expresses the thermodynamic function (free energy function, heat content function) as an additive function of the CH_3, CH_2, CH, and C groups for each molecule plus the constants G_{steric} and G_σ. Owing to the additive nature of these thermodynamic functions, group values can be established from the total heat content and total free energy functions of molecules containing these groups. Summation of the increments for the groups contributing to the molecular structure gives the desired thermodynamic property. Corrections for differences in symmetry factors are considered separately from these increments in Franklin's approach. Numerical values of these increments are tabulated up to 1500° K.

The correlation method of *Souders, Matthews, and Hurd* presents structural group contributions to the heat capacity, heat content, and entropy for the vibrational and characteristic internal rotational contributions. Inspection of the structural formula guides the selection of the increments most appropriate for the environment within the molecule, and these, summed with the translational, external rotational, contributions for the system as a whole, lead to the desired thermodynamic property. The correlation method, furthermore, leads directly to a value for the entropy of formation as distinct from the entropy of the molecule under consideration. Numerical values of the increments in heat content and heat capacity are tabulated to 3000° F, and for the entropy of formation and heat of formation to 2000° K.

The approach of *Van Krevelen and Chermin* was to develop a method based on group increments which gives the result directly in a temperature dependent equation form for the free energy of formation of the molecule in question. The preceding methods necessitate interpolation procedures to calculate the free energy change at any temperature other than those for which the numerical values were compiled. Van Krevelen and Chermin expressed the temperature dependence of the structural group contribution to the free energy by the function:

$$\Delta G^\circ_{f(\text{group})} = A + 10^{-2} BT \tag{5.3}$$

where A and $10^{-2}B$ are, by approximation, equal to the heats and entropies of formation. It is possible to describe the temperature dependence up to 1500° K by two linear equations, the first covering the range 300–600° K, and the second 600–1500° K. The values of the constants A and B for these equations are reported for the various structural groups for hydrocarbons and nonhydrocarbons. The equation for the free energy of formation of a complex molecule is readily obtained by summation of the contributions for structural groups of the molecule.

3. Construction of the Correlations

Following the treatment in the preceding section, the construction of the correlation is discussed separately for each of the methods considered.

Andersen, Beyer, and Watson. The values of the heats of formation, entropies, and heat capacity constants were divided into group contributions by assuming each compound to be composed of a basic or parent group, the latter being modified subsequently as already mentioned by group substitution to give the molecule under consideration. The parent groups and corresponding thermodynamic data compiled by these investigators are given in Table 17, Part II, and can readily be extended to other types of systems. For example, extension to the nitrogen heterocyclics is possible since the thermodynamic data for many of the appropriate compounds, e.g., pyridine, picoline, collidines, and quinolines, are now established. The method differs from the approach of Parks and Huffman in that the thermodynamic increment was correlated with a structural modification corresponding to replacement of CH_3 groups rather than single hydrogen atoms, wherever possible. Thus the contributions for the NO_2 or CN groups were gained by comparison of the difference in thermodynamic properties of nitromethane and methyl cyanide with ethane, rather than methane. The comparison is made with as many of the corresponding homologous compounds in each series as possible, since symmetry factors more frequently would contribute to deviations from simple additivity, especially with the lower members. Corrections for deviations from additivity arising from differences in symmetry numbers are not discussed by these investigators. The first substitution on a parent molecule is of necessity the replacement of a hydrogen atom by a methyl group, and is termed a primary CH_3 substitution. The compilation of thermodynamic increments corresponding to this structural modification is given in Table 18, Part II. When more than one hydrogen atom is replaced in a methyl group (branch chains), the replacement was designated as a secondary CH_3 substitution. The group increments for these are summarized in Table 19, Part II, in which the nature of the carbon atom (A) and its nearest neighbor (B) are defined as atoms of types 1, 2, 3, and 4, if the number of hydrogen atoms on each is 3, 2, 1, and 0, respectively. A carbon atom in an aromatic ring (e.g., benzene, naphthalene) is designated as a type 5 atom in this Table. The contributions for unsaturated bonds and corrections for conjugation effects, as given in Table 20, Part II, were similarly gained by comparison of the appropriate olefinic or acetylenic compound with the corresponding hydrogenated systems. The increment thus corresponds to a simple replacement of a single bond by the multiple bond. The specificity of this increment was again recognized by taking into account the nature (type number) of the two carbon atoms, (A and B) between which the bond is found. The group increments for the nonhydrocarbon group substitutions are found in Table 21, Part II.

Sufficient information to distinguish between the increments derived from precisely established thermodynamic data and approximated values is not given in the original publications (1944) of these investigators.[33] For the heats of formation, the principle sources of data were those of Bichowsky and Rossini,[37] Kharasch,[38] and the, at that time current, publications of Rossini and co-workers, with corrections calculated or approximated as necessary, to obtain values for the ideal gas at 25° C and one atmosphere pressure. The principal sources of the entropy data were from the contributions appearing in the literature based on direct calorimetric third law measurements, or statistical thermodynamic calculations. Entropies, like the heats of formation, were converted to the ideal gas state at 25° C and one atmosphere pressure, using estimates where the necessary experimental data were not known. It was judged by these investigators that the errors introduced by these approximations were less than the experimental errors in the determination of heats of formation and entropies. To calculate the heat capacity constants for the temperature dependent equation, $C_p = a + bT + cT^2$, the estima-

[37] F. R. Bichowsky and F. D. Rossini, "The Thermochemistry of the Chemical Substances," Reinhold, New York, 1936.
[38] M. S. Kharasch, *J. Research Natl. Bur. Standards* **2**, 359 (1929).

tion methods of Dobratz[39] and Stull and Mayfield,[40a] based on statistical thermodynamic concepts and generalized bond vibrational frequencies, were used to obtain the necessary values where high temperature data were not available. Group contributions were derived on the thus estimated values for homologues containing 2, 3, or 4 carbon atoms, since it was recognized that these methods were less reliable for the lowest members in any series of compounds.

A number of the values in the original tables of Andersen et al.[33] were modified subsequently by Brown[40b] using the data compilations on hydrocarbons by Rossini et al.[1] Tables 17–21, Part II in this edition have been revised to include the group increments recommended by Brown.

Franklin. The group values for the free energy function, $(G°-H_0°)$, were calculated after first subtracting an increment corresponding to the symmetry correction, $-R \ln \sigma$, from this function for the molecule in question. The steric factor term, G_{steric}, was assumed negligible in compiling these data. To account for constant deviations from simple additivity found in certain molecular forms, Franklin compiled a table of *correction* factors. Thus for each ethyl side chain, a factor of 0.8 kcal./mole was proposed to take into account the deviation from additivity of this increment over the whole temperature range from 298° K to 1500° K. Inspection of the basic equations defining the heat and free energy of formation:

$$\Delta H_f° = \Delta H_0° + (H° - H_0°) - \sum (H° - H_0°)_{elements} \qquad (5.4)$$

$$\Delta G_f° = \Delta H_0° + (G_0 - H_0°) - \sum (G° - H_0°)_{elements} \qquad (5.5)$$

where $\Delta H_0°$ is the zero point heat of reaction, shows that the method of group increments should apply to these properties since it is satisfactory for $(H° - H_0°)$ and $(G° - H_0°)$. The structural groups and numerical values of the increments of Franklin for $(H° - H_0°)$, $\Delta H_f°$, $(G° - H_0°)$ and $\Delta G°$ for the hydrocarbons only over the temperature range up to 1500° K are found in Tables 22, 23, 24, and 25, Part II, respectively. Most of the data were based on the accurate compilation of the properties of hydrocarbons by the American Petroleum Institute[41] then available.

While the basic paraffinic structural increments for normal and branch chain hydrocarbons need little comment, it is of interest to examine the extension of this method to cyclic and aromatic hydrocarbons, and conjugated olefinic systems. To eliminate the need for "corrections" for resonance energy effects on the simple structural increments derived from the paraffinic hydrocarbons devoid of resonance, separate values were calculated for characteristic resonating groups. Thus the groups:

[39] C. J. Dobratz, *Ind. Eng. Chem.* **33**, 759 (1941).

[40a] D. R. Stull and F. D. Mayfield, *Ind. Eng. Chem.* **35**, 639 (1943); D. R. Stull, *ibid.* **35**, 1301 (1943).

[40b] J. M. Brown, Univ. Wisc. Dept. Chem. Eng. Spec. Probs. Proj. (June, 1953).

[41] "Collection, Analysis, and Calculation of Data on the Properties of Hydrocarbons," Am. Petroleum Inst. Research Project 44, 1947.

5. THE METHODS OF GROUP CONTRIBUTIONS

$$\diagup\hspace{-0.5em}\diagdown CH, \quad \diagup\hspace{-0.5em}\diagdown C-, \quad \text{and} \quad \diagup\hspace{-0.5em}\diagdown C\leftrightarrow$$

designate a single CH group in a benzene ring, a carbon atom in an aromatic nucleus bonded to an alkyl or nonaromatic group, and a carbon atom at the point of juncture of the condensed aromatic nuclei. For straight chain and cyclic polyenes, additional values for the groups:

$$\leftrightarrow CH_2 \quad \text{and} \quad \leftrightarrow C\diagup\hspace{-0.5em}\diagdown^H$$

likewise have been tabulated. Data were not available to obtain the value for the [↔C<] group which is required to complete the description of resonance systems.

The basic structural groups for the olefinics proposed by Franklin are of the type:

$$\diagdown\hspace{-0.5em}_H C=CH_2, \quad \diagdown\hspace{-0.5em} C=C\diagup\hspace{-0.5em}^H_{\diagdown}, \quad \diagdown\hspace{-0.5em} C=C\diagup\hspace{-0.5em}_{\diagdown}$$

rather than the apparently more simple arrangements:

$$=CH_2, \quad =C\diagup\hspace{-0.5em}^{}_H, \quad =C\diagup\hspace{-0.5em}_{\diagdown}$$

since the former take into account the complicating factors of hyperconjugation, *cis-trans* isomerism, and thus proved to be more strictly additive. Values of the [=CH₂] increment were calculated and are tabulated for use, however, only to calculate values for the [=C<$_H$] and [=C<], if such data are required. The values for the nonresonance acetylenic groups:

$$—C\equiv \quad \text{and} \quad HC\equiv$$

appear satisfactory. Extension to acetylenic groups in resonance was not possible at that time owing to the lack of data.

The series of additive increments designated as: cyclo C₆ ring, cyclo C₅ ring, cyclo C₄ ring, and cyclo C₃ ring are correction factors, accounting for deviations from additivity when group constants calculated from paraffins are used to estimate the thermodynamic properties of cycloparaffins. The deviations may be attributed in part to the effect of ring strain, and are possibly due to the simplifying assumptions used to achieve the paraffinic group increments.

A compilation of structural groups and numerical values of the increments for nonhydrocarbons is given by Franklin as listed in Table 26, Part II. From the limited data available, it appears that the method of Franklin can readily be extended to organic compounds other than pure hydrocarbons.

As already mentioned, for the main part the hydrocarbon structural group increments were developed from the tables of data of the American Petroleum Institute at that date. The values for the $\diagup\hspace{-0.5em}\diagdown C\leftrightarrow$ group, however, were calculated from the earlier data of Parks and Huffman for naphthalene and estimated heat capacities.[39] Similarly the increments for the C_4 ring were based on a graphical interpolation based on the values C_3, C_5, and C_6. The increments for the nonhydrocarbon groups were estimated from the early data of Parks and Huffman.[4] International Critical Tables,[3] and a compilation by Aston.[42] Accordingly the values for these increments must be recognized as less well established.

Souders, Matthews, and Hurd. For any polyatomic molecule in the ideal gas phase at one atmosphere pressure, the heat capacity may be expressed by the equation:

$$C^\circ_{p\text{ (comp.)}} = \Sigma\, C^\circ_{p\text{ vib.}} + \Sigma\, C^\circ_{\text{i.r.}} + A \qquad (5.6)$$

where the constant A is $3.5\,R$ and $4\,R$ for the translational and external rotational contribution of a linear and a nonlinear system, respectively. The summations cover all the vibrational group contributions and the internal rotational contributions for the molecule under consideration. Souders, Matthews, and Hurd calculated values of the structural group increments for the vibrational and internal rotational contributions separately, all data being based on the values computed by statistical thermodynamic methods for the simpler polyatomic hydrocarbons. The values of the heat content contributions were obtained by graphical integration of the heat capacity over the temperature range. As mentioned earlier, the values of the structural group increments were calculated in units of B.T.U. per lb. mole up to 3000° F. These data are found in Tables 27, 28, 29, and 30, Part II. To preserve precision in the calculations, more significant figures were carried than are warranted by the absolute accuracy of the values.

The number of structural groups is relatively small in comparison to those of the preceding method where a number of "corrections" were tabulated from resonance effects. Such effects, in the present work, have been included in the values for the internal rotational group increments. Thus, although the internal rotational effect is zero for a triple bond, the finite value tabulated by these investigators is the "correction" due to the conjugation effect for such groups.

Two types of groups were considered in constructing the entropy and heat of formation correlations. The *type I* groups, $\Delta S^\circ_{f\,298}$, $\Delta H^\circ_{f\,298}$, were dependent only on the structural constants of the compounds in the standard state (298.1° K). The *type II* groups, designated as $[\Delta S^\circ_{fT} - \Delta S^\circ_{f\,298}]$ and $[\Delta H^\circ_{fT} - \Delta H^\circ_{f\,298}]$, and summed with the contribution of the type I group increments, extended the standard state values to higher temperatures.

The values of the entropy of formation $(\Delta S^\circ_{f\,298})$ in the standard state were calculated from the equation:

$$\Delta S^\circ_{f\,298} = S^\circ_{298} - 1.36\, m - 15.615\, n \qquad (5.7)$$

[42] J. G. Aston, "Thermodynamic Data on Hydrocarbons," Standard Oil Development Co., 1944.

in which $S°$ is the entropy of the compound, and m and n are the number of carbon and hydrogen atoms, respectively, in the molecule. The values of the constants in (5.7) are the entropy of solid graphite and one half the virtual entropy of hydrogen at 298.16° K. The entropies of the compounds used to construct the structural type I groups were calculated when reliable experimental data were lacking, by statistical thermodynamics and included the hindered rotational contributions. The structural groups developed for the heat content and heat capacity correlations were found satisfactory for estimates of the standard entropy of formation, if conjugation and symmetry effects were taken into account. The numerical values for the type I group contributions for the entropy of formation and heat of formation, and the conjugation, adjacency, and symmetry corrections, compiled by these investigators are given in Tables 31 and 32, Part II, respectively. It should be noted that the units for the entropy and heat of formation are cal./deg.mole and kcal./mole, respectively, in contrast to the preceding tables where the B.T.U.-Fahrenheit scales were used.

The symmetry contributions $(-R \ln \sigma)$ are those from the theoretical development of entropy. Since the CH_3 group increment already includes the effect of the internal symmetry of this group ($\sigma = 3$), the symmetry number used in these correlations is that of the molecule treated as a rigid rotator with no free internal rotation. Thus the symmetry numbers of ethane and the branch paraffins, 2-methylpropane, and 2-methylbutane, here would be 2, 3, and 1, respectively. In cases like 3-methylhexane where there is optical isomerism an additional correction to the entropy equal to $+ R \ln 2$ is necessary to take into account the increased number of orientations now possible within this system. Symmetry does not contribute correspondingly to the heat of formation. The increments of conjugation and adjacency in Table 32, Part II, were proposed to enable more accurate estimates to be achieved for the entropy and heats of formation using the group increments based on the more simple paraffinic systems, i.e., without conjugation or adjacency of certain groups. The number of adjacency group contributions is greater for the heat of formation than for the entropy of formation.

The temperature dependent *type II* groups can be defined in terms of heat capacities by the equations:

$$[\Delta S°_{fT} - \Delta S°_{f298}] = \int_{298}^{T} \frac{C°_{p(\text{comp.})}}{T} dT - \frac{n}{2} \int_{298}^{T} \frac{C°_{p(H_2)}}{T} dT - m \int_{298}^{T} \frac{C°_{p(C)}}{T} dT \quad (5.8)$$

and

$$[\Delta H°_{fT} - \Delta H°_{f298}] = \int_{289}^{T} C°_{p(\text{comp.})} dT - \frac{n}{2} \int_{298}^{T} C°_{p(H_2)} dT - m \int_{298}^{T} C°_{p(C)} dT \quad (5.9)$$

using the heat capacities of the compound, hydrogen, and graphite at zero pressure and where m and n are the number of carbon and hydrogen atoms. From the definition of the heat capacity of the compound as the sum of translational, rotational, vibrational, and internal rotational contributions, as expressed in (5.6), and the above two equations, it can be readily shown that for *each* vibrational group and *each* internal rotational degree of freedom, the following equations apply:

$$(\Delta S^\circ_{fT} - \Delta S^\circ_{f298})_{\text{vib.}} = \int_{298}^{T} \frac{C^\circ_{p(\text{vib.})}}{T} dT - \frac{n'}{2} \int_{298}^{T} \frac{C^\circ_{p(\text{H}_2)}}{T} dT - m' \int_{298}^{T} \frac{C^\circ_{p(\text{C})}}{T} dT \quad (5.10)$$

$$(\Delta H^\circ_{fT} - \Delta H^\circ_{f298})_{\text{vib.}} = \int_{298}^{T} C^\circ_{p(\text{vib.})} dT - \frac{n'}{2} \int_{298}^{T} C^\circ_{p(\text{H}_2)} dT - m' \int_{298}^{T} C^\circ_{p(\text{C})} dT \quad (5.11)$$

$$(\Delta S^\circ_{fT} - \Delta S^\circ_{f298})_{\text{i.r.}} = \int_{298}^{T} \frac{C^\circ_{p(\text{i.r.})}}{T} dT \quad (5.12)$$

$$(\Delta H^\circ_{fT} - \Delta H^\circ_{f298})_{\text{i.r.}} = \int_{298}^{T} C^\circ_{p(\text{i.r.})} dT \quad (5.13)$$

in which n' and m' are the number of hydrogens and carbons in the vibrational group. For each group Souders, Matthews, and Hurd evaluated the vibrational and internal rotational contributions by graphical methods using the heat capacity data already calculated. The numerical values for the *type II* structural group increments compiled by these investigators in this manner are given in Tables 33–36, Part II.

The method of *Van Krevelen and Chermin* correlates the structural group contributions with the free energy of formation as a linear function of temperature, i.e.,

$$\Delta G^\circ_{f(\text{group})} = A + B \times 10^{-2} T \quad (5.14)$$

where A and B are constants characteristic for each group increment. The group contributions were developed from a similar temperature dependent equation for the free energy of formation of reference compounds, assuming in each case, after Franklin, that the latter may be expressed by:

$$\Delta G^\circ_{f(\text{comp.})} = \sum \text{contribution of composing groups} + \sum \text{corrections for deviations} + RT \ln \sigma \quad (5.15)$$

where the last term is the symmetry correction. The structural groups and correction factors and the numerical values of the corresponding constants A and B are given in Tables 37–40, Part II. Two sets of values are given for these parameters, for the temperature ranges 300–600° K, and 600–1500° K, since it was not possible to cover the entire range with one linear temperature function.

The free energy of formation for the reference compound was based on the statistical thermodynamic functions in most cases. The group contributions for the hydrocarbon structural increments and in addition for some of the cyclic paraffins and aromatics, were based on the publications of the

American Petroleum Research Institute Project 44,[41] or recent literature values. As a starting point the [CH$_2$] increment from the A.P.I. collection of data was selected. From this, and reference to the additive principle expressed in (5.15), the values for the alkane, alkene, and alkyne group contributions were developed. Like Franklin, these investigators found it impossible to calculate contributions for the simple =CH$_2$ group having sufficient constancy, and accordingly, developed correlations for the more complex increments (e.g., [H$_2$C=C$\angle$H], [>C=C<], . . .) which seemed more well behaved. The alkyne groups were calculated with the initial assumption that the [HC≡] contribution was equal to half the value of $\Delta G_f°$ for acetylene after subtraction of the symmetry number contribution. Structural increments for characteristic resonance groups in alkenes, e.g.,

[↔CH$_2$], and aromatic rings, e.g., [—C$\overset{\nearrow}{\underset{\searrow}{}}$], were developed in this method in preference to the earlier correlations[34,35] correcting separately for conjugation effects. A series of group increment corrections and branch chain factors were likewise developed to account for the influence of these effects in the use of the basic structural increments obtained above. The numerical values for ring formation for the C$_4$ system were gained by interpolation, and that for the oxygen heterocyclics, from the values for ethylene oxide and tetrahydrofuran.

Van Krevelen and Chermin report increments for some twenty-two non-hydrocarbon structural groups containing oxygen, nitrogen, and sulphur (Table 39, Part II). The compilation is more comprehensive than in the preceding two methods.[33,34] A point differing from the previous investigators is with reference to the [OH] group contributions. The practice in the other methods is to use several values, the exact value to be used depending on whether the hydroxyl group appeared as a primary, secondary, tertiary, or aromatic substituent. The present investigators list only one [OH] increment, the positional effect having already been taken into account by the more specific nature of the groups for the hydrocarbons. While the scope of the nonhydrocarbon groups is larger than embraced by the previous methods, in many instances the group contributions were developed from very little data, owing to the limited information available. To illustrate, for example, the following group increments were based on the data for only one reference compound: [—CHO] from acetaldehyde, [—CO] from acetone, [—COOH] from acetic acid, [O$\overset{\nearrow}{\underset{\searrow}{}}$] from furane, [N$\overset{\nearrow}{\underset{\searrow}{}}$] from pyridine, [—S—] from dimethyl sulfide, [—F] from trifluoroethane. Until more data are available, the values for structural increments for the nonhydrocarbon groups in this Table, like those of the preceding methods, should be regarded, at best, as first approximations not capable of leading to the precision possible within the family of pure hydrocarbons.

The data in Table 40, Part II, compiled by Van Krevelen and Chermin, extend the scope of this method to the use of simple organic compounds that cannot be treated by the preceding tables of group contributions (e.g., C_2H_6, $(CN)_2$, CH_3OH, ...), and some inorganic substances (e.g., NO_2, SO_2, CO_2, ...) in important gas phase reactions in the organic field.

4. Calculation of Thermodynamic Properties

While the general procedure for the calculation of thermodynamic properties by the above group methods is self-evident from the basic principle underlying this approach, the methods differ in the details of procedure. To achieve the best results, the following suggestions have been made by the various investigators to guide the estimation in practice of thermodynamic properties for a complex molecule. All tables of numerical data and group contributions referred to in this section are found in Part II.

Following the procedure recommended by *Andersen, and his associates*,[33] a *parent* group is selected from Table 17, from which the desired complex molecule is to be built by appropriate group modifications. If a choice of parent groups is open, that group with the largest entropy contribution is selected. The "build-up" to the final structure should be planned to minimize the number of steps (i.e., group substitutions). From the parent group, the carbon skeleton of the final structure is next developed by use of the data in Tables 18 and 19. The replacement of the first hydrogen on the parent group by a CH_3 is designated as a primary methyl substitution. The replacement of a second hydrogen on any carbon atom is a secondary methyl substitution. In building the carbon skeleton, if branching is involved, the longest side chain is developed first. The secondary methyl substitutions, as discussed in the preceding section on this correlation, must be specified according to the type of carbon to which the substituent is added (atom A), and the adjacent carbon atom (atom B). Numerical designations, 1, 2, 3, 4, and 5 corresponding to the skeletal carbons found as CH_3, CH_2, CH, C, and C in benzene, respectively, are used to classify the carbon atoms as *types*. Where a choice of B carbon atoms is possible, the increment corresponding to a substitution where B has the highest type number should be selected. If the final structure contains atoms other than carbon and hydrogen, e.g., O, N, S, . . . , secondary methyl groups are substituted in those positions on the carbon skeleton which are to be occupied by the nonhydrocarbon groups. Next the contributions from Table 20 are used as required for multiple bonds on the final structure. The last step is the replacement of the CH_3 groups by the nonhydrocarbon groups as listed in Table 21, to complete the structure of the final molecule if it is a nonhydrocarbon.

Summation of the numerical values corresponding to these group increments gives the desired thermodynamic data. Extension of the scope of this method can readily be achieved by additional small parent groups, e.g.,

5. THE METHODS OF GROUP CONTRIBUTIONS

hydrazine, pyridine, ..., as precise thermodynamic data for such become available.

Andersen, Beyer, and Watson, suggest that for the smaller polyatomic systems, the methods based on statistical thermodynamics and generalized bond frequencies be employed to achieve best estimates. The method of group contributions is recommended to extend the results of such calculations to larger molecules for which the direct calculations are uncertain and time consuming. It is estimated that the limits of error are ± 4 kcal./mole and ± 2 e.u. for heats of formation and entropies, respectively, calculated from these tables as a guide, and best values for the increments. The data on heat capacities are more uncertain, the estimated results deviating from better experimental data by ± 5 per cent at moderate temperatures. The tables of increments, however, are specific, and make it possible to take into account the differences between isomers.

Example 5.1. Estimate $\Delta H°_{f\,298(g)}$ and $S°_{298(g)}$ for the isomers, 1-pentyne, 2-pentyne, and 3-methyl-1-butyne.

(a) 1-Pentyne, $CH{\equiv}C{-}CH_2{-}CH_2{-}CH_3$

			$\Delta H°_{f\,298}$ (kcal./mole)	$S°_{298}$ (e.u.)
Parent group (CH$_4$, Table 17)			− 17.9	44.5
Primary CH$_3$ (Table 18)			− 2.2	10.4
Secondary CH$_2$ groups (Table 19)				
	A	B		
	1	1	− 4.5	9.8
	1	2	− 5.2	9.2
	1	2	− 5.2	9.2
Multiple bond				
	1	2	69.1	− 7.8
			+ 34.1	75.3

(b) The predicted values for the other two isomers, by a similar procedure, using Tables 17–19, Part II, are:

	$\Delta H°_{f\,298}(g)$ (kcal./mole)	$S°_{298}(g)$ (e.u.)
2-Pentyne, $CH_3{-}C{\equiv}C{-}CH_2{-}CH_3$	30.1	76.8
3-Methyl-1-butyne, $CH{\equiv}C{-}CH({-}CH_3){-}CH_3$	32.7	71.9

The results for the three isomers illustrate that the correlations of structural modifications are quite specific and make possible a distinction between isomeric structures. The thermodynamic properties for these three pentynes

70   I. METHODS OF ESTIMATION

were reported by Rossini and associates after the group correlations had been developed. Comparison of these data with the estimated values, are shown below:

	$\Delta H^\circ_{298(g)}$ (kcal./mole)		$S^\circ_{298(g)}$ (e.u.)	
	Literature[43]	Estimated	Literature[43]	Estimated
1-Pentyne	34.50	34.1	79.10	75.3
2-Pentyne	30.80	30.1	79.30	76.8
3-Methyl-1-butyne	32.60	32.7	76.23	71.9

The estimated heats of formation are in good accord with the literature values, the agreement falling in the range of 0.3 to 2 per cent. The difference between the estimated values and literature data for the entropies is somewhat larger, the agreement being in the range of 3 to 6 per cent. The estimated values are consistently lower than the literature values in this case.

Example 5.2. Estimate the value of $S^\circ_{298(g)}$ for 2,2,3,3-tetramethylbutane. (ANSWER: 93.4 and 94.7 e.u., lit. value,[2] 93.06 e.u.).

It is to noted that two values for $S^\circ_{298(g)}$ of 2,2,3,3-tetramethylbutane, i.e., 93.4 and 94.7 e.u., are obtained, depending on the order in which the secondary CH_3 group increments are taken. In the present instance, the order, 1-1, 1-2, 2-2, 2-3, 3-3, and 3-4, where the numbers refer to the type numbers of carbon atoms A and B, respectively, in each case, gives the value in closest agreement with the literature value.

Example 5.3. Estimate the entropy of 1,2-dibromoethane at 298° K and one atmosphere pressure, in the ideal gas state. (ANSWER: 79.9 e.u., lit. value,[33] 78.7 e.u.)

The principle of *Franklin's*[34] method can be summarized by the equation:

$$G(X) = \sum \begin{array}{c}\text{contributions of structural}\\ \text{groups of complex molecule}\end{array} + RT \ln \sigma + \sum \begin{array}{c}\text{correction}\\ \text{factors}\end{array} \qquad (5.15a)$$

where G can be $(G^\circ - H_0^\circ)$, $(H^\circ - H_0^\circ)$, ΔG_f°, and ΔH_f° for the complex molecule X. The symmetry number enters only in the estimation of the free energy properties. Selection of the increments appropriate for the final structure and summation of the corresponding numerical values from Tables 22–26, leads directly to an estimate of the value for the property desired. For complex hydrocarbon molecules, it is judged that in most cases the confidence to within one kcal. can be placed in the results thus calculated. For other types of compounds the results must be regarded as much more approximate owing to the tentative nature of the values for the nonhydrocarbon group increments.

[43] D. D. Wagman, J. E. Kilpatrick, K. S. Pitzer, and F. D. Rossini, *J. Research Natl. Bur. Standards* **35**, 67 (1945).

5. THE METHODS OF GROUP CONTRIBUTIONS

Example 5.4. Estimate the values for $\Delta G_f°$ and $\Delta H_f°$ for 1-methyl-3-ethylbenzene at 298.1 and 1000° K.

(a) 1-Methyl-3-ethylbenzene ($\sigma = 2$); (see Tables 23, 25, Part II).

	$\Delta H_f°$ (kcal./mole)		$\Delta G_f°$ (kcal./mole)	
	298°	1000°	298°	1000°
4[⇗CH⇘]	13.20	9.68	19.36	38.24
2[⇗C—⇘]	11.14	11.70	17.52	29.52
2 [CH₃]	− 20.24	− 25.28	− 8.28	24.76
1 [CH₂]	− 4.93	− 5.94	2.05	19.58
Correction factor	0.6	0.6	−0.7	− 0.7
$RT \ln 2$	—	—	0.42	1.38
	− 0.23	− 9.24	30.37	112.78
Lit. values[44]	− 0.460	− 8.94	30.217	111.14

Comparison shows that estimated results are in excellent agreement with the literature values.

Example 5.5. Estimate the values of $\Delta G_f°$ and $\Delta H_f°$ at 298.1° K for 3-methyl-1-butyne.

(a) 3-Methyl-1-butyne ($\sigma = 2$); (see Tables 23, 25, Part II).

	$\Delta H_{298}°$ (kcal./mole)	$\Delta G_{298}°$ (kcal./mole)
1 [CH≡]	27.10	24.8
1 [—C≡]	27.34	25.65
1 [CH]	− 1.09	7.46
2 [CH₃]	− 20.24	− 8.28
$RT \ln 2$	—	0.42
Adjacent C and CH groups	2.5	3.6
	34.61	53.45
Lit. values[43]	32.60	49.12

The estimated values for both $\Delta H_f°$ and $\Delta G_f°$ are about 6 per cent higher than the literature values.

Example 5.6. Calculate $\Delta G_f°$ at 1000° K and one atmosphere pressure for (a) 1,3,5-trimethyl benzene and (b) 2,2,3,3-tetramethylbutane. (ANSWER: (a) 113.68; lit. value[1] 113.05; (b) 154.06; lit. value 155.47 kcal./mole). Comparison of the predicted values in this case with those reported by Rossini and co-workers[1] shows that the results are in very close accord.

[44] W. J. Taylor, D. D. Wagman, M. G. Williams, K. S. Pitzer, and F. D. Rossini, *J. Research Natl. Bur. Standards* **37**, 95 (1946).

In the method of *Souders, Matthews, and Hurd*,[35] selection of the appropriate group contributions is also guided by inspection of the structural formula. In this method, the vibrational group contributions to heat capacity and heat content are separately evaluated. Summing these together with the translational plus external rotational contribution for the whole molecule gives the desired data, as expressed by the equations:

$$(C_p^\circ)_X = \Sigma C_{p(vib.)}^\circ + \Sigma C_{p(i.r.)}^\circ + A \qquad (5.6)$$

and

$$(H^\circ - H_0^\circ)_X = \Sigma (H^\circ - H_0^\circ)_{vib.} + \Sigma (H^\circ - H_0^\circ)_{i.r.} + AT \qquad (5.16)$$

where the constant A is $3.5\ R$ and $4\ R$ for linear and nonlinear polyatomic molecules, as indicated in the earlier discussion, and X is the complex molecule for which the property is desired. The numerical values for the vibrational contributions to the heat content and heat capacity are found in Tables 27, 28, and 29, 30 respectively. It is suggested that a simple schematic drawing of the molecule under consideration is a helpful guide to selecting the most appropriate internal rotational contribution from the various choices possible in the Tables. One internal rotational contribution is made by each C—C single bond unless there is no lateral extension at either end of the bond, i.e., the —C≡CH group would give no i.r. contribution.

For the calculation of the entropy of formation and heat of formation by this method, two types of group increments have been compiled as already explained, type I for calculation of $S_{f\,298}^\circ$ and $\Delta H_{f\,298}^\circ$, and type II for extension of the estimates at 298° by means of the increments, $[S_{fT}^\circ - S_{f\,298}^\circ]$ and $[\Delta H_{fT}^\circ - \Delta H_{f\,298}^\circ]$ to higher temperatures.

The numerical values for the group contributions to ΔS_f° and ΔH_f° at 298.16° K are given in Tables 31, and 32. The desired property in each case is obtained simply by summation of the contributions of the structural groups, conjugations, adjacencies, symmetry, and optical isomerism. The symmetry correction and optical isomerism do not contribute to the heat of formation. It should be noted that two HC=groups are listed in Table 32. Use of [HC=]$_{trans}$ is advised only when the structure:

$$\begin{array}{c}R\\ \diagdown\\ H\end{array}C{=}C\begin{array}{c}H\\ \diagup\\ R\end{array}$$

in which R is an alkyl group, is concerned. It follows from Eq. 5.8, 5.9, and 5.6, that the type II group contribution for any polyatomic molecule is given by the expressions:

$$[\Delta S_{fT}^\circ - \Delta S_{f\,298}^\circ] = \Sigma [\Delta S_{fT}^\circ - \Delta S_{f\,298}^\circ]_{vib.} +$$

and

$$\Sigma [\Delta S_{fT}^\circ - \Delta S_{f\,298}^\circ]_{i.r.} + \int_{298}^{T} \frac{A}{T} dT \qquad (5.17)$$

$$[\Delta H_{fT}^\circ - \Delta H_{f\,298}^\circ] = \Sigma [\Delta H_{fT}^\circ - \Delta H_{f\,298}^\circ]_{vib.} +$$

$$\Sigma [\Delta H_{fT}^\circ - \Delta H_{f\,298}^\circ]_{i.r.} + \int_{298}^{T} A\ dT \qquad (5.18)$$

5. THE METHODS OF GROUP CONTRIBUTIONS

in which the summations are over all the vibrational and internal rotational group contributions, and A has the significance as in (5.6). Algebraic addition of the values gained for $\Delta S_f°$ and $\Delta H_f°$ at 298.16° K to the above equations give an estimate of the entropy of formation and heat of formation at higher temperatures. The numerical contributions for the type II group increments are given in Tables 33, 34, 35, and 36. The same considerations in the selections of the proper internal rotational contribution as discussed for estimates of the heat content and heat capacity apply here.

The method gives quite precise results. Standard entropies of formation and heats of formation predicted by this method are believed to be within ± 1.0 e.u. and ± 0.5 kcal. per mole of the correct value regardless of the complexity of the hydrocarbon molecule. The effect of temperature on each of these properties predicted by this method has an uncertainty less than 3 per cent. A rather detailed picture of the molecular structure is required in the use of this method. Its scope is limited to hydrocarbons only since increments for structural groups other than carbon-hydrogen were not developed.

Example 5.7. Estimate the $\Delta S_f°$ and $\Delta H_f°$ for 1,3,5-trimethylbenzene at 298° and 900° K.

(a) 1,3,5-Trimethylbenzene at 298° K ($\sigma = 6$); (see Tables 31 and 32, Part II).

	$\Delta H_{f298}°$ (kcal./mole)	$\Delta S_{f298}°$ (e.u.)
3 [—CH$_3$]	− 30.15	− 58.98
3 [—Ċ=]arom.	16.44	− 28.47
3 [HĊ=]arom.	9.99	− 16.29
—R ln 6	—	− 3.56
	− 3.72	− 107.30
Lit. values[1]	− 3.84	− 107.47

(b) At 900° K (Tables 33–36, Part II).

	$(H_{f900}° - H_{298}°)$ (kcal./mole)	$(S_{f900}° - S_{f298}°)$ (e.u.)
3 [—CH$_3$]	− 15.834	− 30.531
3 [HĊ=]arom.	− 4.683	− 9.141
3 [—Ċ=]arom.	5.274	9.597
3 V$_a$ i.r.	2.814	5.406
4 R (900°−298°)	4.782	—
4 R ln (900°/298°)	—	8.778
	− 7.647	− 15.891

The entropy of formation and heat of formation at 900° K, are therefore:

$\Delta S_{f900}° = -15.891 - 107.30 = -123.19$ e.u. (cf. lit.,[1] − 125.65)

$\Delta H_{f900}° = -7.647 - 3.72 = -11.38$ kcal./mole. (cf. lit.,[1] − 12.643).

Example 5.8. Calculate the entropy of formation (ΔS_f°) and heat of formation (ΔH_f°) for 2,2,3-trimethylbutane at 298.1° K and one atmosphere pressure in the ideal gas state. (ANSWER: −167.84 e.u.; −48.58 kcal./mole; lit. value,[1] −167.76 e.u.; −48.96 kcal./mole).

Comparison of the values estimated in the preceding examples with those reported by Rossini and co-workers[1] shows that the results are in excellent agreement.

The basic equation:

$$\Delta G_{f\text{(comp.)}}^\circ = \Sigma \text{ contributions of groups} + \Sigma \text{ corrections} + RT \ln \sigma \quad (5.15)$$

is used in the method of *Van Krevelen and Chermin*[36] to calculate a temperature dependent equation for the free energy of formation of the type:

$$\Delta G_{f\text{comp.}}^\circ = A + B \times 10^{-2} T \quad (5.19)$$

from an inspection of the structural formula for the compound and a summation of the contributions corresponding to the appropriate groups. The numerical values for the structural groups are found in Tables 37, 38, and 39. As already mentioned, the data in Table 40 extended the scope of the method to some molecules whose properties cannot be calculated by the preceding tables. The method presents a more direct approach for the estimation of ΔG_f° as a function of temperature than the preceding methods. An agreement of $\pm$ 0.6 kcal. per mole for ΔG_f° between estimated and observed values for a number of compounds was reported by these investigators. For complex molecules closely related structurally to the reference compounds used to develop the increments, a mean deviation of the estimated value of this magnitude can be expected.

Example 5.9. Estimate the free energy of formation for 3-methyl-1-butyne at 1000° K.

(a) 3-Methyl-1-butyne ($\sigma = 2$); (see Tables 37 and 38, Part II).

	ΔG_f° (600° – 1500° K) (kcal./mole)
2 [CH$_3$]	− 24.620 + 4.872 × 10^{-2} T
1 [⟩CH—]	− 0.705 + 2.910 × 10^{-2} T
1 [—C≡]	26.555 − 0.550 × 10^{-2} T
1 [CH≡]	26.700 − 0.704 × 10^{-2} T
RT ln 2	+ 0.138 × 10^{-2} T
ΔG_{fT}° =	27.930 + 6.666 × 10^{-2} T
ΔG_{f1000}° =	94.59
Lit. value[43] =	92.93

Example 5.10. Estimate the value of ΔG_f° for 1-*trans*-4-dimethylcyclohexane at 500° K and one atmosphere pressure in the ideal gas state. (ANSWER: 44.3 kcal/mole; lit. value[1] 44.35).

Example 5.11. Estimate the value of $\Delta G_f°$ for vinyl chloride at 600° K and one atmosphere pressure in the ideal gas state (ANSWER: 10.5 kcal./mole; lit. value,[45] 9.4).

From the agreement of the values calculated in the preceding examples with the literature values, the use of this method as a direct and relatively simple procedure seems well justified for estimates of the free energy of formation.

5. Nonhydrocarbon Groups

Because of the extensive thermodynamic information already available, reliable computation of the thermodynamic functions by incremental methods for hydrocarbons is possible. It should be recognized that the group increments for the nonhydrocarbon groups have been compiled on the basis of a much smaller amount of existing data; computations for nonhydrocarbons based on these increments (Part II, Tables 21, 26, and 39) are at best only first approximations. The increments do represent the general nature of the change in the thermodynamic values, but the precise values are more subject to change than the corresponding data for the hydrocarbon group. An illustration of this is found in the thermodynamic values for the [SH] group increment. As part of the program of work initiated by Huffman and at present in progress to meet the need for experimental thermodynamic data for the first members in the homologous series of nonhydrocarbons, the thermodynamic properties of ethanethiol and pentanethiol have been recently published.[32,46] Comparison of the accurate experimental entropies for these thiols and the corresponding hydrocarbons, e.g.,

	$S°_{298 \cdot 1}(g)$ (e.u.)		$S°_{298 \cdot 1}(g)$ (e.u.)
$CH_3-CH_2-CH_3$	64.51	$CH_3(CH_2)_4CH_3$	92.93
CH_3-CH_2-SH	77.07	$CH_3(CH_2)_4SH$	99.18
[[SH] for [CH$_3$]]	6.26		6.25

shows that the structural modification corresponding to the replacement of the [CH$_3$] group by an [SH] group is accompanied by an increase of 6.25 e.u. in the polyatomic system. This is to be compared with the value of 5.2 (Table 21, Part II) which was proposed earlier on the basis of the existing data prior to the work above.

No attempt to estimate the probable errors or to revise the values of the tables of the nonhydrocarbon increments (see above, Part II) has been made. This is best accomplished in practice for each system, by reference to the most recent data to gain the estimate of the limits of accuracy in the final values.

[45] F. W. Giauque and J. D. Kemp, *J. Chem. Phys.* **6**, 40 (1938).
[46] J. P. McCullough, D. W. Scott, H. L. Finke, M. E. Gross, K. D. Williamson, R. E. Pennington, G. Waddington, and H. M. Huffman, *J. Am. Chem. Soc.* **74**, 2801 (1952).

Addendum

Benson and Buss. Benson and Buss[46a] have reported a general limiting law for construction of the additivity rules used for the estimation of molecular properties. The generalized law is as follows.

If one considers the following two disproportionation reactions:

$$RNR + SNS \rightleftharpoons 2\ SNR \tag{5.20}$$

$$RNN'R + SNN'S \rightleftharpoons RNN'S + SNN'R \tag{5.21}$$

where R and S are groups (or atoms) bonded to common molecular frameworks (N, and NN', the latter being a designation for an unsymmetrical framework), then, for molecular properties such as C_p, S, and H, the limiting law states that $\Delta C_p \rightarrow 0$, $\Delta H \rightarrow 0$, and $\Delta S \rightarrow \Delta S_\sigma = R \ln K_\sigma$ for either of these two reactions as the separation of R and S becomes large. Here the symmetry number, σ, is the total symmetry number [cf. Eq. 6.2].

This principle has its analogy in solution chemistry where the contributions to various physicochemical properties by several solute species become additive as the concentration approaches zero (i.e., at infinite dilution). In applying this approach to molecular systems, Benson and Buss pointed out that specific atoms or groups of a molecule may be looked upon as "solute" species whose interactions grow negligibly small as their separation within the molecule increases. In the limiting case these should contribute additively to many of the molecular properties. The zero order, first order, and second order approximations of this general law of additivity were shown by Benson and Buss to reduce to three cases, respectively: (i) additivity of atomic properties, (ii) additivity of bond properties, and (iii) additivity of group properties.

(i) *Atomic Properties.* For the case when the common molecular framework, N or NN', vanishes the disproportionation reaction becomes:

$$R_2 + S_2 \rightarrow 2RS \tag{5.22}$$

and this limiting law is shown to reduce to the principle of the additivity of atomic properties. This is the "zero order" approximation of the general limiting law. The partial atomic contributions for the estimation of $C_p°$ and $S°$, as reported by Benson and Buss, are in Table 43, Part II.

These atomic contributions may be used to estimate $C_p°$ and $S°$ for relatively simple molecular systems such as CH_4, CH_2Cl_2, CO_2, NO_2 as well as relatively larger species, e.g., hexamethylethane, acetone, butene-2, and n-octane. Only the number of each type of atom present in the molecule is important, i.e., the atomic additivity increments in Table 43 are independent of the internal bonding of the constituent atoms.

[46a] S. W. Benson and J. H. Buss, *J. Chem. Phys.* **29**, 546 (1958).

Example 5.12. Estimate the values of $C_p°$ and $S°$ for acetone using the Benson and Buss rule of additivity of atomic properties:

(a) $C_p° = 2(3.75) + 6(0.85) + 3.40 = 16.0$ (cf lit., 15.7)

(b) $S° = 2(-32.6) + 6(2.10) + 8.8 - R \ln 18 = 63.9$ (cf lit., 63.7).

Example 5.13. By the rule of additivity of atomic properties, predict the values of $S°$ for (i) formic acid, (ii) formaldehyde, (iii) cyanogen chloride, (iv) dimethylamine, (v) dibromosilane. (ANSWER: (i) 62.8; lit., 60.1. (ii) 52.6; lit., 52.3. (iii) 56.6; lit., 56.3. (iv) 64.0; lit., 65.3. (v) 73.9; lit., 73.9 e.u.)

Benson and Buss indicate that the values for $C_p°$ and $S°$ thus estimated are accurate to ± 2 cal./mole °K for most species. The estimates may be poorer for very simple H containing molecules, such as NH_3 and CH_4, and also for heavily substituted molecules, such as neopentane. The table of partial atomic contributions should not be used for cyclic molecules, such as benzene.

(ii) *Bond Properties.* The first order approximation of the general limiting law is the case where the molecular framework, N, reduces to a single atom (e.g., O, S) or a partially substituted group (e.g., CH_2, NH, C=O).

It was shown by these investigators that the general limiting law then reduces to the principle of additivity of bond properties. The contributions developed by Benson and Buss for the first order approximation are in Table 44, Part II, for $C_p°$, $S°$, and $\Delta H_f°$, respectively. As in the previous case ring structures, such as benzene and alicyclic rings, are irreducible units and the bond contributions (Table 44, Part II) are not to be used for such systems. It should also be noted that the vinyl group is to be treated as a quadrivalent unit while the carbonyl group corresponds to a divalent unit in the application of the first order approximation of the general limiting law of additivity. The bond increments for such systems as the aromatics, alicyclic rings, acetylenes, hydrazines, and nitrates were not included in Table 44 by Benson and Buss owing to the scarcity of thermodynamic data. As will be seen in the second order approximation, these omitted systems fit readily in the principle of group additivities.

Example 5.14. Calculate the values of $C_p°$, $S°$, and $\Delta H_f°$ for ethylbenzene from the partial bond contributions of Benson and Buss (Table 44, Part II).

(a) $C_p° = 5(3.0) + 4.5 + 1.98 + 5(1.74) = 30.2$ (cf lit., 29.7)

(b) $S° = 5(11.7) - 17.4 - 16.4 + 5(12.90) - R \ln 6 = 85.6$ (cf lit., 85.0)

(c) $H_f° = 5(3.25) + 7.25 + 2.73 - 5(3.83) = 7.08$ (cf lit., 7.1)

Example 5.15. Predict the values for $\Delta H_f°$ by the rule of additivity of bond properties for the following: (i) m-xylene, (ii) formaldehyde, (iii) ethylacetate, (iv) dimethylamine, (v) dichloromethane. (ANSWER: −4.5; lit., −4.9. (ii) −27.8; lit., −27.8. (iii) −105.8; lit., −106.6. (iv) −7.0; lit., −6.6. (v) −22.5; lit., −21.5 kcal./mole.)

The estimates of $C_p°$ and $S°$ are generally accurate to about ±1 cal./mole °K, while the values of $\Delta H_f°$, about ±2 kcal./mole; for heavily branched molecules the limits of error may be larger. In terms of structure sensitivity, $\Delta H_f°$ is found to be by far the most sensitive of the thermodynamic properties. Gross deviations from this rule for the $\Delta H_f°$ for peroxides and NO_2 compounds were noted.

(iii) *Group Properties.* When the common molecular framework (N, or NN') is increased to embrace two atoms or structural elements which impose a separation of about 3–5 Å between the attached groups (i.e., such elements would be benzene, cyclohexane or other ring systems, or conjugated systems of unsaturated bonds), the second order approximation of the general limiting law applies. The structural changes are in adjacent atoms rather than in the same atom, and the disproportionation leaves unchanged the nearest neighbors of the atom or group being switched. It was shown that in this case the generalized principle reduces to the rule of additivity of group properties. Ring systems and unsaturated centers again must be treated as irreducible structural entities. The partial group contributions of Benson and Buss are given in Tables 45–47, Part II, respectively, for hydrocarbons, halogen compounds, and a variety of others grouped as miscellaneous.

In this method, a group is defined as a polyvalent atom held together with its ligands. At least one of these ligands must itself be a polyvalent species or the "group" cannot engage in the disproportionations of the type given by Eqs. (5.20) and (5.21). The method of "Group Equations" (Chapter 6) can be shown to be equivalent to this rule of group additivity.

Example 5.16. Estimate the entropy for 3-methyl pentene-2 using the Benson and Buss partial group contributions.

The contributing groups in CH_3—CH=$C(CH_3)CH_2CH_3$ are:

$2[\mathbf{C}$—$(C_d)(H)_3] + [\mathbf{C}_d$—$(C_d)(H)] + [\mathbf{C}_d$—$(C)_2] + [\mathbf{C}$—$(C)(H)_3] + [\mathbf{C}$—$(C_d)(C)(H)_2]$.

From Table 45 the following contributions are:

$[\mathbf{C}_d$—$(C_d)(H)(C)] + [\mathbf{C}$—$(C_d)(H)_3] =$	38.5	
$[\mathbf{C}_d$—$(C_d)(C)_2] + 2[\mathbf{C}$—$(C_d)(H)_3] =$	48.7	
$[\mathbf{C}$—$(C_d)(C)(H)_2]$—$[\mathbf{C}$—$(C_d)(H)_3] =$		−20.9
$[\mathbf{C}$—$(C)(H)_3] =$	30.4	
Correction for bulky *cis*-groups =	+0.6	
	118.2 −20.9 =	97.3
Symmetry contribution:— $R \ln \sigma = -R \ln 27$ (3 methyl groups)		−6.5
	$S° =$	90.8 cal./mole °K

Example 5.17. Using the law of group contributions, calculate the entropy for $CCl_3 \cdot CH_2CO \cdot OCH(CH_3)CH_2Br$ (start from the carbonyl group, and note that corrections for symmetry ($\sigma = 9$), optical activity, and mixing of isomers must be included). (ANSWER: 137.7 e.u.)

Example 5.18. Estimate the values of $\Delta H_f°$ for the following: (i) 2,3-dimethylbutene-1. (ii) 1,2-dibromoethane. (iii) ethylacetate. (iv) glycerol. (ANSWER: (i) -15.6; lit., -14.8. (ii) -10.4; lit., -9.3. (iii) -106.6; lit., -106.6. (iv) -137.3; lit., -139.3 kcal./mole.)

When isomerism is due to a changed geometrical relationship between identical groups (e.g., 2-methylhexane and 3-methylhexane), the isomers are indistinguishable by the scheme of group additivity; when the isomers arise from differences in the groups present in the molecules (e.g., n-butane and isobutane), the isomerism can be distinguished.

For hydrocarbons, the estimates of $C_p°$ and $S°$ are generally accurate to within ± 0.3 cal./mole °K; for very heavily substituted compounds, such as hexamethylbenzene, larger deviations (± 1.5 cal./mole °K) are to be expected. The values for $\Delta H_f°$ are usually within ± 0.4 kcal./mole, but may be in error by as much as ± 3 kcal./mole for heavily substituted compounds. For compounds containing O, N, S, and halogen, the estimates by this principle were found to fall generally within the limits of the experimental data or of the statistical calculations.

For t-C_4H_9OOH, for example, for which $\Delta H_f° = -52$ kcal./mole, an error of 6 kcal./mole would be caused if the hydroperoxide contained 0.2 wt. % of H_2O. The principle of group additivity predicts a value of -59 kcal./mole, and it is suggested[46a] that the latter estimated value is likely closer to the true value than that from the combustion data. Heat of combustion data for halogen compounds may be subject to errors as much as ± 5 kcal./mole. In developing the present compilation for the halogen compounds, Benson and Buss gave most weight to $\Delta H_f°$ data from heats of reaction or equilibrium studies where the errors are not so large.

The method of disproportionation and the preceding additivity rules can also be used directly in estimating bond dissociation energies and radical properties.

Bryant. A modification of the method of Andersen, Beyer, and Watson has been reported by Bryant[46b] for heats of formation of organic fluorine compounds. Multiple substitutions of fluorine for methyl were not considered in the original work and this problem was investigated by Bryant. The results, summarized in Tables 48 and 49, Part II, present an approach that applies well for hydrocarbons and perfluorocarbons. The asterisks indicate changes in the values originally proposed for these increments.

[46b] W. M. D. Bryant, *J. Polymer Sci.* **56**, 277 (1962).

It is recommended that all substitutions of a given type are made before going to those of a higher numerical type in the estimation procedure. Thus straight chain contributions are developed before those for single branches per carbon atom, and these in turn precede those for the second branch per carbon.

Example 5.19. Estimate the heat of formation of hexafluoroethane (1 atm., 298° K).

Base group CH_4	-17.9 kcal.
Primary CH_3 substitution	-2.4
Secondary CH_3 substitution (1,1)	-4.5
Secondary CH_3 substitution (1,2)	-5.0
Secondary CH_3 substitution (2,2)	-6.5
Secondary CH_3 substitution (2,3)	-6.5
Secondary CH_3 substitution (3,3)	-6.5
Secondary CH_3 substitution (3,4)	-5.0
	-54.3 kcal.
Trisubstituted fluorine (6 × 43.7)	-262.2
$C_2F_6(g) : \Delta H^\circ_{f\,298\cdot15} =$	-316.5 kcal.

The experimental heat of formation of hexafluoroethane by Kirkbride and Davidson[46c] of -303 kcal. is believed to be numerically too small by 10–20 kcal.[46d]

Where the carbon skeleton is only partially fluorinated and where fluorine shares a carbon atom with other halogens, it appears that additional small enthalpy increments may be involved. The uncertainty in the heats of formation for fluorocarbons is estimated as ± 0.5–1 kcal. For hydrocarbons, this procedure gave results, when applied to some 25 normal and branch-chain hydrocarbons, that differed from the values given in Circular C461, National Bureau of Standards, generally by about 0.2 kcal. and not more than 0.6 kcal.

It should be noted that for systems where data are scarce, one should select a reliable heat of formation of a compound of a closely related structure and proceed to the desired structure and heat of formation estimate by suitable incremental modifications of this "parent" compound; the error introduced by using a reliable heat of formation at this advanced stage in the estimate should normally be much less than in the calculation beginning with methane.

Using this revised group contribution procedure leads to the heats of formation as summarized in Table 5.1, column two.

These results lead to the generalized expression for $\Delta H^\circ_{f\,298°\,K}$ of gaseous normal perfluoroalkanes, where n is greater than 1.0:

$$C_nF_{2n+2(g)}; \quad \Delta H^\circ_{f\,298°K} = 94.5(n-2) + 316.5 \text{ kcal./mole} \qquad (5.23)$$

[46c] F. W. Kirkbride and F. G. Davidson, *Nature* **174**, 79 (1954).

[46d] J. B. Farmer, I. H. S. Henderson, F. P. Lossing, and D. G. H. Marsden, *J. Chem. Phys.* **24**, 352 (1956).

5. THE METHODS OF GROUP CONTRIBUTIONS

The values for CF_4, the branched fluorocarbons, and perfluoroethylene were gained from assessments of recent experimental data for these and related hydrocarbons and fluorocarbons. The molecular entropies (see Table 5.1, column 3) were calculated from:

$$C_nF_{2n+2(g)}; \quad S^\circ_{298} = 46.61 + 16.066n \quad (5.24)$$

for the n-fluorocarbons with 2 or more C atoms. This expression was deduced from the experimentally established data for CF_4 and C_7F_{16} as parent compounds and the principle of Parks and Huffman.[4] Values for the branched

TABLE 5.1
THERMODYNAMIC PROPERTIES OF GASEOUS FLUOROCARBONS[46b]

Fluorocarbon	$\Delta H_{298.15}$ (cal./mole)	$S_{298.15}$ (e.u.)	$-\Delta G_{298.15}$ (rounded off), (cal./mole)
CF_4	217,800	62.48	207,040
C_2F_6	316,500	78.5	295,600
C_3F_8	411,000	94.6	380,000
$n\text{-}C_4F_{10}$	505,500	110.7	464,400
$n\text{-}C_5F_{12}$	600,000	126.7	548,800
$n\text{-}C_6F_{14}$	694,500	142.8	633,200
$n\text{-}C_7F_{16}$	789,000	158.9	717,600
$n\text{-}C_8F_{18}$	883,500	174.9	802,000
$n\text{-}C_9F_{20}$	978,000	191.0	886,400
$n\text{-}C_{10}F_{22}$	1,072,500	207.1	970,800
$n\text{-}C_{11}F_{24}$	1,167,000	223.1	1,055,200
$n\text{-}C_{12}F_{26}$	1,261,500	239.2	1,139,600
$n\text{-}C_{16}F_{34}$	1,639,500	303.5	1,477,200
$n\text{-}C_{24}F_{50}$	2,395,500	432.0	2,152,300
$i\text{-}C_4F_{10}$	515,600	107	473,400
$i\text{-}C_5F_{12}$	610,100	125	558,400
$\text{Neo-}C_5F_{12}$	624,600	117	570,500
C_2F_4	151,900	71.70	143,480
C_3F_6	258,750	87.8	240,200
$n\text{-}C_{12}F_{24}(-1)$	1,109,300	232.4	999,800

fluorocarbons were estimated assuming the increments were the same as for the n- and branched hydrocarbons; the value for perfluoropropylene was estimated from the data of C_2F_4 as parent. The free energies of formation were calculated from the well-established relation:

$$\Delta G = \Delta H - T\Delta S \quad (1.3)$$

and a knowledge of the standard entropies for graphite and diatomic fluorine, 1.3609 and 48.6 e.u., respectively.

The values for ΔH°_{f298} for the fluorocarbon free radical species were calculated from the bond-dissociation energies:

I. METHODS OF ESTIMATION

ESTIMATED BOND STRENGTHS IN FLUOROCARBONS[46b]

		D (kcal./mole)
CF_4	$\rightarrow CF_3\cdot + F\cdot$	124.0
$n\text{-}C_nF_{2n+2}$	$\rightarrow CF_{2n+1}\cdot + F\cdot$	123.4
$i\text{-}C_3F_8$	$\rightarrow i\text{-}C_3F_7\cdot + F\cdot$	110.0
$(CF_3)_3CF$	$\rightarrow (CF_3)_3C\cdot + F\cdot$	99.0
$CF_2\!=\!CF_2$	$\rightarrow CF_2\!=\!CF\cdot + F\cdot$	125.0

and the heats of formation of the parent fluorocarbons and monoatomic fluorine. Values of the above bond strengths were estimated by Bryant from correlations of experimentally established values of D with bond distances for hydrocarbons, partially fluorinated compounds, and fluorocarbons. Entropies of the free radical species were calculated from the parent fluorocarbons by the expression:

$$C_nF_{2n+1(g)}; \quad S^\circ_{298} = 44.91 + 16.066n \tag{5.25}$$

where n is greater than, or equal to, two. The n-fluorocarbons were used as

TABLE 5.2
THERMODYNAMIC PROPERTIES OF GASEOUS FLUOROCARBON RADICALS[46b]

Free radical	$-\Delta H^\circ_{f\,298.15}$ gaseous fluorocarbon radical (cal./mole)	$S^\circ_{298.15}$ (e.u.)	$-\Delta G^\circ_{f\,298.15}$ (rounded off) (cal./mole)
$CF_3\cdot$	112,700	62.35	109,150
$C_2F_5\cdot$	212,000	77.0	197,900
$n\text{-}C_3F_7\cdot$	306,500	93.1	282,300
$n\text{-}C_4F_9\cdot$	401,000	109.2	366,700
$n\text{-}C_5F_{11}\cdot$	495,500	125.2	451,100
$n\text{-}C_6F_{13}\cdot$	590,000	141.3	535,500
$n\text{-}C_7F_{15}\cdot$	684,500	157.4	619,900
$n\text{-}C_8F_{17}\cdot$	779,000	173.4	704,300
$n\text{-}C_9F_{19}\cdot$	873,500	189.6	788,700
$n\text{-}C_{10}F_{21}\cdot$	968,000	205.6	873,100
$n\text{-}C_{11}F_{23}\cdot$	1,062,500	221.6	957,500
$n\text{-}C_{12}F_{25}\cdot$	1,157,000	237.7	1,041,900
$n\text{-}C_{16}F_{33}\cdot$	1,535,000	302.0	1,397,500
$n\text{-}C_{24}F_{49}\cdot$	2,291,000	430.5	2,054,600
$i\text{-}C_3F_7\cdot$	319,900	91.7	295,300
$(CF_3)_2CFCF_2\cdot$	411,100	108.7	376,700
$(CF_3)_3C\cdot$	435,500	103.5	399,500
$(CF_3)_2CFCF_2CF_2\cdot$	505,600	122.5	460,400
$(CF_3)_2CCF_2\cdot$	520,100	116.6	473,100
$CF_2\!=\!CF\cdot$	45,800	70.2	44,200

parents in deducing this expression, taking note of the entropy changes that arise in the formation of a free radical species (i.e., the changes due to the loss of an F atom, due to changes in symmetry, and due to transition from a singlet to a doublet state). The value for the vinyl radical was gained from perfluoroethylene as parent; the entropy of $CF_3\cdot$ was calculated from molecular data; and the branched fluorocarbon radicals by modification of Eq. (5.25) on an individual basis. These results are in Table 5.2. Free energies of formation for the free radical species were calculated in the conventional manner.

Comparison with the corresponding hydrocarbons, the thermodynamic data at 298.1°K indicate that the fluorocarbon systems become more stable with increasing molecular weight, while hydrocarbons less stable. The problem of an increase in temperature on these properties remains an area for attention.

Ciola. A group increment method for the direct estimation of reaction equilibrium constants, rather than $\Delta G°$, $\Delta H°$, and $\Delta S°$, has been advanced by Ciola.[46e] Thus from the fundamental relation

$$\Delta G° = -RT \ln K \tag{1.1}$$

and the principle of additivity for group increments, it readily follows that the equilibrium constant for a chemical process can be expressed directly as a function of group contributions. For example, in the dehydrogenation of alcohols:

$$CH_3CH_2OH \rightarrow CH_3CHO + H_2 \tag{5.26}$$

the equilibrium constant in this manner is given by:

$$\log K = \log K_{f\text{CHO}} + \log K_{f\text{CH}_3} - \log K_{f\text{CH}_3} - \log K_{f\text{CH}_2\text{OH}} + \delta_{\text{EtOH}}$$
$$+ \delta_{\text{AcH}} + RT \log \frac{\sigma_{\text{EtOH}}}{\sigma_{\text{AcH}}} \tag{5.27}$$

where σ and δ are the symmetry numbers for the respective molecules and empirical correction factors. Inspection shows that these are unity and zero for the compounds in this reaction; similarly it will be noted that the $\log K$ contributions for the CH_3 group increments cancel. Thus the expression reduces to:

$$\log K = \log K_{f\text{CHO}} - \log K_{\text{CH}_2\text{OH}} \tag{5.28}$$

and this is sufficient for the prediction of the desired equilibrium data. The group increment values developed by Ciola are in Tables 50–61, Part II; the method of application is similar to the increment methods of Chapter 5.

Expression (5.28) is a general relation, applicable to the dehydrogenation of a variety of any alcohols (providing the symmetry number ratio equals

[46e] R. Ciola, *Ind. Eng. Chem.* **49**, 1789 (1957).

unity and the correction factors are zero or the same for the reactants and products). The values of log K for this generalized case are:

$T(°K)$	300	400	500	600	700	800	900	1000
log K	-5.96	-2.93	-1.09	$+0.14$	1.03	1.70	2.12	2.62

A further illustration is the hydrogenation and dehydrogenation of olefins. Thus for the four hydrogenation reactions:

$$\text{vinylbenzene} + H_2 \rightarrow \text{ethylbenzene} \tag{5.29}$$

$$\text{propylene} + H_2 \rightarrow \text{propane} \tag{5.30}$$

$$\text{heptene-1} + H_2 \rightarrow \text{heptane} \tag{5.31}$$

$$\text{3-methyl pentene-1} + H_2 \rightarrow \text{3-methylpentane} \tag{5.32}$$

the expressions for the equilibrium constant as a function of group equations reduce to:

$$\log K = \log K_{f\text{CH}_3} + \log K_{f\text{CH}_2} - \log K_{f(-\text{CH}=\text{CH}_2)} + \log \frac{\sigma_1}{\sigma_2} \tag{5.33}$$

A comparison of the values for log K thus predicted for these four hydrogenations with the values calculated from the data of Rossini et al.[1] is as follows:

Log K Calculated by Group Contribution Compared with That Calculated from Rossini et al.[1]

Reaction	(5.29)		(5.30)		(5.31)		(5.32)	
°K	Groups	Rossini et al.	Groups	Rossini et al.	Groups	Rossini et al.	Groups	Rossini et al.
300	15.363	15.453	15.062	15.002	15.062	15.262	15.062	15.197
400	9.931	9.290	9.631	9.517	9.631	9.830	9.631	9.665
500	6.588	6.142	6.288	6.220	6.288	6.484	6.288	6.176
600	4.335	4.015	4.035	3.991	4.035	4.226	4.035	3.789
700	2.890	2.483	2.590	2.394	2.590	2.332	2.590	2.074
800	1.474	1.329	1.374	1.191	1.329	1.372	1.329	0.750
900	0.510	0.424	0.210	0.254	0.424	0.406	0.424	-0.305
1000	-0.268	-0.302	-0.564	-0.493	-0.302	-0.365	-0.302	-1.162

In the application of this approach, the thermodynamic feasibility of a reaction is predicted from the values of log K. When log K is less than zero ($\Delta G° > 0$), the reaction is either improbable or feasible only under drastic conditions. Since log $K_{f(\text{group increment})}$ for various groups is inaccurate owing to uncertainties in the thermodynamic properties of the parent compounds, an arbitrary error of 100 °K is attributed[46e] to the temperature estimated for log $K \neq 0$.

Morgan and Lielmezs. The estimation of the thermodynamic properties for the n-halogenated hydrocarbons, n-$C_mH_{2m+1}X$, where $X = F$, Cl, Br, or I, has been reported by Morgan and Lielmezs.[46f] The group increments are used in the linear relation:

$$G(C_mH_{2m+1}X) = K_T + [CH_2](n - 3) \qquad (5.34)$$

where G can be $-(G° - H_0°/T)$, $(H° - H_0°)/T$, $C_p°$, or $S°$, and K_T is the corresponding property for the appropriate n-propyl halide. Values for the latter are given in Table 62, Part II, as reported by Morgan and Lielmezs. In the above expression n is the number of carbon atoms in a straight unbranched chain issuing from the functional group. The group increments for the methylene contributions are in Table 15, Part II. The thermodynamic properties for the above "parent" n-alkyl halides (Table 62) were calculated by the methods of statistical thermodynamics, using the spectroscopic vibrational data and taking due recognition of the hindered rotor contributions. Values for the methyl and ethyl halides thus calculated are also reported by Morgan and Lielmezs.[46f]

It is suggested that the accuracy of estimates with these increments (Table 62) for the iodides is $\pm 2\%$, while for the fluorides, chlorides, and bromides it is ± 1 to 2%.

[46f] J. P. Morgan and J. Lielmezs, *Ind. Eng. Chem. (Fundamentals)* **4**, 383 (1965).

CHAPTER 6

The Method of Group Equations

1. Introduction

In the method of group equations the thermodynamic functions are estimated for a desired molecule by referring it to other compounds containing the same numbers and types of groups. Thus the properties for n-pentane would be calculated from a knowledge of the data for n-butane, propane, and ethane, e.g.,

$$G(n\text{-pentane}) = G(n\text{-butane}) + G(\text{propane}) - G(\text{ethane}) \qquad (6.1)$$

where G is the free energy content, the heat content, or some desired thermodynamic function. Equations such as (6.1) in which there is an identity of groups, are termed group equations. The method of group equations has been developed and applied by Rossini, Pitzer, and their associates[44,47,48] in compiling the tables of the American Petroleum Institute[1] for the thermodynamic properties of hydrocarbons. The symmetry number enters into the statistical formulas for the free energy function (Table 2.1), the entropy [i.e., $(H° - H_0°)/T - (G° - H_0°)/T$], and the free energy of formation, but not into the expressions for heat content, heat capacity, and heat of formation. In the application of the method of group equations to calculation of rather accurate data for hydrocarbons, these investigators modified the basic equation (6.1) to correct for symmetry factors and internal rotation.

2. Symmetry Number

The problem of the symmetry of a polyatomic molecule and its contribution to the thermodynamic properties of the system has already been discussed (Chapter 2). Inspection of the equations of statistical thermodynamics (Table 2.1) shows that the term $-R \ln \sigma$ in the rotational contribution, for the system treated as a rigid rotator-simple vibrator, is introduced to account for the effect of symmetry. In this expression, the symmetry number, σ, is defined as the number of indistinguishable positions into which the molecule can be turned by rigid rotations. From the discussion of the

---

[47] J. E. Kilpatrick, E. J. Prosen, K. S. Pitzer, and F. D. Rossini, *J. Research Natl. Bur. Standards* **36**, 559 (1946).

[48] J. E. Kilpatrick, C. W. Beckett, E. J. Prosen, K. S. Pitzer, and F. D. Rossini, *J. Research Natl. Bur. Standards* **42**, 225 (1949).

6. THE METHOD OF GROUP EQUATIONS

concept of internal rotation (Chapter 2) it is apparent that a still further symmetry is conferred upon the system if it is assumed that rotation about the C—C single bond is possible. Thus for each terminal rotor, e.g., CH_3, a symmetry number, n, describes the number of indistinguishable positions achieved by rotation of such a group relative to the rest of the system. The *total symmetry number*, s, may be defined by:

$$n^a \times \sigma = s \qquad (6.2)$$

where a is the number of terminal rotors in the structure. The value of $n = 1$ is assumed when $a = 0$.

The influence of the symmetry number of the molecule on the additivity of thermodynamic properties was recognized in the methods of Franklin, Souders, Matthews, and Hurd, and Van Krevelen and Chermin, and expressed by equation (5.15), i.e.,

$$G_{(comp.)} = \Sigma G \begin{pmatrix} \text{contributing} \\ \text{groups} \end{pmatrix} + \Sigma G \begin{pmatrix} \text{correction for} \\ \text{deviations} \end{pmatrix} + R \ln \sigma$$

The symmetry number of the system treated as a rigid rotator-simple vibrator is used since the symmetry of the methyl group is already included in the CH_3 group increment. In the method of group equations, the contribution conferred by the total symmetry number s, and not σ, must be considered since by virtue of this approach, the principle of additivity is applicable only if there is a balance in σ and n on both sides of the group equation.

Inspection of the structural formula enables a ready calculation of the total symmetry number. This is illustrated by the following examples:

Compound	σ	a	n	s
Propane	2	2	3	18
Butene-1	1	1	3	3
Benzene	12	0	1	12
o-Xylene and m-xylene	2	2	3	18
p-Xylene	4	2	3	36
sym-Trimethylbenzene	6	3	3	162

The importance of achieving a balance in the symmetry contributions, based on s, as well as an identity of groups in the method of group equations was recognized by Rossini, Pitzer, and associates in the application of this approach to the calculation of the very precise thermodynamic properties of hydrocarbons.[1] Thus the group equation 6.1 for cis-2-pentene, providing cognizance is taken of difference in the symmetry numbers, would be expressed as:

$$G(cis\text{-}2\text{-pentene}) = G(1\text{-butene}) + G(cis\text{-}2\text{-butene}) - G(\text{propane}) + R \ln 2 \qquad (6.3)$$

where the additional term $+ R \ln 2$ is required to bring the symmetry of the right hand side of the equations into correspondence with that of the desired compound. The term $R \ln s$, as indicated earlier, applies only to the estimates of $-(G° - H_0°)/T$ and $S°$, and not to the heat content, heat capacity, and heat of formation.

3. Internal Rotation

In the application of the method of group equations to calculation of a thermodynamic property of a long chain molecule, the assumption that the barriers hindering internal rotation in the final structure are the same as in the compounds of the group equation being used in the calculation is implicit in the method. Justification for this is found in the theoretical treatment by Pitzer (Chapter 3) for the long chain paraffins. Based on the above assumption, the calculated thermodynamic properties were well in accord with the experimental data.

For the most precise results with the method of group equations, corrections for differences in potential barriers restricting internal rotation should be taken into account. This may be illustrated in the estimation of the thermodynamic properties for cis-2-pentene. Taking into account the correction for symmetry numbers, the equation:

$$G(\text{cis-2-pentene}) = G(\text{1-butene}) + G(\text{cis-2-butene}) - G(\text{propylene}) + R \ln 2 \quad (6.3)$$

would lead to the desired estimates for this compound. The internal rotation in cis-2-pentene differs from the components above in that it involves the rotation of an ethyl group attached to a nonterminal doubly bonded atom. The group equation (6.3) results in net restriction on the internal rotation of the ethyl group that is too small. Inspection of this problem led Kilpatrick and co-workers[47] to propose the following barrier to describe the internal rotation of the ethyl group in cis-2-pentene:

$$V = \tfrac{1}{2} (2400) (1 - \sin 3\theta) \text{ cal./mole} \quad (6.4)$$

for $0° < \theta < 240°$, and

$$V = \infty \quad (6.5)$$

for $240° < \theta < 360°$. Thus, as illustrated in Fig. 6.1.a., the rotational potential in cis-2-pentene becomes infinitely great from 240° to 360°. A potential barrier of the type:

$$V = \tfrac{1}{2} (2400) (1 - \sin 3\theta) \text{ cal./mole} \quad (6.6a)$$

for $0° < \theta < 240°$, and

$$V = 2400 \text{ cal./mole} \quad (6.6b)$$

for $240° < \theta < 360°$, i.e., the rotational potential is constant (Fig. 6.1.b), describes the restricted rotation on the ethyl group in 1-butene. Accordingly in the estimates of the thermodynamic properties for *cis*-2-pentene from the above components, Eq. (6.4) was modified to correct for the greater net

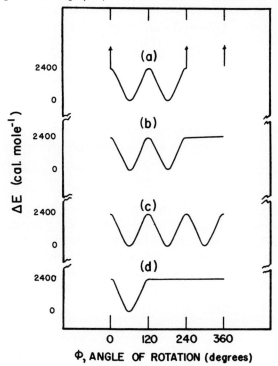

FIG. 6.1. Potential barriers restricting internal rotation in monoolefin hydrocarbons.[47] (a) Barrier for ethyl group in *cis*-2-pentene; (b) barrier for ethyl group in *cis*-1-butene; (c) barrier for isopropyl group in 3-methyl-1-butene, and (d) barrier for ethyl group in 2-methyl-1-butene.

restriction in rotation of the ethyl group for $240° < \theta < 360°$. The expression:

$$G(cis\text{-2-pentene}) = G^*(1\text{-butene}) + G(cis\text{-2-butene} - \text{propylene}) + R \ln 2 +$$
$$\text{(internal rotation of Eq. (6.4), (6.5), for ethyl group)} \quad (6.7)$$

was used where the asterisk on 1-butene indicates that the value of G now does not include the contribution arising from the internal rotation of the ethyl group in this molecule. For calculating the contribution to the given thermodynamic property in such cases, the classical partition function may be used since the moment of inertia of the rotating group is considerably greater than that of a methyl group.

In addition to the preceding barriers, a symmetrical threefold barrier for values of from 0° to 360°, i.e.,

$$V = \tfrac{1}{2}(2400)(1 - \sin 3\theta) \quad (6.8)$$

and a barrier of the type permitting only one position of minimum energy in a rotation of 360°, were considered in calculating the thermodynamic properties of the pentenes.[47] The latter are described by the expressions:

$$V = \tfrac{1}{2}(2400)(1 - \sin 3\theta); \quad 0° < \theta < 120° \tag{6.9a}$$

and

$$V = 2400 \text{ cal./mole}; \quad 120° < \theta < 360° \tag{6.9b}$$

The nature of these two potential functions is illustrated in Fig. 6.1, c and d, respectively. Thus the thermodynamic properties for 2-methyl-1-butene and 3-methyl-1-butene were calculated[47] by use of the group equations:

$$G(\text{2-methyl-1-butene}) = G^*(\text{1-butene}) + G(\text{isobutene} - \text{propylene}) + R \ln 2 +$$
$$(\text{i.r., Eqs. 9a, b, for ethyl group}) \tag{6.10}$$

and

$$G(\text{3-methyl-1-butene}) = G^*(\text{1-butene}) + G(\text{isobutane} - \text{propane}) + R \ln \tfrac{3}{2} +$$
$$(\text{i.r., Eq. 6.8 for isopropyl group}) \tag{6.11}$$

in which the $G^*(\text{1-butene})$ indicates that the contribution for internal rotation is not included in this value for 1-butene. For subsequent calculations for higher olefins in which the nature of the skeletal rotation adjacent to the alkyl group is essentially the same as that in 1-butene or the pentenes, no explicit correction is necessary if the appropriate olefinic butene or pentene is used in the group equations. For precise calculations with the method of group equations, cognizance must be taken, where necessary, of differences in the potential barriers restricting internal rotation as well as in the symmetry numbers.

4. Calculation of Thermodynamic Properties

The application of this method in practice may be most readily illustrated by consideration of some of the group equations developed by Rossini, Pitzer, and associates to calculate the thermodynamic properties for the higher hydrocarbons in the various homologous series. Based on the statistical thermodynamic functions of the lower hydrocarbons in each series, the data for

$$\frac{-(G° - H_0°)}{T}, \frac{(H° - H_0°)}{T}, C_p°, \text{ and } S°$$

are readily calculated by this method over a wide range of temperatures.

Mono-olefin Hydrocarbons. In addition to the examples just considered, the following illustrate further calculations of the thermodynamics for mono-olefins:

$$G(\text{1-hexene}) = G(\text{1-butene}) + G(n\text{-pentane}) - G(\text{propane}) \tag{6.12}$$

$$G(cis\text{-3-hexene}) = G(cis\text{-2-pentene}) + G(cis\text{-2-pentene}) -$$
$$G(cis\text{-2-butene}) - R \ln 4 \tag{6.13}$$

$$G(\text{2,3-dimethyl-2-butene}) = G(cis\text{-2-butene}) + G(trans\text{-2-butene}) +$$
$$G(\text{isobutene}) + R \ln 2 - 2G(\text{propylene}) \tag{6.14}$$

where only in (6.12) is there a balance in the symmetry numbers without the additional correction term.

It is to be noted that use of higher members of any homologous series rather than the very first members is applied whenever it is possible. Thus the increment [—CH_2—CH_2—] was gained from the use of pentane and propane, rather than butane and ethane in group equation (6.12). Use of the former is recommended quite generally in calculations with this method since the corrections arising from differences in symmetry numbers and in the potential barriers restricting internal rotation are thus minimized. The very first members frequently possess much greater symmetry (e.g., ethane, $s = 54$) and have a hindered internal rotational contribution of a more simple nature than the higher members of that homologous series.

An indication of the precision of the method is found in the data for 2-methyl-2-butene. The thermodynamic functions were calculated by the group equation:

$$G(\text{2-methyl-2-butene}) = \tfrac{2}{3}G(cis\text{-2-butene} + trans\text{-2-butene} + \text{isobutene}) -$$
$$G(\text{propylene}) + R \ln 4 \qquad (6.15)$$

The agreement between the value, 80.90 e.u., predicted from the above calculations for S°_{298} and the experimentally established entropy, 80.82, which subsequently became available,[49] leaves little to be desired. Comparison of the heat capacity shows that the calculated value is about one per cent higher than the experimental value.

Acetylene Hydrocarbons. The thermodynamic properties for the acetylene hydrocarbons to C_4 were calculated by statistical thermodynamic methods. The method of group equations was used for the higher members in this series. The following equations illustrate the method as applied in practice:

$$G(n\text{-propylacetylene}) = G(\text{ethylacetylene}) + G(n\text{-butane}) - G(\text{propane}) \qquad (6.16)$$

$$G(\text{isopropylacetylene}) = G(\text{ethylacetylene}) + G(\text{isobutane}) -$$
$$G(\text{propane}) + R \ln \tfrac{3}{2} \qquad (6.17)$$

where in (6.17) a correction of $R \ln \tfrac{3}{2}$ must be included in the entropy and negative of the free energy function because of the difference in the symmetry numbers.

Diolefin Hydrocarbons. For the pentadienes, the method of group equations has been applied[48] since the available thermal, spectroscopic, and other molecular data were insufficient or too uncertain for satisfactory treatment by statistical thermodynamics. The following equations, using reference compounds for which the data are accurately known, illustrate the calculations for some of the isomeric pentadienes:

$$G(\text{1,2-pentadiene}) = G(\text{1-butene}) + G(\text{1,2-butadiene}) - G(\text{propylene}) \qquad (6.18)$$

$$G(\text{2,3-pentadiene}) = 2G(\text{1,2-butadiene}) - G(\text{propadiene}) - R \ln 8 \qquad (6.19)$$

[49] S. S. Todd, G. D. Oliver, and H. M. Huffman, *J. Am. Chem. Soc.* **69**, 1519 (1947).

$$G(\text{3-methyl-1,2-butadiene}) = G(\text{isobutene}) + G(\text{1,2-butadiene}) - G(\text{propylene}) \quad (6.20)$$

$$G(\text{2-methyl-1,3-butadiene}) = G(\text{isobutene}) + G(\text{1,3-butadiene}) -$$
$$G(\text{propylene}) + R \ln 4 - 0.3 \quad (6.21)$$

In the above, the thermodynamic functions for 1,2-butadiene had been calculated by the equation:

$$G(\text{1,2-butadiene}) = G(\text{allene}) + G(\text{propylene}) - G(\text{ethylene}) \quad (6.22)$$

prior to any experimental work on this compound. Confirmation of the increment method is seen in the close agreement of S°_{298} calculated as above, 69.7 e.u., and the value based on the subsequent experimental data,[50] 70.03 e.u. The empirical constant term (–0.3) in Eq. 6.21 for isoprene was added in order to have better agreement with the experimental value at 298.16° K. Thus, without this term, the predicted value for S°_{298} is 75.7 e.u. Compared with the experimental result, 75.2 ± 0.3 e.u., it was judged that while the predicted difference (0.5 e.u.) was well within the expected uncertainties of the method, the additional correction (–0.3) in the group increment equation, which brings the values into still closer agreement, would give better extrapolated values at higher temperatures. It was estimated that, for the pentadienes, the overall uncertainty in the calculated values of the thermodynamic functions was about one cal. per deg. mole at 298.16° K, and greater at higher temperatures. The values for 1,4-pentadiene, calculated by:

$$G(\text{1,4-pentadiene}) = 2G(\text{1-butene}) - G(\text{propane}) - R \ln 4 \quad (6.23)$$

were judged[48] to be particularly uncertain, by the above standards, in this series.

Alkylbenzene Hydrocarbons. The functions for *n*-propylbenzene and the higher normal alkylbenzenes were calculated by the method of group equations from the thermodynamic properties for the normal paraffin hydrocarbons and the lower alkyl benzenes. The following formulas illustrate the extension of this method to this series of compounds:

$$G(\text{1-methyl-3-ethylbenzene}) = G(\text{1,3-dimethylbenzene}) +$$
$$G(\text{ethylbenzene}) - G(\text{toluene}) + R \ln 2 \quad (6.24)$$

$$G(\text{isopropylbenzene}) = G(\text{ethylbenzene}) + G(\text{isobutane}) -$$
$$G(\text{propane}) + R \ln \tfrac{3}{2} \quad (6.25)$$

For the normal alkyl benzenes, C_nH_{2n-6}, where $n \geq 9$, the group equation:

$$G(C_nH_{2n-6}) = G(C_8H_{10}) + G(C_{n-5}H_{2n-8}) - G(C_3H_8) \quad (6.26)$$

[50] J. G. Aston and G. J. Szasz, *J. Am. Chem. Soc.* **69**, 3108 (1947).

6. THE METHOD OF GROUP EQUATIONS

was used, when the values were based on the data for ethylbenzene (C_8H_{10}) and the appropriate normal paraffins. Corrections for changes in the barriers restricting internal rotation were not estimated in view of the uncertainty of the values calculated by the method in the present case.

Example 6.1. (a) From a knowledge of the thermodynamic properties of propane, propylene, and 2,2-dimethylbutane, develop a group equation for estimation of the thermodynamic properties of 3,3-dimethyl-1-butene. (b) Calculate the free energy function and heat of formation at 600° K for 3,3-dimethyl-1-butene.

SOLUTION:

(a) Inspection of the structural formulas suggests that the statistical thermodynamic properties and heat of formation for 3,3-dimethyl-1-butene may be estimated by the equation:

$$G(H_2C{=}CH{-}\underset{\underset{CH_3}{|}}{\overset{\overset{CH_3}{|}}{C}}{-}CH_3) = G(CH_2{=}CH{-}CH_3) + G(CH_3{-}\underset{\underset{CH_3}{|}}{\overset{\overset{CH_3}{|}}{C}}{-}CH_2{-}CH_3) - G(CH_3{-}CH_2{-}CH_3)$$

since there is an identity of groups. The total symmetry numbers for the compounds in the order of the above equation are: $3^3 \times 1 = 27$, $3^1 \times 1 = 3$, $3^4 \times 1 = 81$, and $3^2 \times 2 = 18$, respectively. Comparison of the symmetry numbers shows that a factor of $-R \ln 2$ is required on the right hand side to balance the difference in symmetries of the compounds. Accordingly the completed group equation would be:

$$G(3,3\text{-dimethyl-1-butene}) = G(\text{propylene}) + G(2,2\text{-dimethylbutane}) - G(\text{propane}) - R \ln 2$$

where the last term $(-R \ln 2)$ is to be included in the estimates of $-(G°-H_0°)/T$ and $S°$ only.

(b) Using the compilation of thermodynamic data by Rossini and associates,[1] the following values for the free energy function and heat of formation at 600° K are predicted.

	$-(G° - H_0°)/T$ cal./deg. mole	$\Delta H_f°$ kcal./mole.
G(propylene)	62.05	1.98
G(2,2-dimethyl-butane)	84.1	-50.40
$-G$(propane)	-62.93	28.66
$-R \ln 2$	-1.38	—
G(3,3-dimethyl-1-butene)	81.9	-19.8
Lit. value[1]	81.8	-19.8

It is of interest to note that the same method of calculation was used by Rossini, Pitzer, and associates[47] for the thermodynamics of the mono-olefin hydrocarbons.

Example 6.2. The following group equations are to be used in the evaluation of the statistical thermodynamic functions. Complete the equations with reference to the correction for differences in symmetry numbers:

(a) G(3-methyl-1-pentene) = G(3-methyl-1-butene) + G(2-methylbutene) − G(isobutane)

(b) G(2-pentyne) = G(1-butyne) + G(2-butyne) − G(propyne)

(c) G(*trans*-1,3-pentadiene) = G(*trans*-2-butene) − G(1,3-butadiene) − G(propylene)

(ANSWER: The following terms are to be added to the right hand side of the group equation in each case: (a) $-R \ln 2/3$, (b) $-R \ln 2$, (c) $+R \ln 4$.)

Nonhydrocarbons. It is noteworthy that this approach is being extended to nonhydrocarbons. Pitzer and Barrow[51] have shown that the values for the higher normal thiols calculated by group equations starting with the data for ethanethiol are in good agreement with the properties calculated from spectroscopic data. The calculation of rather accurate data for nonhydrocarbon species awaits precisely established calorimetric and spectroscopic data for the lower members in each family of compounds. A program of such studies under the sponsorship of the American Petroleum Institute is in progress.[52]

Example 6.3. Calculate the statistical thermodynamic functions for succinonitrile at 298.1° K and one atmosphere pressure using the method of group equations.

SOLUTION:

Inspection of the structural formulas leads to the following group equation, based on the thermodynamic properties of propionitrile,[20] butane,[1] and propane,[1]:

G(succinonitrile) = $2G$(propionitrile) + G(*n*-butane) − $2G$(propane) − $R \ln 4$

Using the statistical thermodynamic properties of the above components given in the literature, the following values (cal./deg. mole) are thus estimated for succinonitrile:

$-(G° - H_0°)/T$, 62.41; $(H° - H_0°)/T$, 15.65; $C_p°$, 22.60; $S°_{298}$, 78.06

[51] K. S. Pitzer and G. M. Barrow, *Ind. Eng. Chem.* **41**, 2737 (1949).
[52] H. L. Finke, D. W. Scott, M. E. Gross, G. Waddington, and H. M. Huffman, *J. Am. Chem. Soc.* **74**, 2804 (1952).

These are to be compared with the values: 62.30, 14.95, 22.57, and 77.25, respectively, calculated by the statistical thermodynamic methods (Chapter 2) for the gauche configuration of succinonitrile.[53]

5. Extension of Thermodynamic Values from the Aliphatic to the Aromatic Series

In the preceding chapter, empirical correction factors for "ring formation" were developed for the extension of the methods of group increments from alicyclic to aromatic systems. This problem has been considered, in the light of the method of group equations, by Bremner and Thomas[54] in some detail.

Some basic group equations for the formation of cyclohexane from aliphatic hydrocarbons, without corrections for changes in symmetry number and internal rotation, would be:

$$3G(n\text{-butane}) = G(\text{cyclohexane}) + 3G(\text{ethane}) \qquad (6.27)$$

$$2G(n\text{-pentane}) = G(\text{cyclohexane}) + 2G(\text{ethane}) \qquad (6.28)$$

$$2G(n\text{-hexane}) = G(\text{cyclohexane}) + 2G(\text{propane}) \qquad (6.29)$$

$$2G(n\text{-heptane}) = G(\text{cyclohexane}) + 2G(n\text{-butane}) \qquad (6.30)$$

and for cyclohexene:

$$G(2\text{-butene}) + 2G(n\text{-butane}) = G(\text{cyclohexene}) + 3G(\text{ethane}) \qquad (6.31)$$

where in each case the form of the group equation has been rearranged to give a balanced equation for the corresponding chemical reaction. Rather than use the above to estimate the thermodynamic properties for cyclohexane and cyclohexene, Bremner and Thomas examined the magnitude of the thermodynamic changes accompanying each process as expressed above in the light of the known thermodynamic properties for each compound. A summary of these results is found in Table 6.1, where the numbers enclosed in brackets are the values revised in the light of the more recent data.[1] The interest in the present discussion lies only in the order of magnitude rather than the exact values of these changes. Thus while the results based on the more recent data differ from the earlier calculations, the orders of magnitude for the changes remain essentially unchanged.

[53] G. J. Janz and W. E. Fitzgerald, unpublished work (1957).
[54] J. G. M. Bremner and G. D. Thomas, *Trans. Faraday Soc.* **43**, 779 (1947).

TABLE 6.1

THERMODYNAMIC CHANGES ACCOMPANYING THE FORMATION OF CYCLOHEXANE AND CYCLOHEXENE FROM SOME ALIPHATIC HYDROCARBONS[54]

Group equation	$\Delta G^\circ_{298.1}$ (kcal./mole)	$\Delta H^\circ_{298.1}$ (kcal./mole)	$\Delta S^\circ_{298.1}$ (cal./deg./mole)
(6.27)	$\begin{cases} -5.0 \\ (-3.7) \end{cases}$	$\begin{matrix} -0.9 \\ (-0.3) \end{matrix}$	$\begin{matrix} 14.2 \\ (13.5) \end{matrix}$
(6.28)	− 4.4	− 0.1	14.7
(6.29)	− 4.0	0.6	15.0
(6.30)	− 4.4	0.5	15.9
(6.31)	− 5.9	− 0.7	17.3

Inspection of the results (Table 6.1) shows that the heat of formation, in the first approximation, is additive even though ring closure occurs. The entropy change is very similar (14–17 e.u.) for the five group equations, and can be understood in the light of changes in restricted internal rotation and symmetry accompanying each process. The equations represent processes in which there is an increase in the entropy of translation, each being of the general type:

$$N(\text{molecules}) = N(\text{molecules}) + 1 \qquad (6.32)$$

Calculation of the increase of translation entropy by the Sackur-Tetrode equation:

$$S^\circ_{\text{trans.}} = \tfrac{3}{2} R \ln M + \tfrac{5}{2} R \ln T - R \ln P - 2.298 \qquad (6.33)$$

gives a value of about 35 e.u. for these processes. From comparison with the observed entropy increase (14–17 e.u.), it follows that the process of cyclization is accompanied by a decrease of about 20 e.u. This decrease can be accounted for as changes in the internal entropies in the light of symmetry and hindered internal rotation concepts.

For butane, cyclohexane (chair structure), and ethane, the symmetry numbers are 2, 6, and 6, respectively, the molecules being considered as rigid rotators. Accordingly an entropy decrease of:

$$R (\ln 6 + 3 \ln 6 - 3 \ln 2) = 10 \text{ e.u.} \qquad (6.34)$$

results as a consequence of the changes in symmetries of the molecules in the group equation for the estimation of the thermodynamic properties of cyclohexane from butane and ethane.

Bremner and Thomas approximate the contributions due to hindered internal rotation as equal to that of an ethane-like system, having a potential energy barrier of about 3 kcal. Neglecting the small vibrational contributions, this gives a value of 1.5 e.u. as the entropy contribution for each degree of internal rotation. Each C—C single bond in the aliphatic series is assumed to contribute one degree of internal rotation. For the above group equation the net change in the number of internal rotations is 6, and this corresponds to a further entropy decrease of 9 e.u. for the over all process.

Similar considerations, applied to the group equation for cyclohexene formation, show that the entropy decrease of 17 e.u., after allowing for the translational effect, may be accounted for by the increase in symmetry, 8 e.u., and the decrease in internal rotations, 9 e.u., accompanying this process.

In extension of thermodynamic values from the aliphatics to the aromatic series two further points, as well as the preceding, arise. The group equation:

$$3G(cis\text{-}2\text{-butene}) = G(\text{benzene}) + 3G(\text{ethane}) \tag{6.35}$$

rearranged to correspond to the formation of benzene by the de-ethanation of cis-2-butene was selected[35] to illustrate these points. Using the thermodynamic data in the literature for each compound, the thermodynamic changes in ΔG_T°, ΔH_T°, and ΔS_T° were calculated from 298.1° to 1000° K.

It was found that the heat of reaction, which corresponds to the resonance energy of benzene, was constant (36.7 ± 0.2 kcal./mole) over the whole temperature range. Calculation of the resonance energy by the method of group equations, in contrast to the conventional procedure based on bond energy equations, permits an estimate of the resonance energy at temperatures other than room temperature. The entropy change for the process (6.35) was found to be constant (13.5 ± 0.1 e.u.) in the above temperature range. As with the alicyclics, the entropy change can be accounted for largely as due to translational, internal rotational, and symmetry effects. For the above group equation, the calculated increase in translational entropy is 33.4 e.u., of which 13.4 e.u. are observed. The changes in symmetry, ($R \ln 12 - 3 R \ln 6 - 3 R \ln 2$) and internal rotations (3×1.5) account for an additional 16 e.u. The small remaining decrease (4 e.u.) is suggested to be due, possibly, to a resonance effect as distinct from unconjugated olefins in the aliphatic series.

The latter point may be illustrated more clearly from a consideration of the group equation:

$$3G(\text{cyclohexene}) = G(\text{benzene}) + 2G(\text{cyclohexane}) \tag{6.36}$$

which differs from the preceding group equation (6.35) in that the entropy changes attributed to changes in translational and internal rotational degrees of freedom are no longer present. The entropy change for this process at 298.1° K calculated from the known thermodynamic properties of the compounds was found to be –9.7 e.u. Comparison of the symmetry numbers of the molecules shows that a decrease of 6 e.u. can be accounted for by this factor. The difference, 3.7 e.u., if significant, may be considered as a decrease due to resonance effects, e.g., changes in the vibrational spectra in the aromatics relative to that of aliphatic olefins, conjugated and unconjugated, and possible in part due to the higher symmetry of benzene than that of cyclohexatriene.

In summary, the extension of the thermodynamic values of the aliphatic series to alicyclic and aromatic compounds by the method of group equations implies a process accompanied by an increase in the entropy of translation. When this effect is allowed for, the entropy decrease that remains may, for the greater part, be attributed to changes in symmetry numbers and internal rotations of the molecules being considered. In the aromatic series, an additional small decrease in entropy remains to be explained (approximately 4 e.u. for benzene). This may be due to entropy changes associated with resonance effects as distinct from unconjugated olefinics in the aliphatic series. The method of group equations seems applicable for this purpose providing due cognizance is taken of the changes in symmetry and internal rotations in the estimation of free energy or entropy properties.

CHAPTER 7

Heat of Formation and Heat Capacity

1. Introduction

The importance of an accurate value of the heat of reaction to obtaining a reliable thermodynamic analysis of a process can hardly be overstressed. The concept of heat of formation is the basic principle underlying the calculation of heats of reaction from thermochemical data. The most important auxiliary data required for computing the variations of the heats of formation, free energies, and entropies, are heat capacities. In the present chapter the interest lies in methods, other than those already described, for the estimation of the heats of formation and heat capacities of organic molecules.

2. Heat of Formation

Since only changes in heat content and not absolute values are measureable in practice, a common standard temperature and a standard state for each substance are selected to define heats of formation. It is customary to assign to each chemical element in its stable modification at the standard state conditions (25° C, one atmosphere), a heat of formation equal to zero. A table of standard heats of formation, $\Delta H°_{f298}$, is thus a compact summary for the standard heats of reaction at 25° C and one atmosphere for all chemical reactions, actual and hypothetical, which may take place among the substances included in the table. The well established methods for heat of formation based on direct measurements of heats of combustion and heats of reaction will not be discussed since these have been most adequately treated in the standard texts of physical chemistry and thermochemistry. The present interest is in the empirical methods of calculation based on correlations of data with regularities in the molecular structures of organic compounds.

3. Bond Energies and Binding Energies

The assumption that the energy of a given bond between two atoms is constant regardless of the other bonds to those atoms is the underlying principle of the various formulas proposed for calculation of the heat of formation for organic compounds. The theoretical implications of this assumption have been considered by Dewar.[55] The heat of formation ($\Delta H_f°$) of a compound may be expressed by the equation:

$$\Delta H_f° = \Delta E_0 + \Delta E_z + \int_0^T C_p \, dT - \int_0^T C_p' dT \tag{7.1}$$

where ΔE_0, ΔE_z, and C_p are the total binding energy, zero-point energy, and specific heat of the molecule, respectively, and C_p' is the sum of the specific heats of the atoms from which it is formed. If $\Delta H_f°$ is an additive function of the bonds, it follows that all the terms in the above equation must be additive. The expression for the heat of formation may thus be written:

$$\Delta H_f° = \sum \Delta E_0^* + \sum \Delta E_z^* + \sum \int_0^T C_p^* dT - \int_0^T C_p' dT \tag{7.2}$$

in which ΔE_0^*, ΔE_z^*, and C_p^* are contributions of the individual bonds to the whole. Not only the binding energy, but also the specific heat and zero-point energy of a molecule, must be divisible into contributions by individual bonds. The success of Pauling's bond energies in predicting heats of formation for a wide range of unconjugated molecules is in accord with the above conclusions.

More recently, Cottrell[56] has investigated the role of the zero-point energy (ΔE_z) and the chemical binding energy (ΔE_0) as expressed in Eq. 7.1 specifically for hydrocarbons. On the basis of a review of existing data on heats of formation for hydrocarbons, Rossini,[57] has suggested that greater stability is imparted to hydrocarbon molecules by the following factors: (a) resonance, (b) minimum departure of the bond angles from the tetrahedral value, (c) maximum compactness of the carbon skeleton, and (d) minimum repulsion between nonadjacent bonded atoms. In a qualitative way these generalizations imply that for any given pair of isomers the branched chain isomer conforming most fully to the above conditions will have the greatest heat of formation. The principle of the increased stability of branched chain isomers is recognized as the "Rossini effect." From an inspection of the zero-point energies of some n- and branched isomeric hydrocarbons, e.g., n-butane, 80.6 kcal/mole; isobutane, 80.2; n-pentane, 98.5; neopentane, 96.8; n-hexane, 116.2; 2,2-dimethyl-butane, 114.5; 1-butene, 65.9; isobutene, 65.5, it is readily apparent that the "Rossini effect" may be accounted for, in part, by zero-point energy differences.

A comparison between the heats of formation at 0° K, and the zero-point energies for a number of n- and branched isomeric compounds is shown in Table 7.1. The third column lists the differences in the heats of formation,

[55] M. J. S. Dewar, *Trans. Faraday Soc.* **42**, 767 (1946).
[56] T. L. Cottrell, *J. Chem. Soc.* p. 1448 (1948).
[57] F. D. Rossini, *Ind. Eng. Chem.* **29**, 1424 (1937); *Chem. Revs.* **27**, 1 (1940).

TABLE 7.1

COMPARISON OF INCREASED STABILITY AND ZERO-POINT ENERGY DIFFERENCES FOR SOME HYDROCARBONS[56]

Compound		$\Delta H_{f\,0°K}^{\circ}$ Difference (B–A) (kcal./mole)	ΔE_z Difference (A–B) (kcal./mole)	Percentage of effect due to zero-point energy
A	B			
n-butane	isobutane	1.27	0.4	31
n-pentane	neopentane	4.03	1.7	42
n-hexane	2,2-dimethylbutane	3.63	1.7	47
1-butene	isobutene	3.49	0.6	17

while the fourth, the zero-point energy differences between each isomeric pair. From the last column it is apparent that in some instances the zero-point energy accounts for almost as much as one half of the "Rossini effect." While the data in Table 7.1 may be regarded as little more than qualitative in significance, they do show that an appreciable part of the increased stability of the branched chain compounds can be attributed to the smaller zero-point vibrational energies of these compounds.

The chemical binding energy (ΔE_0) for a hydrocarbon is calculated by subtracting the zero-point vibrational energy and its heat content relative to the molecule at 0°K from the total heat of formation (cf. Eq. 7.1). It was shown[56] that this may be divided into "binding-energy terms" per bond, which are additive, and may be used for estimations of steric strain and resonance energies. The concept of binding energies for estimates of steric strain seems somewhat more sensitive than the conventional method. It was noted by Cottrell that the observed binding energy of cyclohexane was less than that calculated from the sum of the "binding-energy terms" for the corresponding bonds. This indication of the relative instability of cyclohexane is not shown by comparison of the heat of formation with that calculated from conventional bond energies. In general, however, the replacement of conventional bond energy terms by "binding-energy terms" has little effect on discussions involving them.

The use of tables of bond energies to reproduce heats of formation is not possible largely owing to deviations from additivity in the preceding points (cf. Eq. 7.2) and because an indisputable value for the energy of formation of gaseous monoatomic carbon has not yet been established. The estimation of heats of formation is expressed by the relation:

$$\Delta H_f^{\circ} = -\sum n_i q_i + \sum \Delta H_A \qquad (7.3)$$

in which $\sum \Delta H_A$ is the sum of the heats of formation of the gaseous atoms produced by complete decomposition of the compound, and n_i and q_i are the number and energies of the bonds in the molecular structure. In practice it is found that there is no such thing as a strictly constant bond energy between two atoms that persists over a varied series of compounds. The values in Table 1, Part II, from a recent compilation,[58] may be taken as average values from which actual bond energies do not deviate too widely. The ultimate value for the heat of sublimation of graphite at the absolute zero, i.e., for the process:

$$C_{(graphite)} = C_{(gas, \, ^3P \, state)} \; (0° \, K) \tag{7.4}$$

has long been in dispute, the values ranging from about 130 to 210 kcal./mole. Gaydon[59] recommends a value of 170.4 kcal./mole in his comprehensive compilation of dissociation energies. Additional support for this value is found in the recent thermodynamic measurements of Brewer, Gilles, and Jenkins.[60] It should be noted that the values in Table 1 are for 0° K. The heat of formation at the absolute zero, $\Delta H_0°$ can be calculated directly by:

$$\Delta H_0° = \Delta H_{fT}° - \Delta(H_T° - H_0°) \tag{1.11}$$

The division of $\Delta H_0°$ into a part for each chemical bond for reference compounds has been used to compile the values in Table 1, Part II.

Example 7.1. Estimate the heat of formation of ethane using bond energy data.

SOLUTION:

The heat of formation may be estimated from the assumption that all C—H bonds are equivalent. According to Eq. 7.3, and Table 1, Part II, therefore:

$$\Delta H_{f0°K}° = -6(98.2) - (80.5) + 6(103.2/2) + 2(170.4) = -19.3 \text{ kcal./mole}$$

(cf. lit. value,[1] $\Delta H_{f0°K}°$, −16.52 kcal./mole).

Example 7.2. Estimate the heats of formation of n-hexane and its isomers, 2-methylpentane, and 2,2-dimethylbutane from bond energy data. (ANSWER: −32.5 kcal./mole. The calculated values using bond energies are all the same. The literature values,[1] based on experimental data, are −30.91, −32.08, and −34.65 kcal./mole at 0° K).

Because of the assumption that all C—H bonds, for example, are the same, the heats of formation calculated for isomeric hydrocarbons from tables of bond energies (e.g. Table 1, Part II) are necessarily the same. The differences in heats of formation observed experimentally for isomeric hydrocarbons are definite deviations from the above simple principle of additivity.

[58] K. S. Pitzer, *J. Am. Chem. Soc.* **70**, 2140 (1948); "Quantum Chemistry", Prentice Hall, New York, 1953.
[59] A. G. Gaydon, "Dissociation Energies and Spectra of Diatomic Molecules," Chapman and Hall, London, 1947.
[60] L. Brewer, P. Gilles, F. A. Jenkins, *J. Chem. Phys.*, **11**, 797 (1948).

4. Heat of Formation from Group Increments

It has been seen in the preceding chapters that the heat of formation of an organic compound is an additive function of characteristic group equivalents. Since accurate heats of formation cannot be estimated by means of bond energies, Rossini[57] recommended the use of empirical equations based on regularities in heats of formation in the various homologous series of organic compounds.

The results in Fig. 7.1 show the deviations from linearity of the heat of formation in relation to the number of carbon atoms for three homologous

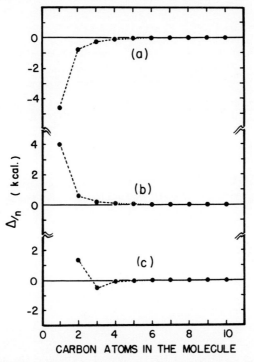

FIG. 7.1. Regularities in heats of formation of some homologous series of organic compounds.[57] (a) n-Hydrocarbons; (b) n-Alkyl Primary Alcohols; (c) n-(Alkene-1) Olefins.

series of gaseous molecules, normal paraffins, normal olefins, and primary normal aliphatic alcohols. The data may be expressed by a quite general equation for the heat of formation:

$$\Delta H_f^\circ = A' + Bm + \delta \tag{7.5}$$

where for each homologous series, $(H-(CH_2)_m-Y)$, A' is a constant characteristic of the end group Y, B is a universal constant for all series, and m is the number of carbon atoms in the normal alkyl radical. The deviation from linearity, δ, for $m > 5$ is zero. The magnitude and the sign of the deviations from linearity for members with $m < 5$ is dependent on the specific interactions of adjacent but nonbonded atoms and deviations from additivity in the zero-point energies. These contributions to the net deviation, as seen in Fig. 7.1, are greatest for the first member of the series, and may be of such sign and magnitude as to produce an irregular trend (cf. olefins, Fig. 7.1).

A summary of the values for the parameters in Eq. 7.5 for several homologous series of hydrocarbons, taken from the work of Prosen, Johnson and Rossini,[61] is given in Table 7.2. The results illustrate the magnitudes of the

TABLE 7.2

Values of the Deviations from Linearity of the Heats of Formation for Several Homologous Series of Hydrocarbons

Series Structures	Normal paraffins Methyl-$(CH_2)_m$-H	Mono olefins (1-alkene) Vinyl-$(CH_2)_m$-H	Normal alkylbenzenes Phenyl-$(CH_2)_m$-H	Normal alkylcyclopentanes Cyclopentyl-$(CH_2)_m$-H	Normal alkylcyclohexanes Cyclohexyl-$(CH_2)_m$-H
A' (kcal./mole)	-15.334	9.740	16.404	-20.516	-31.246
B (kcal./mole) per CH_2 group	-4.926	-4.926	-4.926	-4.926	-4.926
δ, in kcal.mole					
$m = 0$	-2.55 ± 0.37	2.76 ± 0.37	3.42 ± 0.38	2.05 ± 0.40	1.82 ± 0.41
$m = 1$	0.02 ± 0.33	0.07 ± 0.33	0.47 ± 0.34	-0.06 ± 0.37	-0.82 ± 0.40
$m = 2$	0.37 ± 0.30	0.39 ± 0.33	0.57 ± 0.33	-0.01 ± 0.36	0.05 ± 0.45
$m = 3$	0.30 ± 0.28	0.04 ± 0.48	0.24 ± 0.30	-0.10 ± 0.37	-0.18 ± 0.37
$m = 4$	0.04 ± 0.23	0.00 ± 0.47	0.00 ± 0.35	0.00 ± 0.39	0.00 ± 0.37
$m = 5$	0.00 ± 0.23	0.00 ± 0.47	0.00 ± 0.35	0.00 ± 0.39	0.00 ± 0.37

deviations from linearity with increasing values of m, and the accuracy possible for estimation of heats of formation by group increments providing precisely established values are available for the lower homologous members. The uncertainties assigned to the values of δ include the uncertainty associated with the extrapolation of the linear part of Eq. 7.5 to the lower values of m.

In Table 7.3 the values of δ for $m = 0$ for several end groups (Y) in addition to those in Table 7.2 are given from the same work. It is seen that the values of δ vary regularly with changes in the number of hydrogen atoms and the number and kind of carbon atoms on the attaching carbon atom of the Y group.

The heats of formation for the branched chain isomers have been calculated using the appropriate n-paraffin hydrocarbon and experimental heats of

[61] E. J. Prosen, W. H. Johnson, and F. D. Rossini, *J. Research Natl. Bur. Standards* **37**, 51 (1946).

isomerization. With sufficiently accurate data available for the lower members in a homologous series this method of group increments makes it possible to

TABLE 7.3

VARIATION IN THE DEVIATION FROM LINEARITY OF THE HEATS OF FORMATION AND STRUCTURE OF THE END GROUP IN HOMOLOGOUS SERIES OF HYDROCARBONS[61]

End group	Value of δ for $m=0$ (kcal./mole)	Bonds associated with the main or attaching carbon atom of the given end group (not including the attaching bond)				
		C—H	C—C (paraffin)	C=C (phenyl)	C=C (olefin)	C≡C (acetylene)
Methyl	-2.55 ± 0.37	3	—	—	—	—
Ethyl	0.02 ± 0.33	2	1	—	—	—
Cyclohexyl	1.82 ± 0.41	1	2	—	—	—
Cyclopentyl	2.05 ± 0.40	1	2	—	—	—
Isopropyl	2.05 ± 0.50	1	2	—	—	—
tert.-Butyl	2.55 ± 0.50	—	3	—	—	—
Vinyl	2.76 ± 0.37	1	—	—	1	—
Phenyl	3.42 ± 0.38	—	—	2	—	—
Isopropenyl	3.68 ± 0.40	—	1	—	1	—
Ethynyl	4.94 ± 0.55	—	—	—	—	1

deduce, without actual measurement, quite precise heats of formation for the higher members. Extension of this approach to organic molecules other than hydrocarbons awaits accurate experimental data for the first few members in the various homologous series.

It should be recalled that in addition to the methods discussed in this section, the methods of group increments (Tables 17–21, 23, 26, 31, 32, 35, and 36, Part II) and group equations considered in previous chapters may be used for the estimation of heats of formation.

Example 7.3. Estimate the heat of formation for 1-hexene at 298.1° K using the methods of group increments and group equations.

SOLUTION:

1-Hexene (CH_2=CH—CH_2—CH_2—CH_2—CH_3).

(a) With Eq. 7.5 and Table 7.2, the value is $\Delta H^{\circ}_{f298(g)} = 9.740 - 4.926(4) + 0 = -9.964$ kcal./mole.

(b) With the method of Andersen, Beyer, and Watson and Tables 17–21, Part II, the value is -10.2 kcal./mole.

(c) The method of Franklin, and Table 23, Part II, gives an estimate of -9.91 kcal./mole for this value.

(d) Using the method of Souders, Màtthews and Hurd, and Table 31, Part II, one obtains -9.82 kcal./mole.

(e) Using the method of group equations, and the heats of formation compiled by Rossini and associates[1] for propane, n-hexane, and propylene, i.e.,

$$\Delta H_f°(\text{1-hexene}) = \Delta H_f°(\text{propylene}) + \Delta H_f°(\text{hexane}) - \Delta H_f°(\text{propane})$$

gives a value of -10.26 kcal./mole.

The values are to be compared with the literature value[1], -9.96 kcal./mole.

5. Gaseous Free Radicals and Ions

The method of group increments recently has been extended by Franklin[62] to the calculation of the heats of formation of gaseous free radicals and carbonium ions. The basic principle is expressed by the equation:

$$\sum \Delta H_{gi} + \Delta H_{g°} + \Delta H_{g^+} + R = \Delta H_f° \qquad (7.6)$$

where ΔH_{gi} is that part of the heat of formation assigned to the ith group, and $\Delta H_{g°}$ and ΔH_{g^+} are, respectively, the group equivalent values for groups to which an unpaired electron and a positive charge are formally assigned. R is the resonance energy of the molecule. The construction of the correlation was similar to that described for molecules, where the value of $\Delta H_{g°}$ and ΔH_{g^+} for a group has to be calculated from the known heat of formation of a radical or ion containing that group. For ions the appearance potentials, measured under electron impact in a mass spectrometer, can be considered as the heats of reaction for the process. The heat of formation of the ion can be calculated from this datum and the known heat of formation for the nonionized fragment. Thus, for ethane, the appearance potential for the $C_2H_5^+$ ion is 12.84 volts or 296.1 kcal./mole. According to the process:

$$e + C_2H_6 \rightarrow C_2H_5^+ + H + 2e - 296.1 \text{ kcal.} \qquad (7.7)$$

a value of ΔH_f for the ion equal to 223.9 kcal./mole is gained, using -20.2 and 52 as the respective heats of formation for C_2H_6 and H. It is recognized that the carbonium ions rearrange very rapidly unless the ion formed has its charge at the most stable position. Thus it is possible that ions such as n-propyl and isobutyl would, respectively, rearrange to the more stable i-propyl and t-butyl isomers during the process of ionization. Accordingly, the nominally primary ions of the above such groups are formed with approximately the same amount of energy. Thus pseudo-equivalents can be employed with primary ions capable of rearrangement.

The group equivalents for radicals and ions, both at 298° K are given in Tables 41 and 42, respectively, Part II. For application in practice, the group equivalents of $\Delta H°_{f\,298}$ in Tables 23 and 26 are required. Franklin has given two additional ring corrections for $\Delta H°_{f\,298}$: thiocyclopropane, 16 kcal./mole; ethylene oxide, 22 kcal./mole.

[62] J. L. Franklin, *J. Chem. Phys.* **21**, 2029 (1953).

With the aid of the group equivalents for the free radicals, it is possible to achieve an estimate for the strength of bonds in certain ring compounds. Thus for cyclohexane, if a single C—C bond is broken symmetrically, the process would be:

$$\text{cyclohexane} = \dot{C}H_2(CH_2)_4\dot{C}H_2 \quad (7.8)$$

According to Eq. 7.6, it follows that:

$$\Delta H_f° [C_6H_{12}] = \Delta H_f [\dot{C}H_2(CH_2)_4\dot{C}H_2] - D(C\text{—}C)$$

$$= 4\Delta H (>CH_2) + 2\Delta H(\text{—}\dot{C}H_2) - D(C\text{—}C) \quad (7.9)$$

where D is the bond dissociation energy in question. Using the group increments (Tables 23, 41), $D(C\text{—}C) = 78$ kcal./mole. The rupture of this bond, which has not been determined, may be expected to be comparable to the bond rupture in n-hexane leading to the formation of two n-$\dot{C}_3H_7$ groups. The strength of this bond has been given[63] as 76 kcal./mole, a value which compares closely with that calculated for cyclohexane by group increments. In Table 7.4 some of the calculated bond strengths in cyclic compounds, compiled by Franklin,[62] are given. It is to be noted that the values correlate qualitatively with the extent of strain in the rings. As the ring size becomes smaller strain effects increase, and the calculated bond strengths decrease. It is pointed out that such values may indeed represent only upper limits since interactions between electrons formally assigned to "nearest neighbor" atoms would reduce the heat of formation of the diradical.

With reference to the carbonium ions, the method of group equivalents similarly is applicable to the estimation of ionization potentials, or heats of formation of ions. Thus the resulting ion from 2-butene would be expected to be:

$$CH_3\text{—}CH=CH\text{—}CH_3 = CH_3\text{—}\underset{H}{\dot{C}}\text{—}\underset{H}{\overset{+}{C}}\text{—}CH_3 \leftrightarrow CH_3\text{—}\underset{H}{\overset{+}{C}}\text{—}\underset{H}{\dot{C}}\text{—}CH_3 \quad (7.10)$$

where the possibility of resonance between two forms is shown. Calculation of the heat of formation of this ion by group increments (Tables 23, 41, and 42) gives a value of 221 kcal./mole. Combined with the heat of formation for 2-butene, the value of 9.6 electron volts is estimated for the ionization potential. This is to be compared with the value of 9.3 e.v. from direct experiment.[64] The method, thus, permits a ready calculation of the ionization potentials of olefins, aromatics and other unsaturates. It does not apply directly to the calculation of the ionization potentials for saturated organic compounds. In the parent ions of such molecules, there is no way of knowing which electrons are removed, so that a formal location of the charge in the

---

[63] M. Szwarc, *Chem. Revs.* **47**, 75 (1950).
[64] R. E. Honig, *J. Chem. Phys.* **16**, 105 (1948).

TABLE 7.4

CALCULATED BOND STRENGTHS IN CYCLIC COMPOUNDS[62]

		D, kcal./mole
△	→ Ċ—C—Ċ	50
□	→ Ċ—C—C—Ċ	60
⬠	→ Ċ—C—C—C—Ċ	72
⬡	→ Ċ—C—C—C—C—Ċ	78
C—C with S bridge	→ Ċ—C—Ṡ	52
	→ Ċ—S—Ċ	50
C—C with O bridge	→ Ċ—C—Ȯ	49
	→ Ċ—O—Ċ	53

molecule is not possible. Consequently group equivalents cannot be calculated for the saturated compounds, but rather only for the unsaturates as noted above. It is further noted[62] that the method must be used with reservation also when strongly electrophilic groups are present in the ion near the position where the charge is formally located.

6. Heat of Combustion by Group Increments

In a review on the status of data and calculations for the heats of formation of simple organic molecules, Rossini[57] has cautioned that practically none of the data before 1937 on organic compounds can be used in thermodynamic calculations where errors in $\Delta H_f°$ of several kcal./mole are significant. This point is illustrated in Fig. 7.2, in which graphical comparisons of the various data on heats of combustion of ethylene and propylene are given. The scale of ordinates gives the heat of combustion at 25° C and a constant pressure of one atmosphere in international kilojoules per mole. A large part of the thermochemical data for complex organic compounds is dependent upon the earlier experimental results. Valuable compilations of these data and references are given by Bichowsky and Rossini[37] for some molecules of 1 and 2 carbon atoms, and Kharasch[38] for some 1500 compounds of all types.

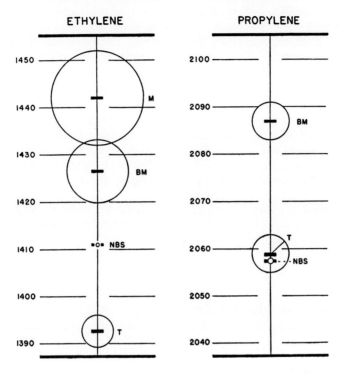

FIG. 7.2. Values of the heats of combustion of ethylene and propylene. The radii of the circles give the estimated uncertainties in the values of: T, Thomsen, 1886; BM, Berthelot and Matignon, 1893; M, Mixter, (1901); and N.B.S., Rossini and Knowlton, 1937.

The observations of Rossini concerning limits of accuracy, accordingly, still pertain, especially for organic compounds other than the hydrocarbons.

An empirical correlation of the heats of combustion has been proposed by Kharasch and Sher,[65] from the above survey of data, based on the electronic concepts of valence theory. With assumption that whenever an organic substance is burned in oxygen the heat generated is due to the interdisplacement of the electrons between the carbon and oxygen atoms, it was calculated that energy equivalent to 26.05 kcal. per electron mole was evolved for the methane and carbon dioxide configurations as standards of reference. This led directly to an empirical equation of the type:

$$-\Delta H_c = 26.05\,N + w \tag{7.11}$$

where $-\Delta H_c$ is the calculated heat of combustion and N is the number of electrons involved, and w is a constant. Since the products of combustion

[65] M. S. Kharasch and B. Sher, *J. Phys. Chem.* **29**, 625 (1925).

for an organic compound are practically always the same, the C atoms ending as CO_2 molecules, and the H and O atoms as H_2O molecules, the additivity rule (Eq. 7.11) implies more or less definite covalent bond energies, independent of the particular molecule in which the bond exists. The parameter, w, is an empirical constant correcting for deviations of the electron distribution from that in the reference position (i.e., the arrangement of electrons around carbon in methane). Thus it was found the values of w were -3.5, $+13$, $+16.5$ kcal./electron mole for the aromatic-alkyl, olefinic, and nitrile bonds respectively.

A summary of equations based on such structural correlations for the calculation of heats of combustion is given in Table 73, Part II, for various distinct types of compounds. The subscripts on the correction factors refer to the corresponding structural correlations in Table 74. The following examples illustrate the application of this method.

Example 7.4. Using the empirical structural correlations of Kharasch,[38] i.e., Table 73, calculate the heat of combustion for (a) p-toluidine, (b) saccharinic acid lactone, (c) o-toluic acid.

SOLUTION:

Inspection of the formula for each compound suggests structural correction factors for use with Eq. 7.11, i.e.,

Compound:	p-Toluidine	Saccharinic acid lactone	o-Toluic acid
Empirical formula	C_7H_9N	$C_6H_{10}O_5$	$C_8H_8O_2$
Structural formula	CH_3—〈 〉—NH_2	(see structure)	〈 〉—COOH, CH_3
N	37	24	36
w	$(6.5 - 3.5 - 3.5)$	$(13 + 3.5 + 6.5 + 13)$	$-2(3.5)$
$-\Delta H_{c18}\ °C(\text{liq.})$ (kcal./mole)	963.3	662.2	930.8
Lit. value (kcal./mole)	958.4	656.6	928.9

Structure for saccharinic acid lactone:

$$\begin{array}{c} CH_3 \\ | \\ OC—COH \\ | \quad\quad | \\ O \quad HCOH \\ \llcorner| \\ —CH \\ | \\ CH_2OH \end{array}$$

The literature values[38] are for $\Delta H_c°$ in the combustion of the compounds as solids (15° C).

Example 7.5. Calculate the heats of combustion for the following compounds: (a) safrole, (b) retenequinone (methylisopropylphenanthraquinone), (c) eugenol acetate. (ANSWER: (a) N, 46; $-\Delta H_c°$ calc. 1246.8, lit. value 1244.1 kcal./mole; (b) N, 84; $\Delta H_c°$ calc. 2187.7, lit. value 2152.4 kcal./mole; (c) N, 56; $\Delta H_c°$ calc. 15043, lit. value 14988 kcal./mole. The literature values[38] are for combustion of the compounds as solids (15° C).

The equations in Table 73 are type formulas only. The values of the structural correction factors as listed in Table 74 are applied once for each such bond in the molecule (cf. Example 7.5 c). The calculated values of $\Delta H_c°$ refer to the combustion of the substance in the liquid or solid state.

Comparison of the values calculated by the Kharasch and Sher method with the observed data established that in most cases the two values agreed within the accuracy of the experimental work (i.e., within 1–2 %). An estimate of the heat of combustion, and thus a value for the heat of formation (not corrected to 0° K) is readily achieved by this exceedingly simple calculation.

The increment in the heat of combustion per CH_2 group added to the chain in various homologous series of organic compounds having n-alkyl chains has been very precisely established in the more recent investigations of Prosen and co-workers.[66, 67] Comparison of experimental values of $-\Delta H_c°$ with calculated values using the equation:

$$-\Delta H_c° = A + Bm \qquad (7.11a)$$

has shown that there are no serious deviations from linearity down to $m = 2$. A summary of these results, and deviations of the experimental values from the values calculated by Eq. 7.11a are found in Table 7.5. The values of B were evaluated by the method of least squares, using all of the available data at the National Bureau of Standards. The values of A were gained by weighting the experimental value of the heat of combustion inversely as the square of its assigned uncertainty for each compound. It follows from this that, within the limits of the experimental uncertainty, the $[CH_2]$ increment is constant beginning with n-propane. This is in accord with the concept that the average bond energy of a particular bond is affected, within experimental error, only by the bonds (or atoms) attached to the atoms of the bond under consideration. Bonds further removed have no significant effect on this bond energy unless major steric effects are present in the system. The values of A and B in Table 7.5 and Eq. 7.11a make possible quite accurate estimates of the heats of combustion for the 1-alkenes, n-alkyl benzenes, n-alkylcyclopentanes, and n-alkylcyclohexanes in the liquid state.

Example 7.6. Estimate the heat of combustion at 25° C for the liquid state for (a) ethylcyclopentane, (b) 1-hexadecene.

[66] E. J. Prosen and F. D. Rossini, *J. Research Natl. Bur. Standards* **34**, 263 (1945).
[67] F. M. Fraser and E. J. Prosen, *J. Research Natl. Bur. Standards* **55**, 329 (1955).

TABLE 7.5

Comparison of Experimental and Calculated Values of Heat of Combustion[67] and the Values of the Constants A and B for the Empirical Equation:[a]

$$-\Delta H_c^\circ = A + Bm \quad (m > 1)$$

Series Structure	1-Alkenes Vinyl—$(CH_2)_m$—H		Normal alkylbenzenes Phenyl—$(CH_2)_m$—H		Normal alkylcyclopentanes Cyclopentyl—$(CH_2)_m$—H		Normal alkylcyclohexanes Cyclohexyl—$(CH_2)_m$—H	
	$-\Delta H_c$ (liq.) expt.	Δ(expt.-calc.)	$-\Delta H_c$ (liq.) expt.	Δ(expt.-calc.)	$-\Delta H_c$ (liq.) expt.	Δ(expt.-calc.)	$-\Delta H_c$ (liq.) expt.	Δ(expt.-calc.)
$m = 0$	—	—	780.98 ± 0.10	(+2.49)	786.54 ± 0.17	(+1.49)	936.88 ± 0.17	(+1.19)
$m = 1$	—	—	934.50 ± 0.12	(−0.23)	941.14 ± 0.18	(−0.15)	1091.13 ± 0.23	(−0.80)
$m = 2$	644.46[b] ± 0.18	+0.10	1091.02 ± 0.17	+0.06	1097.50 ± 0.22	−0.03	1248.23 ± 0.35	+0.06
$m = 3$	800.61 ± 0.26	+0.01	1247.19 ± 0.16	−0.02	1253.74 ± 0.28	−0.03	1404.34 ± 0.27	−0.07
$m = 4$	956.72 ± 0.33	−0.12	1403.46 ± 0.27	+0.01	—	—	1560.78 ± 0.29	+0.13
$m = 5$	1112.87 ± 0.38	−0.21	—	—	—	—	—	—
$m = 10$	—	—	2340.58 ± 0.42	−0.31	2347.54 ± 0.46	+0.09	2497.90 ± 0.43	−0.19
$m = 14$	2519.17 ± 0.44	−0.07	—	—	—	—	—	—
A	—	331.88	—	778.49	—	785.05	—	935.69
B	—	156.24	—	156.24	—	156.24	—	156.24

[a] Units, kcal./mole   [b] At saturation pressure.

SOLUTION:

According to Eq. 7.11a and Table 7.5, the heats of combustion for alkyl cyclopentanes and 1-alkenes may be calculated by:

(a) $-\Delta H_c^\circ = 785.05 + 156.24\, m$

and

(b) $-\Delta H_c^\circ = 331.88 + 156.24\, m$

respectively, where $m > 1$. For ethylcyclopentane $m = 2$, and 1-hexadecene, $m = 14$. Thus the heats of combustion for ethylcyclopentane and 1-hexadecene are predicted as 1097.53 and 2519.24 kcal./mole (cf. lit.[67] value 1097.50 and 2519.17 kcal./mole). The agreement of the calculated with the observed values is well within the limits of experimental error in each case.

7. Heat of Vaporization†

The relation between the vapor pressure and temperature is expressed by the Clapeyron equation:

$$\frac{dp^0}{dT} = \frac{\Delta H_v}{T_0(V_g - V_l)} \tag{7.12}$$

where, for a pure liquid, p^0 is the vapor pressure at temperature T_0, ΔH_v is the molal latent heat of vaporization, and V_g, V_l are the molal volumes of the saturated vapor and liquid respectively. The equation is exact, but limited in practical application because the necessary data are frequently lacking.

An accurate value of ΔH_v can be gained from a modified form of the Clapeyron equation[7] (7.12), i.e.,

$$\Delta H_v = 2.30259 \left[1 - \frac{PV}{zRT}\right] \frac{(z\, BRT^2)}{(C' + T)^2} \tag{7.13}$$

in which z is the compressibility factor, i.e.,

$$z = \frac{PV_g}{RT},$$

and B and C' are the constants of the Antoine vapor pressure equation, i.e.,

$$\log p^0 = A - \frac{B}{C' + T} \tag{7.14}$$

the temperature being expressed on the absolute scale. Thus for the precise calculation of the heat of vaporization a knowledge of the compressibility factor and the constants B and C' from vapor pressure data are required. An approximate form of (7.12) has been derived by Watson[68] for estimation of ΔH_v at the normal boiling point, i.e.,

$$\Delta H_v = 0.95\, RB \left(\frac{T_B}{T_B - 43}\right)^2 \tag{7.15}$$

[68] K. M. Watson, *Ind. Eng. Chem.* **35**, 398 (1943).

† An authoritative survey of the various equations proposed in the past 20 years, with illustrations of calculation procedures for estimates of heats of vaporization with less than 5% error, is given by S. H. Fishtine, *Ind. Eng. Chem.* **55**, 20, 47 (1963).

where 0.95 is an average value of the compressibility factor. The constant B may be evaluated from a knowledge of the vapor pressure at any two temperatures by the relation:

$$B = \ln p_2/p_1 \bigg/ \left(\frac{1}{T_1-43}\right) - \left(\frac{1}{T_2-43}\right) \tag{7.16}$$

Best results are obtained if T_1 and T_2 are relatively close to the boiling point.

At comparatively low temperatures and pressures, where it may be assumed that V_l is negligible relative to V_g, and the vapor obeys the ideal gas law, the differential form of the Clausius-Clapeyron equation:

$$\frac{dp^0}{dT} = \frac{\Delta H_v}{RT^2} p^0 \tag{7.17}$$

and the integrated form, if ΔH is assumed constant over a small temperature interval, i.e.,

$$\ln (p^0{}_1/p^0{}_2) = \frac{\Delta H_v}{R} \left[\frac{T_1-T_2}{T_1 T_2}\right] \tag{7.18}$$

may be used to calculate values of the heat of vaporization.

One of the oldest and simplest relations for estimation of the heat of vaporization at the normal boiling point is *Trouton's Rule*, which states that the molal entropy of vaporization of a liquid at its normal boiling point is 21 e.u., i.e.,

$$\frac{\Delta H_v}{T_B} = 21 \tag{7.19}$$

This rule has long been recognized as inexact, and that 21 is but an average value at best. Typical departures from the rule may be illustrated by the following values for Trouton's rule constant:

aniline, 21.1; ethyl acetate, 22.0; ethyl alcohol, 26.8; water, 26.0; octane, 20.4

Several empirical modifications of Trouton's original rule have been proposed, e.g., two of the most notable are:

Bingham's equation:

$$\frac{\Delta H_v}{T} = 17 + 0.011\, T_B \tag{7.20}$$

and Nernst's equation:

$$\frac{\Delta H_v}{T} = 9.5 \log T_B - 0.007\, T_B \tag{7.20a}$$

Neither of these have been found applicable over wide ranges of boiling points, Bingham's equation predicting results that are too high, and the Nernst equation predicting values that actually pass through a maximum, both at the higher boiling temperatures. The Kistiakowsky equation:

$$\frac{\Delta H_v}{T} = 8.75 + 4.571 \log T \tag{7.21}$$

has been deduced from the third law of thermodynamics and has been offered[69] as a fundamental law for the latent heat of vaporization at one atmosphere. This relation is applicable to nonpolar liquids only, an agreement within 3 per cent being claimed. Hildebrand[70] pointed out that $(\Delta H_v/T)$ is a constant at the temperatures which correspond to a constant molal volume of saturated vapor, and that these temperatures are not necessarily the normal boiling points. A convenient method for use of the *Hildebrand principle* has been described by Lewis and Weber[71] and McAdams and Morrell.[72] The values of Trouton's ratio $(\Delta H_v/T)$ are plotted against $\log(p/T)$ (since p/T is proportional to the vapor concentration for ideal gases). Two lines, both approximately linear, can be thus drawn, one for polar, and the other for nonpolar liquids. Thus ΔH_v can be predicted at various temperatures. The method breaks down at higher temperatures and pressures as the critical region is approached. A comparison of the calculated and observed values for ΔH_v using this principle and the Kistiakowsky equation is as follows:

Liquid	$(C_2H_5)_2O$		C_6H_5Cl	$n-C_8H_{16}$
T° K	273	453	404	398
$(\Delta H_v/T)_a$	23.7	4.8	20.7	21.0
$(\Delta H_v/T)_b$	24.2	4.9	20.7	20.6
$(\Delta H_v/T)_c$	25.0	4.9	21.5	20.4

where the subscripts *a*, *b*, and *c*, indicate the values calculated on the basis of the Hildebrand principle, the Kistiakowsky equation, and the literature data.[3] The Kistiakowsky equation was used to calculate ΔH_v at the normal boiling point, T_B and the heats of vaporization at the other temperatures above were estimated from an expression developed by Watson,[73] i.e.,

$$\frac{\Delta H_{v_1}}{\Delta H_{v_2}} = \left(\frac{1 - T_{r1}}{1 - T_{r2}}\right)^{0.38} \tag{7.22}$$

where T_{r1} and T_{r2} are the reduced temperatures corresponding to T_1 and T_2. Thus only a knowledge of the critical temperature is required for calculation of the complete range of heats of vaporization. The above relation (7.22) applies to both polar and nonpolar liquids with a claimed accuracy of 5 per cent except in the immediate vicinity of the critical region.

Othmer[74] has derived the expression:

$$\frac{\Delta H_{v_1}}{\Delta H_{v_2}} = \frac{T_{c_1} \log p_{r_1}}{T_{c_2} \log p_{r_2}} \tag{7.23}$$

[69] W. Kistiakowsky, *Z. physik. Chem.* **B 107**, 65 (1923).
[70] J. H. Hildebrand, *J. Am. Chem. Soc.* **37**, 970 (1915).
[71] W. K. Lewis and H. Weber, *Ind. Eng. Chem.* **14**, 485 (1922).
[72] W. H. McAdams and J. Morrell, *Ind. Eng. Chem.* **16**, 375 (1924).
[73] K. M. Watson, *Ind. Eng. Chem.* **23**, 362 (1931).
[74] D. F. Othmer, *Ind. Eng. Chem.* **32**, 841 (1940); **34**, 1072 (1942).

from the Clausius-Clapeyron equation, which relates the ΔH_v with the critical temperature and reduced pressure, when the heat of vaporization of a reference substance is known. Thus from a knowledge of the normal boiling point, critical pressure and temperature, ΔH_v is readily calculated. Good graphical methods have been described by Meissner[75] and Gordon,[76] based on the reduced pressures and temperatures of reference substances. The methods give estimated values of ΔH_v showing average agreements within 5 percent with the experimental data for a wide variety of liquids, polar and nonpolar.

Methods for the estimation of the three critical constants, have been developed by Meissner and Redding.[77] For polar compounds, or where the liquid density is unknown, the following equations were proposed to estimate the critical temperature:

(a) compounds with $T_B < 235°\,K$ and all elements:

$$T_c = 1.70\, T_B - 2.0 \tag{7.24}$$

(b) compounds with $T_B > 235°\,K$,

(i) that have halogens or sulfur:

$$T_c = 1.41\, T_B + 66 - 11F \tag{7.25}$$

where F is the number of fluorine atoms in the molecule,

(ii) that are aromatics or naphthenes free of halogens and sulfur:

$$T_c = 1.41\, T_B + 66 - r(0.383\, T_B - 93) \tag{7.26}$$

where r is the ratio of noncyclic C atoms to the total number of C atoms in the compound,

(iii) other than *(i)* and *(ii)*:

$$T_c = 1.027\, T_B + 159 \tag{7.27}$$

These equations have not been tried for liquids with the normal $T_B > 600°\,K$. The estimated values are claimed to give agreement within 5 per cent with experimental values for nearly all compounds with the exception of water. The critical pressure[77] may be estimated from a knowledge of T_c and v_c (the critical volume) by the equation:

$$p_c = \frac{20.8\, T_c}{(v_c - 8)} = \frac{20.8\, T_c}{(M/\rho_c - 8)} \text{ (atm.)} \tag{7.28}$$

[75] H. P. Meissner, *Ind. Eng. Chem.* **33**, 1440 (1941).

[76] D. G. Gordon, Ph. D. thesis, Univ. of Wisconsin, 1942; see also O. A. Hougen and K. M. Watson, "Chemical Process Principles," Vol. I, Wiley, New York, 1943.

[77] H. P. Meissner and E. R. Redding, *Ind. Eng. Chem.* **34**, 521 (1942).

7. HEAT OF FORMATION AND HEAT CAPACITY

which has been found to give values that are in accord with the experimental values to within 10 per cent. The critical volume required in the preceding equation is given by the relation:

$$v_c = (0.377\,\overline{P} + 11.0)^{1.25} \text{ (cc./mole)} \tag{7.29}$$

where $\overline{P}$ is the value of the *parachor* for the compound. The latter property is defined by the well-known relation:

$$\overline{P} = \frac{M\sigma^{0.25}}{(d_l - d_g)} \tag{7.30}$$

where σ is the surface tension and d_l, d_g are the densities of the liquid and vapor (gm./cc.), respectively. The parachor is principally an additive property for organic compounds and can be estimated quite accurately by the summation of the appropriate atomic and structural contributions. A summary of the atomic and structural parachor values is given in Table 75, Part II.

Since much of the earlier heat of combustion data is available only for one state, and since most of the preceding thermodynamic methods refer to the substance in the ideal gas state, a need for evaluation of heats of vaporization frequently arises. The preceding discussion, while not complete in coverage, is sufficient to illustrate some of the various approaches, theoretical and empirical, to this problem.

Example 7.7. Estimate the heat of formation, $\Delta H^\circ_{f\,298^\circ(g)}$, for triethylamine, $(C_2H_5)_3N$, b.p. 89.5° C.

SOLUTION:

The heat of combustion as a liquid at 18° C can be estimated from the empirical structural correlation of Kharasch. This value can be used to obtain the heat of formation as a liquid (18° C). The latter may be corrected to the ideal gas state from a knowledge of the heat of vaporization. The methods discussed in the preceding section are applied to estimate this value. The steps in the calculation of $\Delta H^\circ_{f\,298(g)}$ are outlined as follows:

(a) Triethylamine $(C_2H_5)_3N$. Estimation of the *heat of combustion*, $-\Delta H^\circ_{c(liq.)18^\circ C}$ by the method of Kharasch is achieved using equation 7.11, and Tables 73, 74. Inspection of the structural formula suggests the following calculation, since $N = 39$:

$$-\Delta H_c^\circ = 26.05(39) + 26_{gg} = 1015.9 + 26 = 1041.9 \text{ kcal./mole}$$

(b) The heat of formation as a liquid is calculated for the process:

Thus $\quad(C_2H_5)_3N_{(l)} + 13\tfrac{1}{2}\,O_{2(g)} = 6\,CO_{2(g)} + 7\tfrac{1}{2}\,H_2O + \tfrac{1}{2}\,N_2$

$$\Delta H^\circ_{f(liq.)} = 6(-94.052) + 7\tfrac{1}{2}(-68.320) - (-1041.9) = -34.81 \text{ kcal./mole}$$

(c) The heat of vaporization is estimated using Watson's estimation equations, 7.15, 7.16. To evaluate the constant B, the critical pressure and

temperature are selected as p_2, T_2, respectively. This requires an estimation of the critical properties of triethylamine.

(i) v_c. (Eq. 7.29, 7.30).
The value of the Parachor (Table 75, Part II) for $(C_2H_5)_3N$ is given by:

$$\overline{P} = 6(4.8) + 15(17.1) + 12.5 = 297.8$$

Accordingly, the critical volume is estimated to be:

$$v_c = [0.377(297.8) + 11.0]^{1.25} = 416 \text{ cc./mole}$$

(ii) T_c. Since triethylamine contains no halogen or sulfur and is neither aromatic nor naphthenic, the critical temperature may be estimated by Eq. 7.27, i.e.,

$$T_c = 1.027(362.5) + 159 = 531° \text{ K}$$

(iii) p_c. (Eq. 7.28) is calculated using the above values of T_c and v_c, i.e.,

$$p_c = \frac{20.8 \ (531)}{(461-8)} = 27.0 \text{ atm.}$$

(iv) B. The value of B may be calculated from Eq. 7.16 using p_c, T_c and T_B, $p = 1$, respectively, i.e.,

$$B = \frac{2.303 \log (27.0/1.0)}{[(1/362.5-43)-(1/531-43)]} = 3080$$

(v) ΔH_{vT_B}. From Eq. 7.15, we obtain:

$$\Delta H_v = (0.95)(1.99)(3080)\left(\frac{362.5}{319.5}\right)^2 = 7500 \text{ cal./mole}$$

(vi) $\Delta H_{v18°C}$. From Eq. 7.22, it follows that:

$$\frac{\Delta H_{v\,18°C}}{7500} = \left(\frac{1-291/531}{1-363/531}\right)^{0.38} = 1.15$$

Thus $\Delta H_{v18°C} = 8.6$ kcal./mole.

(d.) $\Delta H°_{f291(g)}$. From the preceding it readily follows that:

$$\Delta H°_{f291(g)} = \Delta H°_{f291(l)} + \Delta H°_{v291} = -34.8 + 8.6 = -26.2 \text{ kcal./mole}$$

The following values for triethylamine may be found in the literature[76]: $\Delta H°_{c291°(liq.)}$, -1037 kcal./mole; T_c, 535.2° K; v_c, 430 cc./mole, p_c, 30.0 atm. The value for $\Delta H°_{f298(g)}$, estimated by the method of group increments (Tables 20, 21, Part II), is -23.2 kcal./mole. Comparison of the estimated values, neglecting the small correction for the difference in temperature, shows that the agreement is all that could be expected in view of the approximations involved in the preceding calculations.

8. Heat Capacity

While the method of calculation based on statistical thermodynamics and spectroscopic data gives very precise results for the heat capacity of polyatomic molecules, the rigorous application of this approach to the more complex organic molecules requires extensive spectroscopic data and mathematical analysis. Mecke[78] proposed a simplification of this approach by the use of generalized vibrational frequencies for the stretching and bending modes of each valence bond. The discussion in this section is concerned with the estimation methods, based on this principle, which have been developed specifically for the calculation of heat capacities of organic molecules.†

9. Method of Generalized Vibrational Assignments

The basic principle of the method is the concept that the many vibrational frequencies associated with a bond may be grouped into two generalized frequencies, ν, and δ, the valence and deformation vibrational frequencies, respectively. The former represents the vibrational contributions acting in the line of vibrating bodies, and the latter the vibrational contributions acting perpendicular to the ν vibrations.

The concept of these generalized vibrational modes was applied by Bennewitz and Rossner[79] to compute the gaseous heat capacities of various organic molecules. The molar heat capacity for a polyatomic nonlinear molecule of the rigid rotator-simple vibrator type may be calculated from the relation:

$$C_v = 1.5R + 1.5R + \sum q_i C_{\nu_i} + \sum q_j C_{\delta_j} \tag{7.31}$$

where C_{ν_i} and C_{δ_j} are the generalized vibrational contributions of the Einstein functions for the valence and deformation modes for the system. The vibrational contribution for a generalized deformation is related to the valence vibrational contribution by the equation:

$$C_{\delta_j} = (\sum q_i C_{\delta_i}) / \sum q_i \tag{7.32}$$

and the number of deformation frequencies is simply given by:

$$\sum q_j = 3n - 6 - \sum q_i \tag{7.33}$$

Accordingly the basic relation (7.31) may be expressed in the form:

$$C_p^\circ = 4R + \sum q_i C_{\nu_i} + \left(\frac{3n - 6 - \sum q_i}{\sum q_i}\right) \sum q_i C_{\delta_i} \tag{7.34}$$

which is the equation of Bennewitz and Rossner. The ν frequencies were evaluated from Raman spectra, and the δ frequencies by a series of ap-

---

[78] H. Mecke, *Z. physik. Chem.* B **16**, 409, 421 (1932).
[79] K. Bennewitz and W. Rossner, *Z. physik. Chem.* B **39**, 126 (1938).
† For the prediction of liquid heat capacities see R. C. Reid and T. K. Sherwood, "Properties of Gases and Liquids." McGraw-Hill, New York, 1958.

proximations from less complex compounds. The generalized vibrational frequencies for carbon-hydrogen-oxygen compounds proposed by these investigators are summarized in Table 76, Part II. The above equation was found to reproduce experimental data within 5 per cent in the temperature region of 400° K. Fugassi and Rudy[80] have recalculated the Einstein functions used in the Bennewitz and Rossner equation in the form to fit the power series equation:

$$(C_p°)_i = a_i + b_i T + c_i T^2 \tag{7.35}$$

to simplify the application of this method in practice. The constants a_i, b_i, c_i for the C_{ν_i} and C_δ vibrational contributions corresponding to the generalized bond frequencies are given in Table 76, Part II.

On the assumption that the organic vapors obey Berthelot's equation of state, these data may be converted to heat capacity data at any pressure, p, by the expression:

$$(C_p°)_p = (C_p°) + R\left[1 + \frac{81}{23} \times \frac{p}{p_c} \times \left(\frac{T_c}{T}\right)^3\right] \tag{7.36}$$

where the critical data can be estimated when necessary, as a first approximation, by the simple relations; $T_c = 1.5\ T_b$ and $\log p_c = 3(T_c/T_b) - 1$, T_b being the normal boiling point. When converting heat capacities from zero to one atmosphere pressure, the increase in heat capacity calculated[40] by (7.36) is given by:

$$C_p = 5.03\ T_c^3/p_c T_b^3 \tag{7.37}$$

The equation of Bennewitz and Rossner was found to be inadequate at lower temperatures ($T < 400°$ K) as more experimental data became available. Thus for propane and butane, the predicted heat capacities were as much as 15 per cent less than the experimental values at 298° K. Dobratz[39] recognized that with free internal rotation in the molecule, this error would be corrected in part. The contribution for a rotational degree of freedom $(R/2)$ is 1 calorie, whereas the average contribution of a deformation vibrational mode at 298° K would be only about 0.5 calorie. To give more accurate results, accordingly, Dobratz modified Eq. 7.34 to the following form:

$$(C_p°)_p = 4R + \frac{aR}{2} + \sum q_i C_{\nu_i} + \left(\frac{3n - 6 - a - \sum q_i}{\sum q_i}\right) \sum q_i C_{\delta_i} \tag{7.38}$$

in which a equals the number of bonds permitting free rotation, C—C or similar. This equation reproduces the data for propane and butane at 298° K to within 5 per cent. A table of bond vibrational contributions to the heat capacity expressed as the three parameter power series (cf. Eq. 7.35) was compiled by Dobratz for the temperature range 300°–700° K, and for organic bonds with halogens, sulfur, and nitrogen as well as oxygen and hydrogen. These data are given in Table 76, Part II.

[80] P. Fugassi and C. E. Rudy, Jr., *Ind. Eng. Chem.* **30**, 1020 (1938).

7. HEAT OF FORMATION AND HEAT CAPACITY

The above method has been re-evaluated in the light of more recent spectroscopic data by Stull and Mayfield.[40a] The bond vibrational frequency assignment, together with the Einstein function contributions by these investigators are given in Table 77, Part II, for the temperature range 250°–1500° K. To increase the accuracy of the method, three types of C=C modes, rather than only one, were considered; namely, a symmetrical and an unsymmetrical type for aliphatics, and an aromatic type vibrational mode. Calculation of acetylenic derivatives was made possible by evaluating the generalized ν and δ frequencies, using the data of methylacetylene and dimethylacetylene for reference. The C—H deformation frequency, based on vibrational assignment for propane, rather than ethane and methane, was assigned a somewhat lower value. The remaining bond frequencies are substantially unchanged. The table of solutions in $C_{vib.}$ to the Einstein function corresponding to one degree of freedom calculated by Stull and Mayfield to extend the method to a wider temperature range is given in Table 48, Part II. The use of Einstein functions rather than the power series type correlations improves the low temperature precision, and extends the upper temperature range. Comparison of the literature data for some 29 hydrocarbons with the values estimated, using the revised vibrational frequencies and more detailed tables of Einstein functions, showed an average difference of ± 4 per cent.

Example 7.8. Estimate the heat capacity of propane at 300° K and 1000° K in the ideal gas state.

SOLUTION:

(a) Eq. 7.38 and Table 77, Part II. From the structural formula for propane:

$a = 2$, $\Sigma q = 10$, $n = 11$, and $\varphi = \dfrac{(3n-6-a-\Sigma q)}{\Sigma q} = 1.5$

From these factors, Eq. 7.38, and Table 77, one obtains:

Bond	q.	φ	300° K	1000° K
C—H, ν	8	—	0.0032	4.3792
C—H, δ	8	1.5	2.1912	18.3756
C—C, ν	2	—	0.7982	3.3696
C—C, δ	2	1.5	4.4940	5.8062
$4R + R$	—	—	9.9350	9.9350
$\Sigma = C^\circ_{pT°K}$ (cal./mole)			17.42	41.87

or *(b)* Eq. 7.38 and Table 76, Part II:

Bond	q	φ	a_i	C_ν contribution $b_i \times 10^3$	$c_i \times 10^6$
C—H	8	—	1.832	− 9.792	13.264
C—C	2	—	− 2.180	12.000	− 6.882
$\Sigma q_i C_\nu$			− 0.348	2.208	6.382
				$C\delta$ contribution	
C—H	8	1.5	− 11.256	46.800	− 16.104
C—C	2	1.5	2.190	10.242	− 7.731
$\varphi \Sigma q_i C_\delta$			− 9.066	57.042	− 23.835

Summing, one obtains:

$C_p^\circ = 0.521 + 59.250 \times 10^{-3}\,T - 17.453 \times 10^{-6}T^2$ since $4R + R = 9.935$.
Thus $C_{p300°K}^\circ = 16.72$ and $C_{p1000°K}^\circ = 42.32$ cal./deg.mole.
or *(c)* Eq. 7.34 and Table 76, Part II:

$$\Sigma q_i C_\nu = 0.212 + 2.208 \times 10^{-3}T + 6.742 \times 10^{-6}T^2$$
$$\varphi \Sigma q_i C_\delta = - 10.275 + 64.648 \times 10^{-3}T - 27.013 \times 10^{-6}T^2$$

where

$$\varphi = \frac{3n - 6 - \Sigma q_i}{\Sigma q_i} = 1.7$$

rather than 1.5 as in *(a)* and *(b)* above. Since $4R = 7.948$, it follows that:

$$C_p^\circ = - 2.115 + 66.856 \times 10^{-3}T - 20.27 \times 10^{-6}T^2$$

Accordingly $C_{p300°}^\circ = 16.12$ and $C_{p1000°}^\circ = 44.47$ cal./deg.mole.

The values are to be compared with 17.66 and 41.83 cal. per deg.mole at 300° and 1000° K, respectively, reported in the literature.[1] The equation of Bennewitz and Rossner, 7.34, does not take into account the internal rotational contribution to heat capacity. The predicted value for propane is thus about 12 per cent less than the observed heat capacity. The internal rotational contribution is taken into account in the modified equation of Dobratz (7.38) and the predicted value is within 3 per cent of the literature value. Using the data calculated by Stull and Mayfield based on more recent spectroscopic data, the best estimated value (within 1–2 per cent of the literature value) is achieved, using Eq. 7.38. The bond vibration contributions by Fugassi and Rudy, and Dobratz, (as the three parameter equation for C_p°, Table 76, Part II) were calculated for the temperature range up to 700° K, whereas the compilation by Stull and Mayfield (Table 78, Part II) was calculated from the Einstein functions up to 1500° K. Comparison of the predicted values at 1000° K gives an indication of the errors to be expected in extrapolations beyond 700° K in the use of Table 76.

Example 7.9. Estimate the heat capacity of benzene at 300° K in the ideal gas state. (ANSWER: (a) Eq. 7.34 and Table 76, Part II; 19.78; (b) Eq. 7.38 and Table 76; 19.78; (c) Eq. 7.38 and Table 77; 19.91; cf. lit.,[2] 19.65 cal./deg.mole. For benzene, in which there is no internal rotational contribution, the estimated heat capacities at the lower temperatures are in good agreement with the observed values (within one per cent), the limits of accuracy being dependent only on the assignment of the generalized frequencies to the valence and deformation vibrational modes of the bonds contributing to the structure in the system.

The method of generalized bond vibrational frequencies is not recommended for the very first members of a homologous series, e.g., methane, ethane, or acetylene, since, owing to the higher symmetry present, the vibrational spectra frequently differ from those of the other members of the series. The procedure for the vibrational assignment does not distinguish between many isomeric structures; e.g., n-butane and iso-butane, so that the precision of the method for such systems is limited accordingly. Since the method assumes free internal rotation about single C—C bonds, the values calculated for temperatures below 250° K are likely to be in error; more so than at higher temperatures. For many organic molecules, the upper limits of temperature must be considered to be about 1100° K. Above 1000–1100° K, electron interaction develops in some molecules which would cause a gradual increase in the heat capacity. The estimated values, accordingly, may be increasingly low above this temperature limit.

10. Temperature Dependence of Heat Capacity

If the heat capacity can be expressed by a purely empirical power series equation:

$$C_p = a + bT + cT^2 + \ldots \quad (1.5)$$

the constants a, b, c, d must be derived from heat capacity data over the range of temperatures of interest with reference to the specific problem under consideration. The algebraic methods of solution, in this case, of three simultaneous equations rest on the elementary and basic procedure of elimination of the constants one by one from different pairs of equations until only one remains, or the more advanced and neater procedure of determinants, in which four determinants of the third order must be solved.

A method based on differential calculus, basically more simple and less laborious than the algebraic methods for obtaining the equations expressing the heat capacity as a power series function of temperature has been described by Janz.[81] Thus for a *three* parameter power series equation as above, the first differential of C_p with respect to T defines the slope of the curve at any point T, i.e.,

$$dC_p/dT = b + 2cT \quad (7.39)$$

The second differential:

$$d^2C_p/dT^2 = 2c \quad (7.40)$$

[81] G. J. Janz, *J. Chem. Educ.* **31**, 72 (1954).

defines the rate of change of slope at any point T. It is evident from Eq. 7.40 that the latter is constant in this case. With these equations the constants a, b, and c are readily obtained from a knowledge of the heat capacities at three temperatures.

To reduce the use of letters and symbols to a minimum, let the following data for methane[1] be used for illustration of this method:

T (°K)	300	600	900
C_p (cal./deg. mole)	8.55	12.55	16.21

The average slope over the temperature range 600–300° K is simply $4.00/300 = 13.3 \times 10^{-3}$. This is also the slope at 450° K, i.e., since the rate of change of slope is constant, the average slope in this temperature interval is:

$$\tfrac{1}{2}\,[(dC_p/dT)_{300°K} + (dC_p/dT)_{600°K}] = b + 2c\,(450°K) \qquad (7.41)$$

or the slope of the curve at $T = 450°$ K. Similarly the average slope over the temperature range 900–600° K is 12.2×10^{-3}, and this, as shown above, is also the slope at 750° K.

The rate of change of slope (d^2C_p/dT^2) from 450° to 750° K is $-(1.1_0 \times 10^{-3})/300 = -3.6_7 \times 10^{-6}$. In this instance, since the slope is decreasing, the rate of change of the slope is negative. Using this value in (7.40) above, c is found to be $-1.8_3 \times 10^{-6}$. Solving the first differential equation for b, and substituting for c, the value for b, obtained quite simply, is $14.9_5 \times 10^{-3}$. These values in Eq. 1.5 give 4.23_1 for the constant a. The equation for the heat capacity as a function of temperature for methane in the temperature interval 300–900° K is given by:

$$C_p = 4.23_1 + 14.9_5 \times 10^{-3}T - 1.8_3 \times 10^{-6}T^2 \qquad (7.42)$$

Thus, at 700° K, the heat capacity calculated by this relation is 13.8_1 cal./deg.mole, and the literature value[2] is 13.88, i.e., the agreement is sufficient for most problems encountered.

7. HEAT OF FORMATION AND HEAT CAPACITY

Addendum

Verma and Doriaswamy. The approach of Verma and Doriaswamy[81a] for heats of formation is to be noted. It is shown that temperature dependence of ΔH_f° for any *group* can be expressed by simple straight-line relations of the form:

$$\Delta H_{f_T}^\circ = A + BT \tag{7.43}$$

and that for the range 300°–1500° K, there are two such straight line portions, the approximate ranges being 300°–850°, 850°–1500°, or 300°–750°, 750°–1500° K for the majority of groups. Values of A and B, tabulated for a variety of hydrocarbons and nonhydrocarbons, are given in Tables 63–71, Part II. The heat of formation for a compound in the form of a temperature-dependent equation can be gained from the relation:

$$\Delta H_f^\circ = \Sigma \text{ contribution of component groups} + \Sigma \text{ correction for deviations} \tag{7.44}$$

The principle is essentially the same as the procedure developed by Van Krevelen and Chermin[36] for ΔG_{f_T} (refer to Chapter 5).

Example 7.10. Find the expression for ΔH_f° as a function of temperature in the range up to 850° K for toluene.

SOLUTION:
For $\Delta H_{f_T}^\circ = A + BT$ use Tables 63–71, Part II, as follows:

	A	$B \times 10^2$
5[HC↗↙]	18.840	−0.835
1[—C↗↙]	5.437	0.037
1[—CH$_3$]	−8.948	−0.436
$\Delta H_{f_T}^\circ$ (toluene)	= 15.329	− 1.234 × 10^{-2} T

Example 7.11. Predict the expression for $\Delta H_f(T)$ for 3-methyl-1-butyne in the range 300°–850° K. (ANSWER: $\Delta H_f^\circ = 35.052 - 0.0787 \times 10^{-2}\ T$.)

[81a] K. K. Verma and L. K. Doriaswamy, *Ind. Eng. Chem. (Fundamentals)* **4**, 389 (1965).

The results of comparisons of the estimates for $\Delta H_f°$ at four temperatures (300°, 600°, 1000°, and 1500° K) for 15 hydrocarbons and 10 nonhydrocarbons are summarized as follows:

	Average error (kcal./mole)	
	Hydrocarbons	Nonhydrocarbons
Franklin[34]	0.31	2.5
Souders et al.[35]	0.34	not applicable
Verma and Doriaswamy[81a]	0.34	0.24

The method of Verma and Doriaswamy appears as reliable as that of Franklin or Souders *et al.* for hydrocarbons, and superior for nonhydrocarbons. The method differs from the previously reported techniques in that the estimate for $\Delta H_f°$ is obtained directly as a temperature-dependent equation.

Waddington et al. Using the heats of formation of a series of completely fluorinated compounds, as determined by combustion calorimetry, and related thermochemical data for structurally similar hydrocarbons, a comparison of the C—F and C—H bond energies has been reported by Waddington *et al.*[81b]

For the four hypothetical reactions it follows from these data that:

			$\Delta H_{f\,298}°$ (kcal./mole)
(a)		$CF_4 + 4H = CH_4 + 4F$	67.7
(b)		$n\text{-}C_7F_{16} + 16H \rightarrow n\text{-}C_7H_{16} + 16F$	213.1
(c)		$C_6F_{11} \cdot CF_3 + 14H \rightarrow C_6H_{11} \cdot CH_3 + 14F$	173.7
(d)		$C_6F_{11} \cdot C_2F_5 + 16H \rightarrow C_6H_{11} \cdot C_2H_5 + 16F$	197.7

The heats of reaction depend primarily on the thermochemical bond energies of the ligands created or destroyed in the substitution of H for F since the molecular conformations for the reactants and products are the same. The heats of reaction for the above four processes, accordingly, may be expressed as follows:

$$4d = 67.7; \quad a + 10b + 3c = 173.7$$

$$10b + 6c = 213.1; \quad a + 12b + 3c = 197.7$$

where **a** now represents the difference in bond energies, $E(C—F) - E(C—H)$, for the CF group relative to the CH group, **b** that for =CF_2 relative to

---

[81b] W. D. Good, D. R. Douslin, D. W. Scott, A. George, J. L. Lacina, J. P. Dawson, and G. Waddington, *J. Phys. Chem.* **63**, 1133 (1959).

=CH_2, **c** that for —CF_3 relative to —CH_3, and **d** the difference for CF_4 relative to CH_4. It follows that $a = 7.2$, $b = 12.0$, $c = 15.5$, and $d = 16.9$ kcal./mole, respectively. These values may be used to estimate the heat of formation of an aliphatic fluorocarbon if the heat of formation of the corresponding hydrocarbon is known.

Laidler. From an analysis of the heats of formation and combustion of organic gases and liquids, and of the heats of vaporization, Laidler[81c] advanced a system for the estimation, at 25°C, of the heats of atomization, heats of formation, and heats of combustion for organic compounds containing C, H, O, and N. To search for systematic additivity effects, the thermochemical data at 25°C were converted into heats of atomization, i.e., the heats required to convert the substances into their constituent atoms, using:

$$C_{graphite} = C_{gaseous} - 171.7 \text{ kcal.}$$

$$\tfrac{1}{2}H_2 = H - 52.09 \text{ kcal.}$$

$$\tfrac{1}{2}O_2 = O - 59.16 \text{ kcal.}$$

$$\tfrac{1}{2}N_2 = N - 113.0 \text{ kcal.}$$

and ΔH_c for $C_{(graphite)}$ and $H_{2(gas)}$ as 94.05 and 68.32 kcal., respectively. It is observed that the heats of atomization of gaseous compounds can be accurately represented by the principle of additivity if the numbers of atoms of various kinds within a molecule are assessed in terms of the different kinds of bonds. The bond contributions thus derived for the calculation of ΔH_a, ΔH_f, and ΔH_v, at 25° C are in Table 72, Part II. The symbols (column 2) indicate the types of bonds used as a basis for this system; these are as follows:

(i) c_1, c_2, and c_3 are the C—C, C=C, and C≡C bond increments, the same in all paraffins.

(ii) p, s, and t are the C—H bond increments for the primary, secondary, and tertiary type bonding in the paraffins.

(iii) p', s' and t' are the C—H primary, secondary, and tertiary C—H bond contributions when these occur at a site next but one to a double bond.

(iv) s_2 and t_2 are the C—H secondary and tertiary C—H bond contributions to sites adjacent to the C=C bond.

(v) t_3 is the C—H bond contribution at sites adjacent to a triple bond.

(vi) The bond contributions for specific groups in the molecule are: a_{1c}, alcohol group; e, ether group; a_{1d}, aldehyde group; k, ketone group; a_c, acid group; a_m, amine group; n_i, nitro group; n_a, nitrate group; n_{m_1} and n_{m_2}, mono- and di-nitramines, respectively.

[81c] K. J. Laidler, *Can. J. Chem.* **34**, 626 (1956).

Example 7.12. What are the heats of combustion and formation for $C_2H_5ONO_2$? (Refer to Table 72)

SOLUTION:

Structure type	Additive contribution	
	$\Delta H_{c(298)}$	$\Delta H_{f(298)}$
1 (c_1)	47.48	−0.45
3 (p)	161.19	11.85
2 (s)	108.92	6.42
1 (n_a)	−4.0	27.5
$C_2H_5ONO_2$:	ΔH_c 313.59	ΔH_f 45.32

The literature values are 313.2 and 45.7 kcal., respectively.

Example 7.13. Predict the heat of combustion for diethylnitramine. (ANSWER: 691.08 kcal.; lit., 692.2 ± 0.9 kcal.)

For benzene derivatives, 42.2 kcal. should be subtracted from the values estimated for the heats of atomization or formation using the bond increments (Table 72); this corrects for the resonance energy effect since the bond increments were derived for the paraffins (i.e., ordinary double bonds).

Heats of vaporization are gained from a comparison of the values for $\Delta H_{f(298)}$ predicted for the liquid and gaseous states, respectively.

Reid and Sherwood. A critical review of the various estimation procedures for a large number of properties of gases and liquids has been given by Reid and Sherwood.[81d] Recommended methods are considered for estimating or correlating the following properties: critical properties; P-V-T relationships; vapor pressures; latent heats of vaporization; $\Delta H_f°$; $\Delta G_f°$; $C_p°$; viscosity; thermal conductivity; diffusion coefficients; and vapor-liquid equilibria.

[81d] R. Reid and T. K. Sherwood, The Properties of Gases and Liquids (Their Estimation and Correlation), 2nd ed. McGraw-Hill, New York, 1966.

CHAPTER 8

Applications of the Thermodynamic Method

1. Introduction

To illustrate the thermodynamic approach for the prediction of reaction equilibria and the feasibility of novel processes, specific applications are considered in this chapter in some detail. The thermodynamics of the hydrogenation of benzene to cyclohexane are discussed in the light of precision data to illustrate that the reaction equilibria can be predicted well within the experimental limits of accuracy. Insight into the thermodynamic stability of the intermediates, the dihydro- and tetrahydro-benzenes is gained from the free energy changes estimated for this hydrogenation per mole of hydrogen added. The thermal dimerization of butadiene to vinylcyclohexene, and the dissociation of the latter to monomeric butadiene are considered thermodynamically. A kinetic rate equation for the latter process is predicted from these data. The problem of thermodynamic or kinetic control in a reaction is illustrated by a discussion of the thermodynamics of the acrylonitrile-butadiene addition reaction. Fundamentals in the catalytic ring closure are discussed from the thermodynamic viewpoint, and in the light of experimental studies. A survey of some applications of the thermodynamic method to various fields is given for reference in further work. Before dealing with the various illustrative examples, a brief comparison of the several comprehensive estimation methods is in order.

2. Comparison of Comprehensive Estimation Methods

A brief summary of the comparison of these methods has been compiled in Table 8.1. The comprehensive nature and type of data that may be estimated from the numerical tabulations are indicated. The remarks on each estimation procedure are made with the assumption that the data presented by these authors at the time of publication are being used. The methods may be divided broadly into two categories, precise and approximate.

The method of statistical thermodynamics permits a precise calculation of the statistical functions providing the required experimental data on molecular parameters and the fundamental vibrational frequency assignments are known. The method is quite general and embraces all types of compounds. It is limited in practice to the more simple polyatomic systems owing to the lack of experimental data and the present limitations of the mathematical methods. The calculation of the fundamental vibrational

TABLE 8.1. COMPARISON OF THE ESTIMATION METHODS

Method	Compounds	Properties estimated	Remarks
Parks and Huffman[4]	All types	$S°_{298}$, $G°_{f298}$	Demonstrated additivity
Pitzer,[5] Person and Pimental,[26] Morgan and Lielmezs[46f]	C, H,[5,26] and halides[46f]	Statistical thermodynamic functions	Accurate estimates; long chain cpds.
Andersen et al.,[33] Bryant[46b]	All types; F[46b]	$\Delta H°_{f298}$, $S°_{298}$, $C°_p = f(T)$	Good approximations
Franklin[34]	All types	$(G° - H°)$, $(H° - H_0°)$, $\Delta H°_f$, $\Delta G°_f$	Improves on Anderson et al. for C, H cpds., but not for others
Souders et al.[35]	C, H cpds.	$C_p°$, $\Delta H°_f$, $\Delta S°_f$	Compares with Franklin for accuracy
Van Krevelin and Chermin,[36] Chermin[123a]	All types	$\Delta G° = A + 10^{-2}BT$	Compares well with Pitzer (C, H), and Anderson et al. (for other cpds.); extended to transition states[123a]
Bremner and Thomas[54]	C, H cpds.	$\Delta G_f°$, $S°$	Aromatic from aliphatic data
Rossini, Pitzer et al.[43,44,47]	C, H cpds.	Statistical thermodynamic functions and $\Delta H_f°$	Used in the A.P.I.–N.B.S. tables[1]
Ciola[46e]	General	$\log K$	At T where $\log K = 0$, an error of 100°K is assigned
Benson and Buss[46a]	All types	$\Delta G°$, $\Delta H°$, $S°$, $C_p°$	Accurate estimates; limiting law approach
Laidler[81c]	C, O, N, S, H	$\Delta H_f°$; $\Delta H°_{comb.}$	Heats of atomization and bond property approach
Waddington et al.[81b]	C—F	$\Delta H_f°$	Limited only by accuracy of parent C, H cpds.
Verma and Doriaswamy[81a]	C, H, O, N, S, X	$\Delta H_f° = A + BT$	Compares with Refs. 34 and 35 for C, H, and superior for other cpds.
Huggins[30]	C, H	$S°$	Entropy of kinking and coiling; long chain cpds.
Kharasch and Sher[65]	All types	$\Delta H°_{comb.\ only}$	Based on survey of data prior to 1925

frequency assignment, based on the experimental spectroscopic data and the theoretical normal coordinate analysis, rapidly becomes exceedingly involved and laborious with increasing complexity in molecular structure.

Using somewhat different statistical methods, Pitzer has been the most successful in treating the problem of calculating the thermodynamic functions for long chain hydrocarbons. Neither the knowledge of the normal coordinates of vibration nor their frequencies are required, but only the masses of the particles and the potential energy as a function of their positions. For the straight chain saturated hydrocarbons the tables of numerical additive values compiled by Pitzer, and revised by Person and Pimentel[26] together with the basic equations permit a ready estimate of precise statistical thermodynamic properties up to 1500° K. Extension of this method to olefins and branched chain aliphatic hydrocarbons is quite limited, the results being of a more approximate nature.

Rossini, Pitzer, and associates, have developed and applied the method of group equations for precise estimates of the statistical thermodynamic properties to all types of hydrocarbons, aliphatic, alicyclic, aromatic, straight, and branched chains. Recognition of the need for a balance in the contributions due to the symmetry factors and restricted internal rotations as well as an identity of groups in the basic group equation made possible the application of the principle of additivity to rather accurate estimates for the complex molecules by reference to the precisely established values for the lower members in each homologous series of hydrocarbons. Extension of this approach to the various types of compounds containing elements other than C and H has been demonstrated, but is quite limited by the lack of data.

The methods based on the inspection of the structural formula for the complex molecule and summation of thermodynamic increments corresponding to the appropriate groups give results more or less approximate in nature. An accuracy of the estimated value within 1 to 10 per cent can readily be achieved, the ultimate probable error being somewhat dependent on skill and experience with the methods as well as the precision of the method and accuracy of the increments. The method and tabulations of Souders and associates probably leads to the best estimates by group increments. The properties estimated are heat capacity, heat content, heat of formation, and entropy of formation in contrast to the preceding methods, but unlike the three other methods based on group increments, the scope is limited to hydrocarbons since increments for the nonhydrocarbon structural groups were not developed. A detailed consideration of the molecular structure is needed to guide the selection of the proper internal rotational contribution, and to allow for conjugations, adjacencies, symmetry, and optical isomerism.

An estimate of the free energy of formation expressed in the form of a simple temperature dependent equation is most directly achieved by the procedure and tables of Van Krevelen and Chermin. This is the only property to be estimated by their method, but the scope of the tables embraces the hydrocarbons in general and a wide variety of compounds containing halogens, O, N, and S. The contribution of symmetry must be separately considered as in the preceding methods since the tables of increments were compiled after subtracting these from the parent molecules. Distinction between isomeric structures is possible since the structural increments are quite specific. The method of Andersen, Beyer, and Watson leads to estimates of the heat of formation and the entropy both at 25° C, and the heat capacity expressed as a three term power series temperature dependent equation. By means of the latter, the heat of formation, entropy, and the free energy at temperatures other than 25° C are readily estimated. Unlike the other methods, the contributions of symmetry are inherent in the values of the group increments rather than given as a separate calculation. While the tables of increments are specific and make possible distinction between isomeric structures, it should be noted that more than one value for a molecule can be achieved, depending on the order in which the group increments are taken. This method and the method of Van Krevelen and Chermin are more comprehensive than the others in covering compounds other than pure hydrocarbons. The group increment method for estimates of the free energy of formation and the heat of formation developed by Franklin is based on an extension of the work by Pitzer, for the long chain aliphatic hydrocarbons, with simplifications. Increments are tabulated for the free energy content and heat content functions, and the free energy and heat of formation, enabling estimates of these properties at various temperatures for all types of hydrocarbons. A limited table of nonhydrocarbon increments was compiled, demonstrating that this method is practical for compounds other than pure hydrocarbons. Franklin's extension to nonhydrocarbons is more limited than the preceding two methods. For pure hydrocarbons, the methods of Souders and associates, Franklin, and Van Krevelen and Chermin may be expected to give somewhat better results than that of Anderson, Beyer, and Watson since the factors contributing to deviations from simple additivity, e.g., symmetry, and steric effects, are specifically corrected for in the former methods.

The work of Parks and Huffman leading to the simple method of structural modification first clearly demonstrated the additive nature of the thermodynamic properties for all types of organic compounds. The increments compiled by these investigators generally are not recommended except for very approximate estimates since the values were based on the earlier data and on relatively few compounds. The method can be applied to achieve quite accurate estimates from relatively simple structural changes in which the modification corresponds closely to that for which the increment was derived from the reference compounds.

A comparison of the principles of each method, and the use in practice, can be gained by reference to the illustrative examples in the preceding chapters. Each approach should be capable of giving almost comparable estimates for the more complex polyatomic systems providing the latest data are used to formulate the group increments or equations.

3. Hydrogenation of Benzene

The thermodynamics of the hydrogenation of benzene have been discussed by Janz.[82] In addition to the over-all hydrogenation to cyclohexane, the process is considered in three steps:

$$\text{(8.1)}$$

For benzene and cyclohexane, the thermodynamic data are well established.[2] The statistical thermodynamic functions[83] and experimental heat capacities[84, 85] for cyclohexene have also been reported.

The heat of combustion of cyclohexene has been recently measured by Pergiel and Prosen.[86] Using these data [$\Delta H_c°$ (25° C) (liquid), -896.20 ± 0.17 kcal. per mole, $\Delta H_f°$ (25° C) (liquid), -9.70 ± 0.19 kcal./mole] and the heat of vaporization[87] corrected to the present definition of the calorie, gives $\Delta H_f°$ (25° C) (gas), -1.72 ± 0.22 kcal. per mole, for cyclohexene. A second value calculated from the heat of hydrogenation of cyclohexene to cyclohexane[88] is -1.2 ± 0.1 kcal. per mole. Of the two values, the former has been used in the present work. For cyclohexadiene no recent heat of combustion data are available. Using the heat of hydrogenation of cyclohexadiene to benzene,[88] the heat of formation for cyclohexadiene, $\Delta H_f°$ (25° C) (gas), 25.6 ± 0.1 kcal. per mole, was obtained.

The entropy of cyclohexadiene was estimated from that of cyclohexene.[83] An increment of 5.3 e.u. accompanies the hydrogenation of one double bond in aliphatic olefins (e.g., cis-butene — 1,3-butadiene, $71.90-66.62 = 5.28$; cis-2-pentene — cis-1,3-pentadiene, $82.76-77.5 = 5.26$). Accordingly the entropy of cyclohexadiene (assuming a symmetry number of 2) is 68.9 e.u.

[82] G. J. Janz, *J. Chem. Phys.* **22**, 751 (1954).
[83] C. W. Beckett, N. K. Freeman, and K. S. Pitzer, *J. Am. Chem. Soc.* **70**, 4227 (1948).
[84] H. M. Huffman, M. Eaton, and G. D. Oliver, *J. Am. Chem. Soc.* **70**, 2991 (1948).
[85] J. B. Montgomery and T. DeVries, *J. Am. Chem. Soc.* **64**, 2375 (1942).
[86] F. Y. Pergiel and E. J. Prosen, National Bureau of Standards Thermochemical Laboratory, private communication, 1954.
[87] J. H. Matthews, *J. Am. Chem. Soc.* **48**, 562 (1926).
[88] G. B. Kistiakowsky, J. R. Ruhoff, H. A. Smith, and W. E. Vaughan, *J. Am. Chem. Soc.* **58**, 137 (1936).

The heat capacities were calculated by the method of Dobratz (Eq. 7.17) using the tables of bonding frequencies reported by Stull and Mayfield, i.e., Table 47, Part II.

TABLE 8.2

THERMODYNAMIC DATA FOR BENZENE, CYCLOHEXADIENE, CYCLOHEXENE, AND CYCLOHEXANE

	ΔH°_{f298} kcal./mole	S°_{298} cal./deg.mole	$C_p^\circ = a + bT^2 + cT^3$ (cal./deg.mole)		
			a	$b \times 10^3$	$c \times 10^6$
Benzene	19.82	64.3	− 3.54	90.37	− 36.67
Cyclohexa-diene-1,3	25.6	68.9	− 1.35	97.6	− 37.3
Cyclohexene	− 1.72	74.2	− 9.14	131.9	− 57.1
Cyclohexane	− 29.43	71.3	− 12.46	143.2	− 54.94

The thermodynamic data for benzene, cyclohexane, cyclohexene, and cyclohexadiene are listed in Table 8.2. The free energy changes were calculated using the modified form of the van't Hoff isochore:

$$\Delta G_T^\circ = I_H + (\Delta a - I_S)T - \Delta aT \ln T - \tfrac{1}{2}\Delta bT^2 - \tfrac{1}{6}\Delta cT^3 \qquad (1.9)$$

where the constants I_H and I_S are evaluated from the data at 25° C, and Δ signifies $(p - r)$ for each property, and are summarized in Table 8.3 and Fig. 8.1. It follows that over the whole temperature range considered, cyclo-

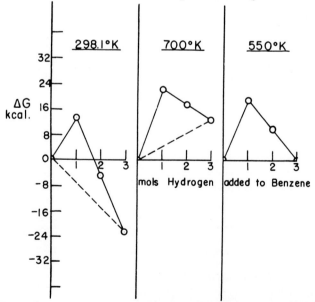

FIG. 8.1. Thermodynamic free energy change in the hydrogenation of benzene per molecule of hydrogen added.

hexadiene is thermodynamically unstable relative to the other three compounds. The earlier prediction of Taylor[88a] at 600° K based on very qualitative calculations is in accord with these results. It has been known experimentally for a long time that cyclohexadiene and cyclohexane react to form benzene and cyclohexane even by passage through heated tubes without any catalysts. It is clear from the thermodynamic conclusions that at moderately high temperatures, cyclohexadiene and cyclohexene could never be isolated, even in trace amounts in the hydrogenation of benzene.

TABLE 8.3

CHANGE IN FREE ENERGY PER MOLECULE OF HYDROGEN ADDED IN THE HYDROGENATION OF BENZENE[b]

Reaction	$T°$ K				
	298.16	400	550	700	1000
a[a]	−23.40	−14.35	0.00	13.62	41.97
b	13.2	15.7	19.6	23.5	31.3
c	−17.9	−14.8	−10.2	−5.5	4.2
d	−18.7	−15.2	−9.4	−4.4	6.5

[a] For the over-all reaction, the free energy tables in reference 1 were used. [b] kcal./mole

A comparison of the results reported by Burrows and Lucarni[89] and Zharkov and Frost[90] by direct measurement in this temperature region with the values calculated using precise thermodynamic data is shown in Fig. 8.2.

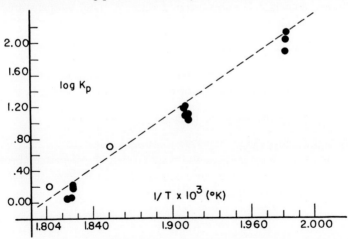

FIG. 8.2. Comparison of experimental data and thermodynamic values for the benzene-cyclohexane hydrogenation reaction at equilibrium. ○ Burrows and Lucarni[89]; ● Zharkov and Frost[90]; - - - Thermodynamic[1].

[88a] H. S. Taylor, *J. Am. Chem. Soc.* **60**, 627 (1938).
[89] G. H. Burrows and C. Lucarni, *J. Am. Chem. Soc.* **49**, 1157 (1927).
[90] V. R. Zharkov and A. V. Frost, *J. Gen. Chem. U.S.S.R.* **2**, 534 (1932).

The free energy change, from the data of Burrows and Lucarni, is −23.43 kcal. per mole at 25° C, whereas the value calculated thermodynamically is −23.40 kcal. per mole. The agreement leaves little to be desired, especially in view of the various difficulties in the direct experimental measurement of equilibrium data.

4. The Thermal Dimerization of Butadiene

The kinetics of the dimerization of butadiene and of the dissociation of vinylcyclohexene:

$$2(CH_2=CH-CH=CH_2) \rightarrow \bigcirc-CH=CH_2 \qquad (8.2)$$

have been re-examined by Duncan and Janz[91] in the light of thermodynamic equilibrium calculations.

Three major kinetic investigations, two on the thermal dimerization of butadiene and one on the depolymerization of dimeric butadiene, have been described in the literature. From 446 to 660° K, Kistiakowsky and Ransom[92] found the dimerization to be second order, proceeding by the rate equation:

$$k_a = 9.20 \times 10^9 \exp(-23{,}690/RT) \text{ cc. mole}^{-1} \text{ sec.}^{-1} \qquad (8.3)$$

It appeared also that with increasing temperature the activation energy for the dimerization increased. The work of Rowley and Steiner[93] in the temperature range 690–925° K substantiated this. At these temperatures the rate expression obtained was:

$$k_a = 1.38 \times 10^{11} \exp(-26{,}800/RT) \text{ cc. mole}^{-1} \text{ sec.}^{-1} \qquad (8.4)$$

Depolymerization of dimeric butadiene was reported by Doumani and associates[94] to proceed by a first order mechanism according to the relation:

$$k_d = 2.35 \times 10^8 \exp(-36{,}000/RT) \text{ sec.}^{-1} \qquad (8.5)$$

in the temperature range 783–977° K. From these data the heat of reaction and equilibrium constants for the reaction can be calculated. From 750 to 950° K the equilibrium constant is given by the ratio of the forward and reverse rates, and in terms of partial pressure is:

$$K_p = 5.87 \times 10^2/2RT \exp(9{,}200/RT) \qquad (8.6)$$

The equilibrium constants and free energy changes for the equilibrium from this expression are as shown in Table 8.4.

[91] N. E. Duncan and G. J. Janz, *J. Chem. Phys.* **20**, 1644 (1952). See also G. J. Janz and M. A. DeCrescente, *J. Phys. Chem.* **63**, 1470 (1959).
[92] G. B. Kistiakowsky and W. W. Ransom, *J. Phys. Chem.* **7**, 725 (1939).
[93] D. Rowley and H. Steiner, *Discussions Faraday Soc.* **10**, 198 (1951).
[94] T. F. Doumani, R. F. Deering, and A. C. McKinnis, *Ind. Eng. Chem.* **39**, 89 (1947).

TABLE 8.4

EQUILIBRIUM DATA PREDICTED FROM EXPERIMENTAL RATE DATA

T (° K)	750	775	800	825	850	875	900	925	950
K_p	0.36	0.26	0.16	0.07	−0.10	−0.09	−0.17	−0.24	−0.30
$\Delta G°$(kcal./mole)	−1.23	−0.92	−0.60	−0.28	0.04	0.36	0.68	1.01	1.34

The temperature dependence of the free energy change is exceedingly small according to these results. The heat of reaction, is predicted as 9.2 kcal. per mole in this temperature range.

Comparison of these data with results from the thermodynamic method shows that only at 800° K is there approximate agreement. The free energy changes for the butadiene-vinylcyclohexene equilibrium were calculated using the van't Hoff isochore in the modified form:

$$\Delta G_T° = I_H + (\Delta a - I_S)T - \Delta aT \ln T - \tfrac{1}{2}\Delta bT^2 - \tfrac{1}{6}\Delta cT^3 \qquad (1.9)$$

where the constants I_H and I_S are evaluated from the data at 298.1° K. For this purpose the following data were used for vinylcyclohexene: $\Delta H°_{f\,298.1°\,K} =$ 16.8 kcal. per mole, $S°_{298.1°\,K} = 96.4$ cal. per deg.mole, $C_p° = 2.22 + 148.2 \times 10^{-3}T - 62.38 \times 10^{-6}T^2$. These data were calculated from reference compounds, ethylbenzene[1], cyclohexane[1], and cyclohexene[46, 95] for which the data are well established, and the method of group increments or contributions. The thermodynamic properties of butadiene have been reported elsewhere.[96] The free energy changes thus calculated for the dimerization of butadiene for temperatures up to 1000° K are given in Table 8.5.

TABLE 8.5

EQUILIBRIUM FREE ENERGY CHANGE PREDICTED FROM THERMODYNAMIC DATA

T (° K)	298.1	300	400	500	600	700	800	900	1000
$\Delta G°$(kcal./mole)	−25.7	−25.6	−21.9	−17.9	−14.7	−11.5	−7.9	−4.7	−1.3

In Fig. 8.3 a comparison is shown of the results from the thermodynamic method and the experimentally established rate equations. The lack of agreement between the two methods is much greater than the errors inherent in the above calculation. The heat of reaction at 25° C calculated from heats of formation is −36.7 ± 0.5 kcal./mole.

[95] M. B. Epstein, K. S. Pitzer, and F. D. Rossini, *J. Research Natl. Bur. Standards* **42**, 379 (1946).

[96] F. Brickwedde, M. Moskow, and J. G. Aston, *J. Research Natl. Bur. Standards* **37**, 263 (1946).

The cause of disagreement observed for the two methods is found to lie in the equation[94] for the thermal dissociation of vinylcyclohexene. The

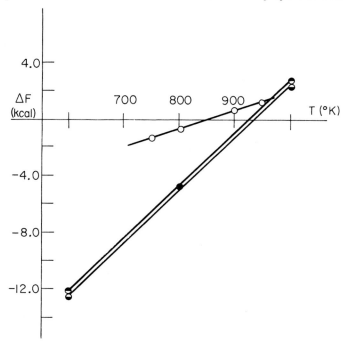

FIG. 8.3. The free energy change-temperature relation for the equilibrium:
$$2(CH_2=CH-CH=CH_2) \rightleftharpoons C_8H_{12}.$$

● Data from experimental rate equations.   ◐ Thermodynamic Calculation.
◐ Data from experimental rate equation for the dimerization and calculated equation for depolymerization.

frequency factor can be calculated using the statistical expression of the theory[97] of reaction rates:

$$a = Ke^2kT/h \, (\exp \Delta S_c^*/R) \tag{8.7}$$

The entropy value for this activated complex has been calculated by Wasserman[98] as $S^\circ_{c600^\circ K} = 120.44$ e.u. per mole, and the entropy of vinylcyclohexene at 600° K is 112.6 e.u. per mole. Thus for the dissociation ΔS_c^* is 7.8 e.u. and the frequency factor is $10^{15.7}$ if the transmission coefficient is taken as approximately unity (true in many first order reactions). The energy of activation for the reverse reaction can be obtained from a knowledge of the heat of reaction at 800° K and the energy of activation for the forward

---

[97] S. Glasstone, K. Laidler, and H. Eyring, "Theory of Rate Processes," McGraw-Hill, New York, 1941.
[98] A. Wassermann, *J. Chem. Soc.* p. 612 (1942).

reaction.[93] Using the data for butadiene,[96] and the heat of formation of vinylcyclohexene corrected to 800° K, the heat of reaction $\Delta E°_{800° K} = 35.0$ kcal. per mole. The value for the latter from experimental kinetic data was 9.2 kcal. It is unlikely that the calculated value is in error by more than 2 or 3 kcal. per mole. The energy of activation for the dissociation reaction is thus 61.8 kcal. per mole. This value is in accord with the order of magnitude for the energy of activation reported[99] for the dissociation of cyclohexene (57.5 kcal.) and that previously estimated[92] for dissociation of vinylcyclohexene (64 kcal.). Accordingly, the rate of dissociation of 3-vinylcyclohexene is given by:

$$k_d = 5.02 \times 10^{15} \exp(-61,800/RT) \text{ sec.}^{-1} \qquad (8.8)$$

The expression for the equilibrium constant for the butadiene-vinylcyclohexene reaction is:

$$K_p = 2.76 \times 10^{-5}/RT \exp(35,000/RT) \qquad (8.9)$$

if the rate equation above is used with the data of Rowley and Steiner. The free energy changes and $\log K_p$ calculated from this equation are in good agreement with the results of the thermodynamic calculation as is shown in Fig. 8.3.

The frequency factor and activation energy reported by Doumani et al.[94] seem to be unusually low for the homogeneous thermal dissociation of vinylcyclohexene and do not lend themselves to equilibrium calculations as seen by the criterion of the thermodynamic methods. The low values may possibly be accounted for by a catalysis or wall effect present but not reported in the investigation.

5. Thermodynamic or Kinetic Control

The Diels-Alder reaction, in which a stable six-membered ring is formed by the 1:4-addition of an unsaturated linkage to a diene, is a well-known process of wide synthetical application. With acrylonitrile, in which both the C=C and C≡N dienophilic groups are present, two competing processes can be foreseen, i.e.,

$$\text{(diene)} + \text{CH}_2\text{=CH-C≡N} \rightarrow \text{(A)} \quad \text{and} \quad \text{(B)} \qquad (8.10)$$

Thus 4-cyanocyclohexene (A) and 3,6-dihydro-2-vinylpyridine (B) would be predicted as products if acrylonitrile adds to the diene by virtue of the

[99] L. Kuchler, *Trans. Faraday Soc.* **35**, 874 (1939); *Nachr. Ges. Wiss. Göttingen, Jahresber. Geschäftsjahr. Math. physik. Kl. Fachgruppen* III, N.F. **1**, 231 (1939).

vinyl- and cyano-groups respectively. The theoretical and experimental aspects of these processes were investigated by Janz and Duncan.[100]

The relative reactivity of these two groups in acrylonitrile can be predicted from thermodynamic and kinetic considerations. A measure of the thermodynamic feasibility of cyanocyclohexene formation (A) and vinylpyridine formation (B) is given by the free energy change for each reaction. This can be calculated at 650° K from a knowledge of the heats of formation and entropies at 298° K, and the heat capacities for these compounds. For this purpose the thermodynamic data for cyanocyclohexene (ΔH°_{298}, 30.9 kcal./mole; S°_{298}, 89.7 e.u. and $C_p = 3.43 + 119 \times 10^{-3}T - 48.4 \times 10^{-6}T^2$) and vinylpyridine (40.3, 78.0, and $-6.01 + 114 \times 10^{-3}T - 45.5 \times 10^{-6}T^2$ respectively) were calculated by the method of group increments from the data for methylcyclohexane[1] and methylpyridine.[101] Using these with the data for hydrogen[1], butadiene[96] and acrylonitrile,[102] the free energy changes for cyanocyclohexene and vinylpyridine formation from acrylonitrile and butadiene at 650° K are -22 and -15 kcal./mole respectively. These results refer to the ideal gaseous state and are necessarily qualitative in nature, but do show that both reactions are thermodynamically promising in this temperature range. In the homogeneous gas phase reaction at 400° C, only cyanocyclohexene was formed. This is understandable in the light of kinetic considerations. The relative rates for these two reactions would be given by the relation:

$$\frac{k_1}{k_2} = \exp\left[(\Delta S^*_{12})/R - (\Delta E_{12})/RT\right] \qquad (8.11)$$

if it is assumed that the rate controlling step in each reaction is the Diels-Alder step, i.e., the cyclization. Sufficient kinetic data in high temperature Diels-Alder associations are available[93,103,104] to estimate the difference in energies of activation (ΔE_{12}) and entropies of activation (ΔS_{12}^*). From the kinetic data for the additions of acrolein[104] and cyanogen[103] to butadiene at elevated temperatures, a minimum value for ΔE_{12} as -11.9 kcal. per mole is obtained. Whereas the reactivity of the C=C bond in acrylonitrile is quite comparable to the same group in acrolein,[104] the C≡N bond in acrylonitrile is judged less reactive than the same bond in cyanogen.[103] The difference in the entropies of the two activated complexes may be calculated from the knowledge that each complex is stereochemically quite similar to the stable cyclic product in such reactions.[93] Thus ΔS_{12}^* is taken equal to the difference in entropies of 3-cyanocyclohexene (96.0) and 2-vinyldihydropyridine (97.1) at 650° K. The latter were calculated from a knowledge of S°_{298} and heat capacities.[105] These data in the above equation give $k_1 = 5800$ (k_2), i.e., the

[100] G. J. Janz and N. E. Duncan, *Nature* **171**, 933 (1953); *J. Am. Chem. Soc.* **75**, 5389 (1953).
[101] P. J. Hawkins and G. J. Janz, *J. Chem. Soc.* p. 1481 (1949).
[102] F. Halverson, R. Stamm, and J. Whalen, *J. Chem. Phys.* **16**, 808 (1948).
[103] P. J. Hawkins and G. J. Janz, *J. Am. Chem. Soc.* **74**, 1790 (1952).
[104] G. B. Kistiakowsky and J. R. Lacher, *J. Am. Chem. Soc.* **58**, 123 (1936).
[105] G. J. Janz, W. J. G. McCulloch, and N. E. Timpane, *Ind. Eng. Chem.* **45**, 1343 (1953).

formation of cyanocyclohexene is predicted to be 5800 times more rapid than vinylpyridine formation. Although thermodynamic calculation had shown both reactions to be very promising in the temperature region of 400° C, from the above results one would expect very little, if any, vinylpyridine in the reaction of acrylonitrile with butadiene in the homogeneous gas phase at atmospheric pressure.

In the homogeneous gas phase at 400° C and short contact time acrylonitrile and butadiene formed cyanocyclohexene in good yields, and gave no detectable amounts of vinylpyridine. In the presence of a catalyst under similar conditions of contact time and temperature, vinylpyridine is formed in addition to cyanocyclohexene. The effect of the catalyst is to lower the energies of activation in these reactions. These results indicate a considerable preferential catalysis of the pyridinic cyclization reaction since the reactivity of the C≡N group is much more nearly that of the C=C group in the presence of the catalyst than in the homogeneous reaction.

Under suitable conditions both the C=C and the C≡N bonds in acrylonitrile add to butadiene, although in the homogeneous gas phase, only the C=C bond of the nitrile reacts appreciably. The experimental results and the thermodynamic free energy calculations are in accord, thus, with the assumption that a kinetic control of the relative rates operates.

6. Ring Closure

A characteristic of the process involved in ring closure of paraffins and olefins is that they are all processes of dehydrogenation. The fundamental principles from the thermodynamic viewpoint governing the formation of aromatics have been considered by Taylor and Turkevich,[106] and Steiner.[107] While these calculations were based on the earlier data of Parks and Huffman[4] and Thomas and co-workers,[108] the results accordingly being only first approximations to the actual equilibria, they do give a schematic picture of the thermodynamic implications of the various processes.

Taylor and Turkevich[106] applied the thermodynamic approach as a guide to evaluate the following processes which may be conceived for n-hexane:

(a) Polymerization
2(Hexane) = Dodecane + H_2

(b) Aromatization
Hexane = Benzene + $4H_2$

(c) Dehydrogenation
Hexane = Hexene + H_2

[106] H. S. Taylor and J. Turkevich, *Trans. Faraday Soc.* **35**, 921 (1939).
[107] H. Steiner, *J. Inst. Petroleum* **33**, 410 (1947).
[108] C. T. Thomas, G. Egloff, and J. C. Morrell, *Ind. Eng. Chem.* **29**, 1266 (1937).

(d) Cyclization
Hexane = Cyclohexane + H_2

(e) Cyclization-Isomerization
Hexane = Methylcyclopentane + H_2

(f) Cracking
Hexane = Propene + Propane

(g) Complete Pyrolysis
Hexane = Carbon + Hydrogen

(h) Disproportionation
2(Hexane) = Pentane + Heptane

The list of these processes can be extended by considering the interactions of the original paraffin with any of the products, i.e., hydrogen, paraffins, olefins, naphthenes, and aromatics. By use of the thermodynamic characteristics of the individual reactions a quantitative analysis of this complex picture, based on the possibility of the various reactions at different temperature intervals, can be achieved.

The standard free energy changes for the various reactions were calculated up to 1000° K. The calculations were extended through the predicted equilibrium constants to determine the extent of decomposition of n-hexane in mole per cent in each case over this temperature range. The results are summarized in Fig. 8.4. From inspection of the temperature dependence of

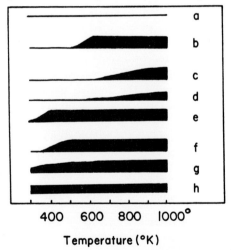

FIG. 8.4. Thermodynamics for various reactions of n-hexane as a function of temperature. The ordinates represent the mole per cent of products from the reaction of one mole of hexane at a pressure of one atmosphere.

the decomposition of hexane into the above various reaction products, the following conclusions may be drawn.

Polymerization of paraffins with the elimination of hydrogen is ruled out on the thermodynamic basis over the whole temperature range considered (up to 1000° K). Cyclization to benzene sets in about 500° K and the equilibrium at one atmosphere pressure is completely on the side of benzene at 600° K. The dehydrogenation to an olefin, and the cyclization to cyclohexane do not become appreciable much below 800° K, but the formation of methyl cyclopentane sets in as low as 350° K. The products of these reactions are all thermodynamically unstable relative to aromatics above 550° K, and may serve as intermediates in the aromatization processes. Cracking reactions, leading to olefins and lower paraffins, disproportionation, and complete pyrolysis to carbon and hydrogen are all predicted to be thermodynamically appreciable in the temperature region of cyclization to an aromatic product. The latter are all characterized by the fact that a carbon-carbon bond must be broken. The choice of alternative paths, cracking or aromatization, cannot be affected by a shift of temperature, but must be carried out by proper choice of catalysts.

The stability of the various intermediates in the conversion of n-hexane to benzene, the dihydro, and tetrahydro intermediates as a function of temperature has been considered in the preceding discussion on the thermodynamics of the hydrogenation of benzene (i.e., Fig. 8.1). The conversion of cyclohexadiene to benzene is predominantly favored. It is also readily seen that the aromatic hydrocarbons are the more stable form at higher temperatures. This fact may be correlated with the resonance phenomena of aromatic compounds. The stabilization of the aromatic ring due to the "resonance energy" reduces to a remarkable extent the energy necessary to produce aromatic compounds.

The latter principle has been illustrated by Steiner[107] in another fashion, by considering the possible formation of benzene from a series of C_6 hydrocarbons, e.g.,

(a) n-Hexane = Benzene + $4H_2$

(b) n-Hexene-1 = Benzene + $3H_2$

(c)   Cyclohexane = Benzene + $3H_2$

(d)   Cyclohexene = Benzene + $2H_2$

(e) n-Hexane = n-Hexene + H_2

(f)   Cyclohexane = Cyclohexene + H_2

From the thermodynamic free energy changes, the temperatures at which these reactions proceed 99 mole per cent to the right hand side were calculated. Thus for the first four processes in that order these were 370°, 120°, 230°, and 310° C, while for the latter two, 870° and 720° C, respectively. It is readily apparent that temperatures some 400°–500° C higher than required for aromatization are necessary to let the latter two reactions go to "completion". The reason for the peculiar characteristics of the equilibria involving aromatics lies undoubtedly in the marked stability due to the resonance energy of the conjugated double bond system in such compounds.

One of the main conclusions to be drawn from thermodynamic considerations is that one should operate at temperatures above 300° C for ring closure and aromatization of a paraffinic hydrocarbon. To reduce the amount of side reactions, e.g., cracking, and complete pyrolysis to carbon, substances that activate the C—H bond rather than rupture the C—C bond are of special interest as catalysts. It is to be expected that the thermodynamic conditions which were shown to apply to the simple aromatics hold generally for the more complicated structures as well, e.g., polycylic and condensed rings. The greater thermal stability shown by many of the highly condensed aromatic ring systems is in accord with this conclusion.

7. Cyclic Additions

As distinct from the ring closure by a cyclization-dehydrogenation process, a number of addition reactions exist in which two unsaturated molecules unite to form a 4, 5, or 6 membered ring. Of these, the diene reaction of Diels-Alder has been studied with particular care. The reaction is of wide synthetical application for fine chemicals and has been proposed as one of the fundamental processes in the high temperature synthesis of aromatic compounds, e.g., benzene, tetralin, and octahydronaphthalene, from low molecular weight intermediates. The basic reaction is the addition of an ethylenic or acetylenic group (i.e., dienophilic group) contained in one reactant across the 1,4-position of a conjugated olefinic system leading to the formation of a six membered cyclic product.

The dimerization of butadiene (Eq. 8.2), and the formation of cyanocyclohexene (Eq. 8.10, A) are examples of basic Diels-Alder processes in which the products may be regarded as derivatives of tetrahydrobenzene. Inspection of the thermodynamic free energy changes predicted for these processes, e.g., Table 8.5 and Fig. 8.3, shows that the cyclic addition is favored up to moderately high temperatures. Comparison with the thermodynamics of the hydrogenation of benzene (i.e., Table 8.3 and Fig. 8.1) gives a further insight into the conditions governing this process. Thus at temperatures higher than 290° C, and in the presence of catalysts favoring dehydrogenation, one would predict that the cyclic adduct would spontaneously aromatize under the conditions of the reaction. In the temperature region of 400° C and in the presence of suitable catalyst surfaces, both ethylbenzene and benzonitrile have been observed as the products of these reactions by Janz and associates.[100,109] If the tetrahydro derivative is desired as the adduct, the thermodynamic conditions show that the reaction should be carried out at relatively low temperatures.

[109] G. J. Janz and R. E. Myers, *J. Am. Chem. Soc.* **75**, 1510 (1953).

The cyclic addition of the nitriles-dienes reaction (cf. Eq. 8.10, B) differs from the preceding reactions in that the addition leads to a derivative of a dihydro rather than a tetrahydro aromatic system. Thus for the addition of benzonitrile to butadiene, the reaction may be formulated:

$$\text{butadiene} + \underset{N}{\overset{C-C_6H_5}{|||}} \underset{3}{\overset{1}{\rightleftharpoons}} [\text{N-C}_6\text{H}_5] \overset{2}{\rightleftharpoons} \text{N-C}_6\text{H}_5 + H_2 \quad (8.12)$$

The over-all reaction observed (3) is a composite of the cyclization (1) and dehydrogenation (2) reactions, leading to the formation of 2-phenylpyridine as product.

The free energy changes and log K_p were calculated by Janz and co-workers[105] using the modified form of the van't Hoff isochore (1.9) for this process since the thermodynamic data for the dihydro intermediate were approximated. The heats of formation and heat capacities for hydrogen, butadiene, benzonitrile, and phenylpyridine were taken from the literature but the thermodynamic properties for 2-phenyldihydropyridine: $\Delta H^°_{f298(g)}$ = 63.3 kcal./mole, $S^°_{298}$ = 99.5 cal./deg. mole, and $C_p^°$ = $-10.36 + 185 \times 10^{-3}T - 71.8 \times 10^{-5}T^2$, were estimated by the methods of group increments from the functions of phenylpyridine and the increments associated with the addition of one hydrogen molecule to pyridine. Using these data the free energy changes shown in Table 8.6 and Fig. 8.5 were calculated.

TABLE 8.6

THERMODYNAMIC FREE ENERGY CHANGE AND LOG K_p FOR FORMATION OF 2-PHENYL-PYRIDINE FROM BUTADIENE AND BENZONITRILE

	T, °K	298.1	300	400	500	600	700	800	900	1000
Reaction 1	$\Delta G^{°a}$	0.4	0.4	5.6	11.0	16.8	23.0	29.5	36.5	44.8
	log K_p	−0.3	−0.3	−0.3	−4.8	−6.1	−7.2	−8.1	−8.9	−9.8
Reaction 2	$\Delta G^°$	−16.8	−16.8	−19.4	−21.9	−24.5	−27.1	−29.7	−32.3	−35.7
	log K_p	12.2	12.2	10.6	9.6	8.9	8.5	8.1	7.9	7.8
Reaction 3	$\Delta G^°$	−16.5	−16.5	−13.7	−10.9	−7.7	−4.1	−0.2	4.2	9.1
	log K_p	12.0	11.9	7.5	4.8	2.8	1.3	0.1	−1.0	−2.0

[a] kcal./mole.

The temperature dependence of each step is shown clearly. In the temperature region of 400° C the over-all reaction is thermodynamically favored because of the very large negative ΔG for the second step. At all temperatures the dihydro intermediate is unstable relative to the pyridine, and in the presence of suitable dehydrogenation conditions should lose hydrogen readily to form the pyridinic product.

The free energy change for the cyclization step may be taken as a first approximation of the free energy of activation for the over-all reaction, if one assumes that the rate controlling step in the reaction mechanism is the initial addition reaction. It is concluded that catalysts promoting cyclization reactions, i.e., activating the C—H bond, should promote the rate of this

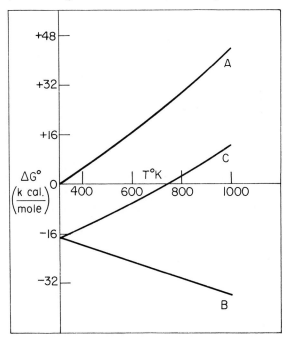

FIG. 8.5. Temperature dependence of free energy changes in cyclization dehydrogenation reaction of benzonitrile and butadiene to form 2-phenylpyridine. A. Cyclization to dihydropyridine; B. Dehydrogenation to phenylpyridine; C. Over-all reaction.

process. A subsequent report[110] has shown that the rate is appreciably increased by such surfaces, e.g., Al_2O_3, and Cr_2O_3–Al_2O_3 catalysts. Wassermann and his co-workers[111] have shown that the Diels-Alder addition reaction, while not highly sensitive to catalysts, is susceptible to a general acid catalysis which even applies to the case where the reactants are both hydrocarbons.

8. Other Applications

The thermodynamic approach has been applied to predict reaction equilibria for the reactions of cyanogen and hydrogen cyanide with ethane and methane,[112] the vapor phase hydration of ethylene,[113] the catalytic

[110] G. J. Janz and W. J. G. McCulloch, *J. Am. Chem. Soc.* **77**, 3014 (1955).
[111] A. Wassermann, *J. Chem. Soc.* p. 618 (1947); W. Rubin, H. Steiner, and A. Wassermann, *ibid.* **1949**, 3046 (1949).
[112] H. W. Thompson, *Trans. Faraday Soc.* **37**, 344 (1941).
[113] R. H. Bliss and B. F. Dodge, *Ind. Eng. Chem.* **29**, 19 (1937).

alkylation of benzene with ethylene,[114] various organic syntheses,[115] the reaction of benzene and ethyl ether,[116] direct cyanogenation of benzene by cyanogen,[117] bromination of toluene,[118] and dehydrogenation, isomerization, alkylation, and cyclization reactions of hydrocarbons,[119] and polymerization reactions.[120-123]

It should be recalled that the free energy changes and equilibrium constants calculated from any but the most precise data are of interest only for predicting the driving forces, i.e., the thermodynamic feasibility, of the processes considered. The degree to which such results may be applied more quantitatively will be dependent on the limitations imposed by the approximation methods and data, and on the experience or skill with which these methods are used. Application of the thermodynamic approach to the many problems of interest in the field of organic compounds will become increasingly easier as accurate experimental data for the simpler members of the various types of organic compounds are established.

Addendum

Rate Constants. Chermin[123a] has indicated the manner in which a correlation between the rate constants and structure of the components in the reaction might be developed for homogeneous gas-phase processes.

From a comparison of the Eyring theory of absolute reaction rates and the Arrhenius rate equation, it follows that:

$$\Delta H^\dagger = E - xRT \tag{8.13}$$

and

$$A = K(RT)^{x-1} e^x \frac{kT}{h} \exp[\Delta S^\dagger/R] \tag{8.14}$$

where E and A are the Arrhenius parameters, and the ($\dagger$) refers to the transition state. Further,

$$\Delta S^\dagger = S^\dagger - \Sigma_i a_i S_i \tag{8.15}$$

and

$$\Delta H^\dagger = \Delta H_f^\dagger - \Sigma_i a_i \, \Delta H_{i_f} \tag{8.16}$$

---

[114] W. A. Pardee and B. F. Dodge, *Ind. Eng. Chem.* **35**, 274 (1943).
[115] R. S. Nyholm, *Australian Chem. Inst. J. Proc.* **14**, 135 (1947).
[116] P. H. Given and D. L. O. Hammick, *J. Chem. Soc.* p. 928 (1947).
[117] G. J. Janz, *J. Am. Chem. Soc.* **74**, 4529 (1952).
[118] S. W. Benson and J. H. Buss, *J. Phys. Chem.*, **61**, 104 (1957).
[119] F. D. Rossini, *in* "Physical Chemistry of Hydrocarbons" (A. Farkas, ed.), Vol. I p. 364, Academic Press, New York, 1950.
[120] F. W. Billmeyer, Jr., "Textbook of Polymer Science," Chap. 9. Wiley (Interscience), New York, 1962.
[121] W. M. D. Bryant, *J. Polymer Sci.* **56**, 277 (1962).
[123] F. S. Dainton and K. J. Ivin, *Nature* **162**, 705 (1948); **172**, 804 (1953).
[123a] H. A. G. Chermin, *Chem. Eng. Sci.* **17**, 705 (1962).

where the subscript i refers to the ith reactant.

For a reaction such as:

$$RH + CH_3 \cdot \rightarrow R \cdots H \cdots CH_3 \rightarrow R \cdot + CH_4 \tag{8.17}$$

where the transition state or activated complex is $R \cdots H \cdots CH_3$, it can readily be shown that:

$$\Delta H^\ddagger = \Delta H_f^\ddagger(\cdots H \cdots CH_3) - [\Delta H_{fCH_3-} + \Delta H_{fCH_3 \cdot}] \tag{8.18}$$

and

$$\Delta S^\ddagger = S^\ddagger(\cdots H—CH_3) - [S_{CH_3-} + S_{CH_3 \cdot}] \tag{8.19}$$

or that the values of the Arrhenius parameters E and A are to a first approximation independent of the number of C atoms in the normal alkane.

This suggests that the thermodynamic properties of the activated complex may be approximated from group increments. To develop the necessary correlations, it would be necessary to start from reasonable assumptions as to the structure, so as to calculate the thermodynamic properties of the activated complex by the methods of statistical thermodynamics for a wide temperature range. These values could then be split into group increments, assuming that the contribution from normal groups are the same as known from normal molecules. In this way the group contributions for the unstable groups, such as, $\cdots H \cdots CH_3$, $\cdots H \cdots CH_2—$, $\cdots H \cdots CH=$, would follow. These being established, the rate constants for reactions previously not investigated can be predicted.

It also follows that for two reactions for which it may be assumed that the unstable parts of the activated complex are identical:

$$A + B \rightleftarrows (\text{activated state}) \rightarrow \text{products} \tag{8.20}$$

$$C + D \rightleftarrows (\text{activated state}) \rightarrow \text{products} \tag{8.21}$$

The ratio of the rate constants is:

$$\frac{k_I}{k_{II}} = \exp\left(\frac{S_A + S_B - S_C - S_D}{R}\right) \exp\left(\frac{\Delta H_{fC} + \Delta H_{fD} - \Delta H_{fA} - \Delta H_{fB}}{RT}\right) \tag{8.22}$$

where A, B and C, D are the reactants in the two processes, respectively. Inspection of this relation shows the rate constants for reactions with analogous mechanisms can readily be predicted from the thermodynamic quantities of the reactants and the rate data for one of the reactions.

PART II
NUMERICAL DATA

II. NUMERICAL DATA

TABLE 1
Bond Energies[a,b]
(kcal. per mole at 0° K)

Elements		Hydrides		Chlorides	
H—H	103.2	H—H	103.2	H—Cl	102.1
Li—Li	26	Li—H	58	Li—Cl	118.5
C—C	80.5(85)	C—H	98.2	C—Cl	78
N—N	37	N—H	92.2	N—Cl	46(?)
O—O	34	O—H	109.4	O—Cl	49
F—F	38	F—H	135	F—Cl	59.0
Na—Na	17.8	Na—H	47	Na—Cl	97.7
Si—Si	(45)	Si—H	76(?)	Si—Cl	87
P—P	(52)	P—H	77	P—Cl	77
S—S	63(?)	S—H	87(?)	S—Cl	65(?)
Cl—Cl	57.1	Cl—H	102.1	Cl—Cl	57.1
K—K	11.8	K—H	42.9	K—Cl	101.4
Cu—Cu	—	Cu—H	62	Cu—Cl	83
Ge—Ge	(42)	Ge—H	—	Ge—Cl	—
As—As	(39)	As—H	56	As—Cl	69
Se—Se	(50)	Se—H	67	Se—Cl	59
Br—Br	45.4(53)	Br—H	86.7	Br—Cl	52.1
Rb—Rb	11.1	Rb—H	39	Rb—Cl	101.0
Ag—Ag	—	Ag—H	53	Ag—Cl	71
Sn—Sn	(35)	Sn—H	—	Sn—Cl	76
Sb—Sb	(42)	Sb—H	—	Sb—Cl	75
Te—Te	(49)	Te—H	59	Te—Cl	—
I—I	35.6(51)	I—H	70.6	I—Cl	49.6
Cs—Cs	10.4	Cs—H	41	Cs—Cl	103

		Multiple Bonds		
	Single		Double	Triple
C—C	80.5		145	198
N—N	37		—	225.1
O—O	34		117.2	—
P—P	(52)		—	116.0
S—S	63(?)		101(?)	—
As—As	(39)		—	90.8
Se—Se	(50)		65	—
Sb—Sb	(42)		—	69
Te—Te	(49)		53	—
C—N	66		—	189
C—O	79		173	—
P—N	—		—	138(?)
S—O	—		120(?)	—
Te—O	—		62.8	—

[a] K. S. Pitzer, *J. Am. Chem. Soc.* **70,** 2140 (1948); "Quantum Chemistry." Prentice Hall, New York, 1953.

[b] For more recent and detailed compilations, refer to: T. L. Cottrell, "The Strengths of Chemical Bonds." Butterworth, London and Washington, D.C., 1954; C. T. Mortimer: "Reaction Heats and Bond Strengths." Pergamon Press, Oxford, 1962.

TABLE 2

Atomic Covalent Radii and Bond Angles[a]

Covalent radius	Angstroms		Angles between bonds
I. Hydrogen	0.30		
II. Carbon			
1. Single bond carbon	0.77		Regular tetrahedral angles, 108° between bonds
2. Double bond carbon	0.67	(for double bonds)	Bonds all lie in one plane
	0.77	(for single bonds)	124°
			112° >C=
			124°
3. Triple bond carbon	0.60	(for triple bond)	Linear
	0.77	(for single bond)	≡C—
4. Benzene carbon	0.695	(for each of the two C—C bonds)	Planar 120°
	0.77	(the bond extending outward)	120° >C— 120°
III. Oxygen			
1. Single bond oxygen	0.66		111° —O⟋
2. Double bond oxygen	0.57		=O
IV. Nitrogen			
1. Amino nitrogen	0.70	(flat pyramid with three bonds making tetrahedral angles with each other)	108° | 108° ⟋N⟍ 108°
2. Nitrate nitrogen	0.65	(double bond)	Planar
	0.70	(single bond)	120 ⟍N⟪ 120
3. Isonitrile nitrogen	0.55	(double bond)	Linear
	0.70	(single bond)	—N=
4. Cyanide nitrogen	0.55		≡N
V. Sulfur			
1. Single bond sulfur	1.04		105° —S⟋
2. Double bond sulfur	0.95		=S
3. Sulfate sulfur	0.95	(double bond)	Tetrahedral angles
	1.04	(single bond)	108° ‖ 108° ⟋S⟹ 108°⟶ | ⟵108°
VI. Sodium	1.81		
VII. Chlorine	0.99		
VIII. Bromine	1.14		
IX. Iodine	1.33		

[a] O. A. Hougen and K. M. Watson, "Chemical Process Principles," Vol. 2, Wiley, New York, 1943.

TABLE 3

A Six-Place Table of the Einstein Functions[a]

$x=\dfrac{h\nu_0{}^b}{kT}$	$\dfrac{x}{e^x-1}$	$\dfrac{x^2 e^x}{(e^x-1)^2}$	$-\ln(1-e^{-x})$	$x=\dfrac{h\nu_0}{kT}$	$\dfrac{x}{e^x-1}$	$\dfrac{x^2 e^x}{(e^x-1)^2}$	$-\ln(1-e^{-x})$
		x at Intervals of 0.005					
0.000	1.000000	1.000000	∞	0.175	0.915051	0.997452	1.829194
0.005	0.997502	0.999998	5.300816	0.180	0.912699	0.997304	1.803449
0.010	0.995008	0.999992	4.610166	0.185	0.910350	0.997153	1.778474
0.015	0.992519	0.999981	4.207196	0.190	0.908007	0.996997	1.754227
0.020	0.990033	0.999967	3.922006	0.195	0.905667	0.996837	1.730672
0.025	0.987552	0.999948	3.701353	0.200	0.903331	0.996673	1.707772
0.030	0.985075	0.999925	3.521520	0.205	0.901000	0.996505	1.685495
0.035	0.982602	0.999898	3.369856	0.210	0.898672	0.996333	1.663811
0.040	0.980133	0.999867	3.238809	0.215	0.896349	0.996157	1.642692
0.045	0.977669	0.999831	3.123508	0.220	0.894030	0.995976	1.622112
0.050	0.975208	0.999792	3.020628	0.225	0.891715	0.995792	1.602046
0.055	0.972752	0.999748	2.927796	0.230	0.889404	0.995603	1.582473
0.060	0.970300	0.999700	2.843261	0.235	0.887098	0.995411	1.563370
0.065	0.967852	0.999648	2.765692	0.240	0.884795	0.995214	1.544717
0.070	0.965408	0.999592	2.694056	0.245	0.882497	0.995013	1.526497
0.075	0.962969	0.999531	2.627533	0.250	0.880203	0.994808	1.508691
0.080	0.960533	0.999467	2.565462	0.255	0.877913	0.994599	1.491284
0.085	0.958102	0.999398	2.507303	0.260	0.875627	0.994386	1.474259
0.090	0.955675	0.999325	2.452608	0.265	0.873345	0.994168	1.457601
0.095	0.953252	0.999248	2.401002	0.270	0.871068	0.993947	1.441298
0.100	0.950833	0.999167	2.352168	0.275	0.868794	0.993722	1.425335
0.105	0.948419	0.999082	2.305836	0.280	0.866525	0.993492	1.409701
0.110	0.946008	0.998992	2.261771	0.285	0.864260	0.993259	1.394384
0.115	0.943602	0.998899	2.219772	0.290	0.861999	0.993021	1.379373
0.120	0.941200	0.998801	2.179664	0.295	0.859742	0.992779	1.364656
0.125	0.938802	0.998699	2.141291	0.300	0.857489	0.992534	1.350226
0.130	0.936408	0.998593	2.104517	0.305	0.855240	0.992284	1.336070
0.135	0.934018	0.998483	2.069221	0.310	0.852996	0.992030	1.322182
0.140	0.931633	0.998368	2.035296	0.315	0.850755	0.991772	1.308552
0.145	0.929251	0.998250	2.002646	0.320	0.848519	0.991510	1.295171
0.150	0.926874	0.998127	1.971183	0.325	0.846287	0.991244	1.282033
0.155	0.924501	0.998000	1.940829	0.330	0.844059	0'990974	1.269129
0.160	0.922132	0.997869	1.911515	0.335	0.841835	0.990700	1.256453
0.165	0.919768	0.997734	1.883176	0.340	0.839615	0.990422	1.243998
0.170	0.917407	0.997595	1.855753	0.345	0.837399	0.990140	1.231756

[a] J. Sherman and R. B. Ewell, *J. Phys. Chem.* **46,** 641 (1942).
[b] Units of ν_0, sec^{-1}.

(Table continued)

TABLE 3 (continued)

$x=\dfrac{h\nu_0}{kT}$	$\dfrac{x}{e^x-1}$	$\dfrac{x^2 e^x}{(e^x-1)^2}$	$-\ln(1-e^{-x})$	$x=\dfrac{h\nu_0}{kT}$	$\dfrac{x}{e^x-1}$	$\dfrac{x^2 e^x}{(e^x-1)^2}$	$-\ln(1-e^{-x})$
0.350	0.835188	0.989854	1.219723	0.550	0.750082	0.975168	0.860264
0.355	0.832980	0.989564	1.207892	0.555	0.748038	0.974722	0.853486
0.360	0.830777	0.989270	1.196257	0.560	0.745998	0.974271	0.846786
0.365	0.828578	0.988971	1.184813	0.565	0.743962	0.973817	0.840164
0.370	0.826382	0.988669	1.173555	0.570	0.741930	0.973359	0.833618
0.375	0.824191	0.988363	1.162477	0.575	0.739901	0.972897	0.827147
0.380	0.822004	0.988053	1.151575	0.580	0.737877	0.972432	0.820750
0.385	0.819822	0.987739	1.140844	0.585	0.735857	0.971963	0.814424
0.390	0.817643	0.987421	1.130279	0.590	0.733841	0.971490	0.808170
0.395	0.815468	0.987099	1.119877	0.595	0.731829	0.971013	0.801986
0.400	0.813298	0.986773	1.109633	0.600	0.729822	0.970532	0.795870
0.405	0.811132	0.986443	1.099543	0.605	0.727818	0.970048	0.789822
0.410	0.808969	0.986109	1.089604	0.610	0.725818	0.969560	0.783840
0.415	0.806811	0.985771	1.079811	0.615	0.723822	0.969068	0.777923
0.420	0.804657	0.985429	1.070161	0.620	0.721830	0.968573	0.772070
0.425	0.802507	0.985083	1.060651	0.625	0.719842	0.968074	0.766280
0.430	0.800361	0.984733	1.051278	0.630	0.717858	0.967571	0.760552
0.435	0.798219	0.984379	1.042037	0.635	0.715878	0.967065	0.754885
0.440	0.796082	0.984022	1.032927	0.640	0.713903	0.966555	0.749278
0.445	0.793948	0.983660	1.023944	0.645	0.711931	0.966041	0.743730
0.450	0.791818	0.983294	1.015085	0.650	0.709963	0.965523	0.738240
0.455	0.789693	0.982925	1.006347	0.655	0.707999	0.965002	0.732807
0.460	0.787571	0.982552	0.997728	0.660	0.706039	0.964477	0.727431
0.465	0.785454	0.982174	0.989225	0.665	0.704083	0.963949	0.722110
0.470	0.783341	0.981793	0.980835	0.670	0.702131	0.963417	0.716843
0.475	0.781232	0.981408	0.972557	0.675	0.700184	0.962881	0.711630
0.480	0.779127	0.981019	0.964388	0.680	0.698240	0.962341	0.706469
0.485	0.777026	0.980626	0.956324	0.685	0.696300	0.961798	0.701361
0.490	0.774929	0.980230	0.948366	0.690	0.694363	0.961252	0.696304
0.495	0.772836	0.979829	0.940509	0.695	0.692432	0.960702	0.691298
0.500	0.770747	0.979425	0.932752	0.700	0.690504	0.960148	0.686341
0.505	0.768662	0.979016	0.925093	0.705	0.688580	0.959591	0.681433
0.510	0.766582	0.978604	0.917530	0.710	0.686660	0.959030	0.676574
0.515	0.764505	0.978188	0.910062	0.715	0.684743	0.958465	0.671762
0.520	0.762432	0.977768	0.902685	0.720	0.682831	0.957897	0.666997
0.525	0.760364	0.977344	0.895399	0.725	0.680923	0.957326	0.662278
0.530	0.758299	0.976917	0.888201	0.730	0.679019	0.956750	0.657604
0.535	0.756239	0.976485	0.881091	0.735	0.677119	0.956172	0.652976
0.540	0.754183	0.976050	0.874965	0.740	0.675222	0.955589	0.648392
0.545	0.752130	0.975611	0.867124	0.745	0.673330	0.955004	0.643851

II. NUMERICAL DATA

$x = \dfrac{h\nu_0}{kT}$	$\dfrac{x}{e^x - 1}$	$\dfrac{x^2 e^x}{(e^x - 1)^2}$	$-\ln(1 - e^{-x})$	$x = \dfrac{h\nu_0}{kT}$	$\dfrac{x}{e^x - 1}$	$\dfrac{x^2 e^x}{(e^x - 1)^2}$	$-\ln(1 - e^{-x})$
0.750	0.671441	0.954415	0.639354	0.950	0.599101	0.928068	0.488968
0.755	0.669557	0.953822	0.634898	0.955	0.597371	0.927342	0.485828
0.760	0.667676	0.953226	0.630485	0.960	0.595646	0.926614	0.482713
0.765	0.665800	0.952626	0.626113	0.965	0.593924	0.925882	0.479623
0.770	0.663927	0.952023	0.621781	0.970	0.592206	0.925147	0.476558
0.775	0.662058	0.951416	0.617490	0.975	0.590491	0.924409	0.473518
0.780	0.660193	0.950806	0.613239	0.980	0.588781	0.923668	0.470502
0.785	0.658332	0.950192	0.609026	0.985	0.587074	0.922924	0.467510
0.790	0.656475	0.949575	0.604852	0.990	0.585363	0.922159	0.464541
0.795	0.654622	0.948955	0.600716	0.995	0.583672	0.921427	0.461597
0.800	0.652773	0.948331	0.596618	1.000	0.581977	0.920674	0.458675
0.805	0.650928	0.947704	0.592556	1.005	0.580285	0.919917	0.455777
0.810	0.649086	0.947073	0.588531	1.010	0.578597	0.919158	0.452901
0.815	0.647249	0.946439	0.584543	1.015	0.576913	0.918396	0.450048
0.820	0.645415	0.945801	0.580590	1.020	0.575233	0.917630	0.447217
0.825	0.643586	0.945161	0.576672	1.025	0.573556	0.916862	0.444408
0.830	0.641760	0.944516	0.572788	1.030	0.571884	0.916091	0.441621
0.835	0.639938	0.943869	0.568939	1.035	0.570215	0.915317	0.438856
0.840	0.638120	0.943218	0.565124	1.040	0.568549	0.914540	0.436112
0.845	0.636306	0.942564	0.561343	1.045	0.566888	0.913759	0.433389
0.850	0.634496	0.941906	0.557594	1.050	0.565230	0.912976	0.430687
0.855	0.632689	0.941245	0.553878	1.055	0.563576	0.912190	0.428006
0.860	0.630887	0.940581	0.550194	1.060	0.561926	0.911401	0.425345
0.865	0.629088	0.939913	0.546542	1.065	0.560279	0.910610	0.422705
0.870	0.627293	0.939242	0.542921	1.070	0.558636	0.909815	0.420084
0.875	0.625503	0.938568	0.539331	1.075	0.556997	0.909017	0.417484
0.880	0.623715	0.937891	0.535772	1.080	0.555361	0.908217	0.414903
0.885	0.621932	0.937210	0.532244	1.085	0.553730	0.907413	0.412341
0.890	0.620153	0.936526	0.528745	1.090	0.552102	0.906607	0.409799
0.895	0.618378	0.935839	0.525276	1.095	0.550477	0.905798	0.407276
0.900	0.616606	0.935148	0.521835	1.100	0.548857	0.904986	0.404772
0.905	0.614838	0.934455	0.518424	1.105	0.547240	0.904171	0.402286
0.910	0.613074	0.933758	0.515041	1.110	0.545627	0.903354	0.399819
0.915	0.611314	0.933058	0.511687	1.115	0.544017	0.902534	0.397371
0.920	0.609558	0.932354	0.508360	1.120	0.542411	0.901710	0.394940
0.925	0.607806	0.931648	0.505061	1.125	0.540809	0.900884	0.392528
0.930	0.606057	0.930938	0.501789	1.130	0.539211	0.900056	0.390133
0.935	0.604312	0.930228	0.498544	1.135	0.537616	0.899224	0.387756
0.940	0.602571	0.929509	0.495326	1.140	0.536025	0.898390	0.385396
0.945	0.600834	0.928790	0.492134	1.145	0.534437	0.897553	0.383054

(Table continued)

TABLE 3 (*continued*)

$x=\dfrac{h\nu_0}{kT}$	$\dfrac{x}{e^x-1}$	$\dfrac{x^2 e^x}{(e^x-1)^2}$	$-\ln(1-e^{-x})$	$x=\dfrac{h\nu_0}{kT}$	$\dfrac{x}{e^x-1}$	$\dfrac{x^2 e^x}{(e^x-1)^2}$	$-\ln(1-e^{-x})$
1.150	0.532853	0.896714	0.380729	1.350	0.472453	0.861024	0.300079
1.155	0.531273	0.895871	0.378420	1.355	0.471016	0.860082	0.298335
1.160	0.529696	0.895026	0.376129	1.360	0.469582	0.859139	0.296603
1.165	0.528124	0.894179	0.373854	1.365	0.468151	0.858193	0.294882
1.170	0.526554	0.893328	0.371596	1.370	0.466724	0.857244	0.293173
1.175	0.524989	0.892475	0.369353	1.375	0.465301	0.856294	0.291475
1.180	0.523427	0.891619	0.367128	1.380	0.463881	0.855341	0.289789
1.185	0.521869	0.890761	0.364918	1.385	0.462464	0.854386	0.288114
1.190	0.520314	0.889900	0.362724	1.390	0.461051	0.853429	0.286450
1.195	0.518763	0.889036	0.360545	1.395	0.459641	0.852470	0.284797
1.200	0.517215	0.888170	0.358382	1.400	0.458235	0.851509	0.283155
1.205	0.515671	0.887301	0.356235	1.405	0.456832	0.850545	0.281524
1.210	0.514131	0.886430	0.354103	1.410	0.455433	0.849580	0.279903
1.215	0.512595	0.885556	0.351986	1.415	0.454037	0.848612	0.278294
1.220	0.511062	0.884679	0.349884	1.420	0.452644	0.847642	0.276695
1.225	0.509532	0.883800	0.347797	1.425	0.451255	0.846670	0.275106
1.230	0.508006	0.882918	0.345724	1.430	0.449870	0.845696	0.273528
1.235	0.506484	0.882034	0.343667	1.435	0.448487	0.844720	0.271960
1.240	0.504965	0.881147	0.341623	1.440	0.447108	0.843742	0.270403
1.245	0.503450	0.880258	0.339594	1.445	0.445733	0.842762	0.268855
1.250	0.501939	0.879366	0.337580	1.450	0.444361	0.841780	0.267318
1.255	0.500431	0.878472	0.335579	1.455	0.442992	0.840795	0.265791
1.260	0.498927	0.877575	0.333592	1.460	0.441627	0.839809	0.264273
1.265	0.497426	0.876676	0.331619	1.465	0.440265	0.838821	0.262766
1.270	0.495929	0.875774	0.329660	1.470	0.438906	0.837831	0.261268
1.275	0.494435	0.874870	0.327714	1.475	0.437551	0.836839	0.259780
1.280	0.492945	0.873964	0.325782	1.480	0.436199	0.835844	0.258301
1.285	0.491458	0.873055	0.323863	1.485	0.434851	0.834848	0.256833
1.290	0.489975	0.872144	0.321957	1.490	0.433506	0.833850	0.255373
1.295	0.488496	0.871230	0.320065	1.495	0.432164	0.832850	0.253923
1.300	0.487020	0.870314	0.318185	1.500	0.430825	0.831849	0.252482
1.305	0.485547	0.869395	0.316318	1.505	0.429490	0.830845	0.251051
1.310	0.484078	0.868474	0.314464	1.510	0.428159	0.829839	0.249629
1.315	0.482613	0.867551	0.312623	1.515	0.426830	0.828832	0.248215
1.320	0.481151	0.866626	0.310794	1.520	0.425505	0.827822	0.246811
1.325	0.479693	0.865698	0.308978	1.525	0.424183	0.826811	0.245416
1.330	0.478238	0.864768	0.307174	1.530	0.422865	0.825798	0.244030
1.335	0.476786	0.863835	0.305382	1.535	0.421550	0.824783	0.242652
1.340	0.475339	0.862900	0.303602	1.540	0.420238	0.823766	0.241283
1.345	0.473894	0.861963	0.301835	1.545	0.418929	0.822748	0.239923

II. NUMERICAL DATA

$x=\dfrac{h\nu_0}{kT}$	$\dfrac{x}{e^x-1}$	$\dfrac{x^2 e^x}{(e^x-1)^2}$	$-\ln(1-e^{-x})$	$x=\dfrac{h\nu_0}{kT}$	$\dfrac{x}{e^x-1}$	$\dfrac{x^2 e^x}{(e^x-1)^2}$	$-\ln(1-e^{-x})$
1.550	0.417624	0.821728	0.238572	1.750	0.368064	0.779584	0.190887
1.555	0.416322	0.820706	0.237229	1.755	0.366890	0.778501	0.189838
1.560	0.415024	0.819682	0.235895	1.760	0.365719	0.777416	0.188796
1.565	0.413728	0.818656	0.234569	1.765	0.364551	0.776330	0.187761
1.570	0.412436	0.817629	0.233251	1.770	0.363386	0.775243	0.186731
1.575	0.411148	0.816600	0.231942	1.775	0.362224	0.774154	0.185707
1.580	0.409862	0.815569	0.230640	1.780	0.361065	0.773064	0.184690
1.585	0.408580	0.814536	0.229348	1.785	0.359910	0.771973	0.183679
1.590	0.407301	0.813502	0.228063	1.790	0.358757	0.770881	0.182674
1.595	0.406025	0.812466	0.226786	1.795	0.357607	0.769788	0.181675
1.600	0.404753	0.811429	0.225517	1.800	0.356461	0.768693	0.180682
1.605	0.403483	0.810390	0.224256	1.805	0.355317	0.767597	0.179694
1.610	0.402217	0.809349	0.223003	1.810	0.354176	0.766500	0.178713
1.615	0.400955	0.808306	0.221758	1.815	0.353039	0.765402	0.177738
1.620	0.399695	0.807262	0.220520	1.820	0.351904	0.764303	0.176770
1.625	0.398439	0.806216	0.219291	1.825	0.350773	0.763202	0.175804
1.630	0.397186	0.805169	0.218068	1.830	0.349645	0.762101	0.174846
1.635	0.395936	0.804120	0.216854	1.835	0.348519	0.760998	0.173893
1.640	0.394689	0.803070	0.215647	1.840	0.347397	0.759894	0.172947
1.645	0.393446	0.802017	0.214447	1.845	0.346277	0.758790	0.172005
1.650	0.392205	0.800964	0.213255	1.850	0.345161	0.757684	0.171070
1.655	0.390968	0.799909	0.212070	1.855	0.344047	0.756577	0.170140
1.660	0.389734	0.798852	0.210893	1.860	0.342937	0.755469	0.169215
1.665	0.388504	0.797794	0.209722	1.865	0.341830	0.754360	0.168296
1.670	0.387276	0.796734	0.208559	1.870	0.340725	0.753249	0.167382
1.675	0.386052	0.795673	0.207403	1.875	0.339624	0.752138	0.166474
1.680	0.384831	0.794610	0.206254	1.880	0.338525	0.751026	0.165571
1.685	0.383613	0.793546	0.205113	1.885	0.337429	0.749913	0.164673
1.690	0.382398	0.792481	0.203978	1.890	0.336337	0.748799	0.163781
1.695	0.381186	0.791413	0.202850	1.895	0.335247	0.747684	0.162894
1.700	0.379978	0.790345	0.201729	1.900	0.334160	0.746568	0.162012
1.705	0.378772	0.789275	0.200615	1.905	0.333076	0.745451	0.161135
1.710	0.377570	0.788204	0.199507	1.910	0.331996	0.744333	0.160263
1.715	0.376371	0.787131	0.198407	1.915	0.330918	0.743214	0.159397
1.720	0.375175	0.786057	0.197313	1.920	0.329843	0.742094	0.158535
1.725	0.373982	0.784982	0.196225	1.925	0.328770	0.740973	0.157679
1.730	0.372792	0.783905	0.195145	1.930	0.327701	0.739852	0.156827
1.735	0.371606	0.782827	0.194071	1.935	0.326635	0.738729	0.155981
1.740	0.370422	0.781747	0.193003	1.940	0.325572	0.737606	0.155139
1.745	0.369242	0.780666	0.191942	1.945	0.324511	0.736482	0.154302

(Table continued)

TABLE 3 (*continued*)

$x = \dfrac{h\nu_0}{kT}$	$\dfrac{x}{e^x - 1}$	$\dfrac{x^2 e^x}{(e^x - 1)^2}$	$-\ln(1 - e^{-x})$	$x = \dfrac{h\nu_0}{kT}$	$\dfrac{x}{e^x - 1}$	$\dfrac{x^2 e^x}{(e^x - 1)^2}$	$-\ln(1 - e^{-x})$
1.950	0.323453	0.735356	0.153471	2.150	0.283459	0.689787	0.123846
1.955	0.322399	0.734231	0.152644	2.155	0.282516	0.688637	0.123189
1.960	0.321347	0.733104	0.151821	2.160	0.281575	0.687486	0.122535
1.965	0.320298	0.731976	0.151004	2.165	0.280637	0.686335	0.121885
1.970	0.319252	0.730848	0.150191	2.170	0.279701	0.685184	0.121239
1.975	0.318209	0.729719	0.149384	2.175	0.278768	0.684032	0.120596
1.980	0.317168	0.728589	0.148580	2.180	0.277838	0.682880	0.119957
1.985	0.316131	0.727458	0.147782	2.185	0.276910	0.681728	0.119321
1.990	0.315096	0.726327	0.146988	2.190	0.275985	0.680575	0.118690
1.995	0.314064	0.725195	0.146198	2.195	0.275063	0.679422	0.118061
2.000	0.313035	0.724062	0.145413	2.200	0.274143	0.678269	0.117436
2.005	0.312009	0.722928	0.144633	2.205	0.273226	0.677115	0.116815
2.010	0.310986	0.721794	0.143857	2.210	0.272311	0.675961	0.116197
2.015	0.309965	0.720659	0.143086	2.215	0.271399	0.674807	0.115583
2.020	0.308948	0.719523	0.142319	2.220	0.270490	0.673652	0.114972
2.025	0.307933	0.718386	0.141556	2.225	0.269583	0.672498	0.114365
2.030	0.306921	0.717249	0.140798	2.230	0.268679	0.671343	0.113761
2.035	0.305911	0.716112	0.140044	2.235	0.267777	0.670187	0.113160
2.040	0.304905	0.714973	0.139295	2.240	0.266879	0.669032	0.112562
2.045	0.303901	0.713834	0.138550	2.245	0.265982	0.667876	0.111968
2.050	0.302900	0.712695	0.137809	2.250	0.265088	0.666721	0.111378
2.055	0.301902	0.711554	0.137072	2.255	0.264197	0.665565	0.110790
2.060	0.300907	0.710413	0.136340	2.260	0.263308	0.664408	0.110206
2.065	0.299914	0.709272	0.135611	2.265	0.262422	0.663252	0.109625
2.070	0.298925	0.708130	0.134887	2.270	0.261539	0.662096	0.109047
2.075	0.297938	0.706987	0.134168	2.275	0.260658	0.660939	0.108473
2.080	0.296953	0.705844	0.133452	2.280	0.259779	0.659782	0.107902
2.085	0.295972	0.704700	0.132740	2.285	0.258903	0.658625	0.107334
2.090	0.294993	0.703556	0.132032	2.290	0.258030	0.657468	0.106769
2.095	0.294017	0.702411	0.131328	2.295	0.257159	0.656311	0.106207
2.100	0.293044	0.701266	0.130629	2.300	0.256291	0.655154	0.105648
2.105	0.292073	0.700120	0.129932	2.305	0.255425	0.653996	0.105092
2.110	0.291105	0.698974	0.129240	2.310	0.254562	0.652839	0.104539
2.115	0.290140	0.697827	0.128553	2.315	0.253701	0.651681	0.103990
2.120	0.289177	0.696680	0.127869	2.320	0.252842	0.650524	0.103444
2.125	0.288218	0.695532	0.127189	2.325	0.251987	0.649366	0.102901
2.130	0.287261	0.694384	0.126512	2.330	0.251133	0.648209	0.102361
2.135	0.286306	0.693235	0.125840	2.335	0.250282	0.647051	0.101824
2.140	0.285355	0.692086	0.125172	2.340	0.249434	0.645893	0.101290
2.145	0.284406	0.690937	0.124507	2.345	0.248588	0.644735	0.100758

II. NUMERICAL DATA

$x = \dfrac{h\nu_0}{kT}$	$\dfrac{x}{e^x - 1}$	$\dfrac{x^2 e^x}{(e^x - 1)^2}$	$-\ln(1 - e^{-x})$	$x = \dfrac{h\nu_0}{kT}$	$\dfrac{x}{e^x - 1}$	$\dfrac{x^2 e^x}{(e^x - 1)^2}$	$-\ln(1 - e^{-x})$
2.350	0.247745	0.643578	0.100229	2.550	0.215972	0.597372	0.081299
2.355	0.246904	0.642420	0.099703	2.555	0.215225	0.596222	0.080876
2.360	0.246065	0.641262	0.099180	2.560	0.214481	0.595072	0.080456
2.365	0.245229	0.640104	0.098660	2.565	0.213738	0.593923	0.080039
2.370	0.244396	0.638947	0.098143	2.570	0.212998	0.592774	0.079623
2.375	0.243564	0.637789	0.097629	2.575	0.212260	0.591625	0.079210
2.380	0.242736	0.636632	0.097118	2.580	0.211525	0.590477	0.078799
2.385	0.241909	0.635474	0.096609	2.585	0.210792	0.589330	0.078390
2.390	0.241086	0.634317	0.096103	2.590	0.210061	0.588183	0.077983
2.395	0.240264	0.633159	0.095600	2.595	0.209332	0.587036	0.077579
2.400	0.239445	0.632002	0.095100	2.600	0.208605	0.585890	0.077177
2.405	0.238628	0.630845	0.094602	2.605	0.207881	0.584744	0.076776
2.410	0.237814	0.629688	0.094108	2.610	0.207158	0.583598	0.076379
2.415	0.237002	0.628531	0.093616	2.615	0.206438	0.582453	0.075983
2.420	0.236193	0.627374	0.093126	2.620	0.205721	0.581309	0.075589
2.425	0.235386	0.626217	0.092640	2.625	0.205005	0.580165	0.075198
2.430	0.234581	0.625061	0.092156	2.630	0.204291	0.579021	0.074808
2.435	0.233779	0.623905	0.091674	2.635	0.203580	0.577878	0.074421
2.440	0.232979	0.622748	0.091195	2.640	0.202871	0.576736	0.074035
2.445	0.232182	0.621592	0.090719	2.645	0.202164	0.575594	0.073652
2.450	0.231386	0.620436	0.090246	2.650	0.201459	0.574452	0.073271
2.455	0.230594	0.619281	0.089775	2.655	0.200756	0.573311	0.072892
2.460	0.229803	0.518125	0.089307	2.660	0.200056	0.572171	0.072515
2.465	0.229015	0.616970	0.088841	2.665	0.199357	0.571031	0.072140
2.470	0.228229	0.615815	0.088378	2.670	0.198661	0.569892	0.071767
2.475	0.227446	0.614660	0.087917	2.675	0.197967	0.568753	0.071396
2.480	0.226665	0.613506	0.087459	2.680	0.197275	0.567615	0.071027
2.485	0.225886	0.612351	0.087003	2.685	0.196585	0.566477	0.070660
2.490	0.225110	0.611197	0.086550	2.690	0.195897	0.565340	0.070295
2.495	0.224336	0.610044	0.086099	2.695	0.195212	0.564203	0.069932
2.500	0.223564	0.608890	0.085650	2.700	0.194528	0.563068	0.069570
2.505	0.222794	0.607737	0.085204	2.705	0.193847	0.561932	0.069211
2.510	0.222027	0.606584	0.084761	2.710	0.193167	0.560798	0.068854
2.515	0.221262	0.605431	0.084320	2.715	0.192490	0.559664	0.068498
2.520	0.220500	0.604279	0.083881	2.720	0.191815	0.558530	0.068145
2.525	0.219739	0.603127	0.083445	2.725	0.191142	0.557397	0.067793
2.530	0.218981	0.601975	0.083011	2.730	0.190471	0.556265	0.067443
2.535	0.218225	0.600824	0.082579	2.735	0.189802	0.555134	0.067095
2.540	0.217472	0.599673	0.082149	2.740	0.189135	0.554003	0.066749
2.545	0.216721	0.598522	0.081723	2.745	0.188470	0.552873	0.066405

(Table continued)

TABLE 3 (*continued*)

$x = \dfrac{h\nu_0}{kT}$	$\dfrac{x}{e^x - 1}$	$\dfrac{x^2 e^x}{(e^x - 1)^2}$	$-\ln(1 - e^{-x})$	$x = \dfrac{h\nu_0}{kT}$	$\dfrac{x}{e^x - 1}$	$\dfrac{x^2 e^x}{(e^x - 1)^2}$	$-\ln(1 - e^{-x})$
2.750	0.187808	0.551743	0.066063	2.875	0.171894	0.523743	0.058070
2.755	0.187147	0.550614	0.065722	2.880	0.171283	0.522633	0.057772
2.760	0.186488	0.549486	0.065383	2.885	0.170674	0.521524	0.057475
2.765	0.185832	0.548359	0.065046	2.890	0.170067	0.520416	0.057180
2.770	0.185177	0.547232	0.064711	2.895	0.169462	0.519309	0.056887
2.770	0.184525	0.546106	0.064378	2.900	0.168858	0.518203	0.056595
2.785	0.183874	0.544980	0.064047	2.905	0.168257	0.517097	0.056305
2.785	0.183226	0.543856	0.063717	2.910	0.167658	0.515993	0.056016
2.790	0.182579	0.542702	0.063389	2.915	0.167060	0.514889	0.055728
2.795	0.181935	0.541609	0.063062	2.920	0.166464	0.513786	0.055442
2.800	0.181293	0.540486	0.062738	2.925	0.165871	0.512785	0.055158
2.805	0.180652	0.539365	0.062415	2.930	0.165279	0.511584	0.054876
2.810	0.180014	0.538244	0.062093	2.935	0.164689	0.510484	0.054594
2.815	0.179377	0.537123	0.061774	2.940	0.164101	0.509385	0.054314
2.820	0.178743	0.536004	0.061456	2.945	0.163514	0.508286	0.054036
2.825	0.178110	0.534885	0.061140	2.950	0.162930	0.507189	0.053759
2.830	0.177480	0.533767	0.060826	2.955	0.162347	0.506093	0.053484
2.835	0.176851	0.532650	0.060513	2.960	0.161767	0.504997	0.053210
2.840	0.176225	0.531534	0.060202	2.965	0.161188	0.503903	0.052937
2.845	0.175600	0.530419	0.059893	2.970	0.160611	0.502810	0.052666
2.850	0.174978	0.529304	0.059585	2.975	0.160036	0.501717	0.052397
2.855	0.174357	0.528190	0.059279	2.980	0.159462	0.500625	0.052128
2.860	0.173738	0.527077	0.058974	2.985	0.158891	0.499535	0.051861
2.865	0.173122	0.525965	0.058671	2.990	0.158321	0.498445	0.051596
2.870	0.172507	0.524853	0.058370	2.995	0.157753	0.497357	0.051332
				3.000	0.157187	0.496269	0.051069

x AT INTERVALS OF 0.01

x	$\dfrac{x}{e^x - 1}$	$\dfrac{x^2 e^x}{(e^x - 1)^2}$	$-\ln(1 - e^{-x})$	x	$\dfrac{x}{e^x - 1}$	$\dfrac{x^2 e^x}{(e^x - 1)^2}$	$-\ln(1 - e^{-x})$
3.00	0.157187	0.496269	0.051069	3.10	0.146241	0.474732	0.046095
3.01	0.156060	0.494097	0.050548	3.11	0.145184	0.472602	0.045626
3.02	0.154941	0.491929	0.050032	3.12	0.144135	0.470476	0.045162
3.03	0.153829	0.489764	0.049522	3.13	0.143092	0.468355	0.044702
3.04	0.152724	0.487604	0.049017	3.14	0.142057	0.466238	0.044247
3.05	0.151626	0.485448	0.048517	3.15	0.141028	0.464125	0.043797
3.06	0.150535	0.483296	0.048022	3.16	0.140005	0.462018	0.043352
3.07	0.149451	0.481149	0.047533	3.17	0.138989	0.459915	0.042911
3.08	0.148374	0.479006	0.047049	3.18	0.137980	0.457816	0.042475
3.09	0.147304	0.476867	0.046570	3.19	0.136978	0.455722	0.042043

$x=\dfrac{h\nu_0}{kT}$	$\dfrac{x}{e^x-1}$	$\dfrac{x^2 e^x}{(e^x-1)^2}$	$-\ln(1-e^{-x})$	$x=\dfrac{h\nu_0}{kT}$	$\dfrac{x}{e^x-1}$	$\dfrac{x^2 e^x}{(e^x-1)^2}$	$-\ln(1-e^{-x})$
3.20	0.135982	0.453633	0.041616	3.60	0.101129	0.374290	0.027704
3.21	0.134993	0.451549	0.041194	3.61	0.100372	0.372419	0.027424
3.22	0.134010	0.449470	0.040775	3.62	0.099621	0.370554	0.027148
3.23	0.133033	0.447395	0.040361	3.63	0.098876	0.368695	0.026874
3.24	0.132063	0.445325	0.039951	3.64	0.098135	0.366841	0.026603
3.25	0.131099	0.443260	0.039546	3.65	0.097399	0.364993	0.026335
3.26	0.130142	0.441200	0.039145	3.66	0.096669	0.363152	0.026069
3.27	0.129191	0.439145	0.038748	3.67	0.095943	0.361316	0.025807
3.28	0.128246	0.437095	0.038355	3.68	0.095222	0.359485	0.025546
3.29	0.127308	0.435050	0.037966	3.69	0.094507	0.357661	0.025829
3.30	0.126376	0.433010	0.037581	3.70	0.093796	0.355843	0.025034
3.31	0.125449	0.430975	0.037200	3.71	0.093090	0.354031	0.024782
3.32	0.124529	0.428945	0.036823	3.72	0.092389	0.352224	0.024532
3.33	0.123616	0.426921	0.036449	3.73	0.091693	0.350424	0.024285
3.34	0.122708	0.424901	0.036080	3.74	0.091002	0.348629	0.024041
3.35	0.121806	0.422887	0.035715	3.75	0.090316	0.346840	0.023799
3.36	0.120910	0.420878	0.035353	3.76	0.089634	0.345058	0.023559
3.37	0.120021	0.418874	0.034995	3.77	0.088957	0.343281	0.023322
3.38	0.119137	0.416876	0.034641	3.78	0.088285	0.341510	0.023087
3.39	0.118259	0.414882	0.034290	3.79	0.087617	0.339746	0.022855
3.40	0.117387	0.412894	0.033943	3.80	0.086954	0.337987	0.022625
3.41	0.116520	0.410912	0.033599	3.81	0.086296	0.336234	0.022397
3.42	0.115660	0.408935	0.033259	3.82	0.085642	0.334488	0.022172
3.43	0.114805	0.406963	0.032923	3.83	0.084993	0.332747	0.021949
3.44	0.113956	0.404996	0.032590	3.84	0.084348	0.331012	0.021728
3.45	0.113113	0.403036	0.032260	3.85	0.083708	0.329284	0.021509
3.46	0.112276	0.401080	0.031934	3.86	0.083073	0.327561	0.021293
3.47	0.111444	0.399130	0.031611	3.87	0.082441	0.325845	0.021079
3.48	0.110618	0.397186	0.031292	3.88	0.081815	0.324135	0.020867
3.49	0.109797	0.395247	0.030976	3.89	0.081192	0.322431	0.020657
3.50	0.108982	0.393313	0.030663	3.90	0.080574	0.320733	0.020450
3.51	0.108172	0.391386	0.030353	3.91	0.079961	0.319041	0.020244
3.52	0.107368	0.389463	0.030046	3.92	0.079352	0.317355	0.020041
3.53	0.106569	0.387547	0.029743	3.93	0.078747	0.315675	0.019839
3.54	0.105776	0.385636	0.029443	3.94	0.078146	0.314001	0.019640
3.55	0.104988	0.383731	0.029145	3.95	0.077549	0.312333	0.019442
3.56	0.104206	0.381831	0.028851	3.96	0.076957	0.310672	0.019247
3.57	0.103429	0.379937	0.028560	3.97	0.076369	0.309017	0.019054
3.58	0.102657	0.378049	0.028272	3.98	0.075785	0.307367	0.018862
3.59	0.101890	0.376167	0.027986	3.99	0.075205	0.305724	0.018673

(Table continued)

TABLE 3 (continued)

$x=\dfrac{h\nu_0}{kT}$	$\dfrac{x}{e^x-1}$	$\dfrac{x^2 e^x}{(e^x-1)^2}$	$-\ln(1-e^{-x})$	$x=\dfrac{h\nu_0}{kT}$	$\dfrac{x}{e^x-1}$	$\dfrac{x^2 e^x}{(e^x-1)^2}$	$-\ln(1-e^{-x})$
4.00	0.074629	0.304087	0.018485	4.40	0.054692	0.243635	0.012353
4.01	0.074058	0.302456	0.018300	4.41	0.054264	0.242248	0.012230
4.02	0.073490	0.300832	0.018116	4.42	0.053839	0.240868	0.012107
4.03	0.072927	0.299213	0.017934	4.43	0.053418	0.239494	0.011986
4.04	0.072367	0.297601	0.017754	4.44	0.052999	0.238125	0.011866
4.05	0.071812	0.295995	0.017576	4.45	0.052584	0.236763	0.011747
4.06	0.071260	0.294394	0.017400	4.46	0.052171	0.235406	0.011630
4.07	0.070713	0.292800	0.017225	4.47	0.051762	0.234056	0.011513
4.08	0.070169	0.291213	0.017052	4.48	0.051356	0.232711	0.011398
4.09	0.069629	0.289631	0.016881	4.49	0.050952	0.231372	0.011284
4.10	0.069093	0.288055	0.016712	4.50	0.050552	0.230040	0.011171
4.11	0.068561	0.286486	0.016544	4.51	0.050155	0.228713	0.011059
4.12	0.068033	0.284923	0.016378	4.52	0.049760	0.227392	0.010949
4.13	0.067508	0.283365	0.016214	4.53	0.049369	0.226077	0.010839
4.14	0.066987	0.281814	0.016051	4.54	0.048980	0.224768	0.010731
4.15	0.066470	0.280270	0.015890	4.55	0.048594	0.223465	0.010623
4.16	0.065957	0.278731	0.015731	4.56	0.048211	0.222168	0.010517
4.17	0.065447	0.277198	0.015573	4.57	0.047831	0.220877	0.010412
4.18	0.064941	0.275672	0.015417	4.58	0.047454	0.219591	0.010308
4.19	0.064439	0.274152	0.015262	4.59	0.047080	0.218312	0.010205
4.20	0.063940	0.272637	0.015109	4.60	0.046708	0.217038	0.010103
4.21	0.063445	0.271129	0.014958	4.61	0.046339	0.215770	0.010002
4.22	0.062954	0.269627	0.014808	4.62	0.045973	0.214508	0.009902
4.23	0.062466	0.268132	0.014659	4.63	0.045609	0.213252	0.009803
4.24	0.061981	0.266642	0.014512	4.64	0.045249	0.212001	0.009705
4.25	0.061500	0.265158	0.014367	4.65	0.044891	0.210757	0.009608
4.26	0.061023	0.263681	0.014223	4.66	0.044535	0.209518	0.009512
4.27	0.060549	0.262210	0.014080	4.67	0.044183	0.208285	0.009416
4.28	0.060078	0.260744	0.013939	4.68	0.043833	0.207057	0.009322
4.29	0.059611	0.259285	0.013800	4.69	0.043485	0.205836	0.009229
4.30	0.059147	0.257832	0.013661	4.70	0.043140	0.204620	0.009138
4.31	0.058687	0.256385	0.013525	4.71	0.042798	0.203410	0.009046
4.32	0.058230	0.254948	0.013389	4.72	0.042458	0.202205	0.008955
4.33	0.057776	0.253509	0.013255	4.73	0.042121	0.201006	0.008866
4.34	0.057326	0.252081	0.013122	4.74	0.041786	0.199813	0.008777
4.35	0.056879	0.250658	0.012991	4.75	0.041454	0.198626	0.008689
4.36	0.056435	0.249241	0.012861	4.76	0.041125	0.197444	0.008603
4.37	0.055994	0.247831	0.012732	4.77	0.040797	0.196268	0.008517
4.38	0.055557	0.246426	0.012604	4.78	0.040473	0.195097	0.008431
4.39	0.055123	0.245027	0.012478	4.79	0.040150	0.193933	0.008347

II. NUMERICAL DATA

$x = \dfrac{h\nu_0}{kT}$	$\dfrac{x}{e^x - 1}$	$\dfrac{x^2 e^x}{(e^x - 1)^2}$	$-\ln(1 - e^{-x})$	$x = \dfrac{h\nu_0}{kT}$	$\dfrac{x}{e^x - 1}$	$\dfrac{x^2 e^x}{(e^x - 1)^2}$	$-\ln(1 - e^{-x})$
4.80	0.039831	0.192773	0.008264	5.20	0.028845	0.150827	0.005532
4.81	0.039513	0.191620	0.008181	5.21	0.028612	0.149885	0.005477
4.82	0.039198	0.190471	0.008099	5.22	0.028380	0.148948	0.005422
4.83	0.038885	0.189329	0.008019	5.23	0.028150	0.148015	0.005368
4.84	0.038575	0.188192	0.007938	5.24	0.027921	0.147087	0.005314
4.85	0.038267	0.187060	0.007859	5.25	0.027695	0.146165	0.005261
4.86	0.037962	0.185934	0.007781	5.26	0.027470	0.145247	0.005209
4.87	0.037658	0.184814	0.007703	5.27	0.027247	0.144334	0.005157
4.88	0.037357	0.183699	0.007626	5.28	0.027026	0.143426	0.005105
4.89	0.037059	0.182589	0.007550	5.29	0.026806	0.142523	0.005054
4.90	0.036762	0.181485	0.007475	5.30	0.026588	0.141624	0.005004
4.91	0.036468	0.180387	0.007400	5.31	0.026372	0.140731	0.004954
4.92	0.036176	0.179294	0.007326	5.32	0.026157	0.139842	0.004905
4.93	0.035886	0.178206	0.007253	5.33	0.025945	0.138958	0.004856
4.94	0.035598	0.177123	0.007181	5.34	0.025733	0.138078	0.004807
4.95	0.035313	0.176046	0.007109	5.35	0.025524	0.137204	0.004759
4.96	0.035030	0.174975	0.007038	5.36	0.025316	0.136334	0.004712
4.97	0.034749	0.173909	0.006968	5.37	0.025110	0.135469	0.004665
4.98	0.034470	0.172848	0.006898	5.38	0.024905	0.134608	0.004618
4.99	0.034193	0.171792	0.006829	5.39	0.024702	0.133752	0.004572
5.00	0.033918	0.170742	0.006761	5.40	0.024500	0.132901	0.004527
5.01	0.033646	0.169697	0.006693	5.41	0.024300	0.132055	0.004482
5.02	0.033375	0.168657	0.006626	5.42	0.024102	0.131213	0.004437
5.03	0.033107	0.167623	0.006560	5.43	0.023905	0.130376	0.004393
5.04	0.032840	0.166594	0.006495	5.44	0.023710	0.129543	0.004349
5.05	0.032576	0.165570	0.006430	5.45	0.023516	0.128715	0.004306
5.06	0.032314	0.164551	0.006366	5.46	0.023324	0.127891	0.004263
5.07	0.032053	0.163537	0.006302	5.47	0.023133	0.127072	0.004220
5.08	0.031795	0.162529	0.006239	5.48	0.022944	0.126257	0.004178
5.09	0.031539	0.161526	0.006177	5.49	0.022756	0.125447	0.004136
5.10	0.031284	0.160528	0.006115	5.50	0.022569	0.124642	0.004095
5.11	0.031032	0.159535	0.006054	5.51	0.022385	0.123840	0.004054
5.12	0.030781	0.158547	0.005994	5.52	0.022201	0.123044	0.004014
5.13	0.030533	0.157564	0.005934	5.53	0.022019	0.122251	0.003974
5.14	0.030286	0.156587	0.005875	5.54	0.021839	0.121463	0.003934
5.15	0.030041	0.155614	0.005816	5.55	0.021660	0.120680	0.003895
5.16	0.029798	0.154647	0.005758	5.56	0.021482	0.119901	0.003856
5.17	0.029557	0.153685	0.005701	5.57	0.021306	0.119126	0.003818
5.18	0.029318	0.152727	0.005644	5.58	0.021131	0.118355	0.003780
5.19	0.029081	0.151775	0.005588	5.59	0.020957	0.117589	0.003742

(Table continued)

TABLE 3 (*continued*)

$x=\dfrac{h\nu_0}{kT}$	$\dfrac{x}{e^x-1}$	$\dfrac{x^2 e^x}{(e^x-1)^2}$	$-\ln(1-e^{-x})$	$x=\dfrac{h\nu_0}{kT}$	$\dfrac{x}{e^x-1}$	$\dfrac{x^2 e^x}{(e^x-1)^2}$	$-\ln(1-e^{-x})$
5.60	0.020785	0.116827	0.003705	6.00	0.014909	0.089679	0.002482
5.61	0.020614	0.116070	0.003668	6.01	0.014785	0.089079	0.002457
5.62	0.020445	0.115317	0.003631	6.02	0.014662	0.088482	0.002433
5.63	0.020276	0.114567	0.003595	6.03	0.014540	0.087888	0.002408
5.64	0.020110	0.113823	0.003559	6.04	0.014419	0.087298	0.002384
5.65	0.019944	0.113082	0.003524	6.05	0.014299	0.086712	0.002361
5.66	0.019780	0.112346	0.003489	6.06	0.014180	0.086129	0.002337
5.67	0.019617	0.111613	0.003454	6.07	0.014061	0.085550	0.002314
5.68	0.019455	0.110885	0.003420	6.08	0.013944	0.084974	0.002291
5.69	0.019295	0.110161	0.003386	6.09	0.013828	0.084402	0.002268
5.70	0.019136	0.109442	0.003352	6.10	0.013712	0.083833	0.002245
5.71	0.018978	0.108726	0.003318	6.11	0.013598	0.083267	0.002223
5.72	0.018822	0.108014	0.003285	6.12	0.013484	0.082705	0.002201
5.73	0.018666	0.107307	0.003252	6.13	0.013372	0.082146	0.002179
5.74	0.018512	0.106603	0.003220	6.14	0.013260	0.081591	0.002157
5.75	0.018359	0.105904	0.003188	6.15	0.013149	0.081039	0.002136
5.76	0.018208	0.105208	0.003156	6.16	0.013039	0.080490	0.002114
5.77	0.018057	0.104517	0.003125	6.17	0.012930	0.079945	0.002093
5.78	0.017908	0.103829	0.003094	6.18	0.012823	0.079410	0.002073
5.79	0.017760	0.103146	0.003063	6.19	0.012714	0.078864	0.002052
5.80	0.017613	0.102466	0.003032	6.20	0.012608	0.078329	0.002031
5.81	0.017467	0.101791	0.003002	6.21	0.012502	0.077797	0.002011
5.82	0.017323	0.101119	0.002972	6.22	0.012398	0.077268	0.001991
5.83	0.017179	0.100451	0.002942	6.23	0.012294	0.076742	0.001971
5.84	0.017037	0.099788	0.002913	6.24	0.012191	0.076220	0.001952
5.85	0.016896	0.099127	0.002884	6.25	0.012089	0.075700	0.001932
5.86	0.016756	0.098471	0.002855	6.26	0.011987	0.075184	0.001913
5.87	0.016617	0.097819	0.002827	6.27	0.011887	0.074671	0.001894
5.88	0.016479	0.097170	0.002799	6.28	0.011787	0.074162	0.001875
5.89	0.016343	0.096526	0.002771	6.29	0.011688	0.073655	0.001856
5.90	0.016207	0.095885	0.002743	6.30	0.011590	0.073151	0.001838
5.91	0.016073	0.095247	0.002716	6.31	0.011493	0.072651	0.001820
5.92	0.015939	0.094614	0.002689	6.32	0.011396	0.072154	0.001802
5.93	0.015807	0.093984	0.002662	6.33	0.011300	0.071659	0.001784
5.94	0.015676	0.093358	0.002635	6.34	0.011205	0.071168	0.001766
5.95	0.015545	0.092736	0.002609	6.35	0.011111	0.070680	0.001748
5.96	0.015416	0.092117	0.002583	6.36	0.011018	0.070195	0.001731
5.97	0.015288	0.091502	0.002558	6.37	0.010925	0.069713	0.001714
5.98	0.015161	0.090891	0.002532	6.38	0.010833	0.069233	0.001697
5.99	0.015035	0.090283	0.002507	6.39	0.010742	0.068757	0.001680

II. NUMERICAL DATA

$x=\dfrac{h\nu_0}{kT}$	$\dfrac{x}{e^x-1}$	$\dfrac{x^2 e^x}{(e^x-1)^2}$	$-\ln(1-e^{-x})$	$x=\dfrac{h\nu_0}{kT}$	$\dfrac{x}{e^x-1}$	$\dfrac{x^2 e^x}{(e^x-1)^2}$	$-\ln(1-e^{-x})$
6.40	0.010652	0.068284	0.001663	6.80	0.007582	0.051616	0.001114
6.41	0.010562	0.067814	0.001646	6.81	0.007518	0.051252	0.001103
6.42	0.010473	0.067347	0.001630	6.82	0.007454	0.050890	0.001092
6.43	0.010385	0.066882	0.001614	6.83	0.007390	0.050530	0.001081
6.44	0.010297	0.066421	0.001598	6.84	0.007327	0.050173	0.001071
6.45	0.010211	0.065316	0.001582	6.85	0.007265	0.049818	0.001060
6.46	0.010124	0.065506	0.001566	6.86	0.007203	0.049465	0.001049
6.47	0.010039	0.065053	0.001550	6.87	0.007142	0.049115	0.001039
6.48	0.009954	0.064603	0.001535	6.88	0.007081	0.048767	0.001029
6.49	0.009870	0.064156	0.001520	6.89	0.007021	0.048421	0.001018
6.50	0.009787	0.063712	0.001505	6.90	0.006961	0.048078	0.001008
6.51	0.009704	0.063270	0.001490	6.91	0.006901	0.047736	0.000998
6.52	0.009623	0.062831	0.001475	6.92	0.006843	0.047397	0.000988
6.53	0.009541	0.062395	0.001460	6.93	0.006784	0.047060	0.000978
6.54	0.009461	0.061962	0.001446	6.94	0.006726	0.046726	0.000969
6.55	0.009381	0.061531	0.001431	6.95	0.006669	0.046393	0.000959
6.56	0.009301	0.061104	0.001417	6.96	0.006612	0.046063	0.000950
6.57	0.009223	0.060678	0.001403	6.97	0.006556	0.045735	0.000940
6.58	0.009145	0.060256	0.001389	6.98	0.006500	0.045409	0.000931
6.59	0.009067	0.059836	0.001375	6.99	0.006444	0.045085	0.000921
6.60	0.008991	0.059419	0.001361	7.00	0.006389	0.044764	0.000912
6.61	0.008915	0.059005	0.001348	7.01	0.006334	0.044444	0.000903
6.62	0.008839	0.058593	0.001334	7.02	0.006280	0.044127	0.000894
6.63	0.008764	0.058184	0.001321	7.03	0.006227	0.043812	0.000885
6.64	0.008690	0.057777	0.001308	7.04	0.006173	0.043498	0.000877
6.65	0.008616	0.057373	0.001295	7.05	0.006121	0.043187	0.000868
6.66	0.008543	0.056972	0.001282	7.06	0.006068	0.042878	0.000859
6.67	0.008471	0.056573	0.001269	7.07	0.006016	0.042571	0.000851
6.68	0.008399	0.056177	0.001257	7.08	0.005965	0.042266	0.000842
6.69	0.008328	0.055783	0.001244	7.09	0.005914	0.041963	0.000834
6.70	0.008257	0.055392	0.001232	7.10	0.005863	0.041662	0.000825
6.71	0.008187	0.055003	0.001219	7.11	0.005813	0.041363	0.000817
6.72	0.008118	0.054617	0.001207	7.12	0.005763	0.041066	0.000809
6.73	0.008049	0.054223	0.001195	7.13	0.005714	0.040771	0.000801
6.74	0.007980	0.053852	0.001183	7.14	0.005665	0.040478	0.000793
6.75	0.007913	0.053473	0.001172	7.15	0.005616	0.040187	0.000785
6.76	0.007845	0.053097	0.001160	7.16	0.005568	0.039898	0.000777
6.77	0.007779	0.052723	0.001148	7.17	0.005520	0.039611	0.000770
6.78	0.007713	0.052352	0.001137	7.18	0.005473	0.039326	0.000762
6.79	0.007647	0.051983	0.001126	7.19	0.005426	0.039042	0.000754

(Table continued)

TABLE 3 (continued)

$x = \dfrac{h\nu_0}{kT}$	$\dfrac{x}{e^x - 1}$	$\dfrac{x^2 e^x}{(e^x - 1)^2}$	$-\ln(1 - e^{-x})$	$x = \dfrac{h\nu_0}{kT}$	$\dfrac{x}{e^x - 1}$	$\dfrac{x^2 e^x}{(e^x - 1)^2}$	$-\ln(1 - e^{-x})$
7.20	0.005379	0.038761	0.000747	7.60	0.003805	0.028935	0.000501
7.21	0.005333	0.038481	0.000739	7.61	0.003772	0.028722	0.000496
7.22	0.005287	0.038204	0.000732	7.62	0.003740	0.028511	0.000491
7.23	0.005242	0.037928	0.000725	7.63	0.003707	0.028301	0.000486
7.24	0.005197	0.037654	0.000718	7.64	0.003675	0.028093	0.000481
7.25	0.005152	0.037382	0.000710	7.65	0.003643	0.027886	0.000476
7.26	0.005108	0.037111	0.000703	7.66	0.003612	0.027680	0.000471
7.27	0.005064	0.036843	0.000696	7.67	0.003581	0.027476	0.000467
7.28	0.005021	0.036576	0.000689	7.68	0.003550	0.027274	0.000462
7.29	0.004978	0.036311	0.000683	7.69	0.003519	0.027072	0.000457
7.30	0.004935	0.036053	0.000676	7.70	0.003488	0.026872	0.000453
7.31	0.004892	0.035787	0.000669	7.71	0.003458	0.026674	0.000448
7.32	0.004850	0.035527	0.000662	7.72	0.003428	0.026477	0.000444
7.33	0.004809	0.035269	0.000656	7.73	0.003398	0.026281	0.000440
7.34	0.004767	0.035013	0.000649	7.74	0.003369	0.026087	0.000435
7.35	0.004726	0.034759	0.000643	7.75	0.003340	0.025894	0.000431
7.36	0.004685	0.034507	0.000636	7.76	0.003311	0.025702	0.000427
7.37	0.004645	0.034256	0.000630	7.77	0.003282	0.025512	0.000422
7.38	0.004605	0.034006	0.000624	7.78	0.003253	0.025323	0.000418
7.39	0.004565	0.033759	0.000618	7.79	0.003225	0.025135	0.000414
7.40	0.004526	0.033513	0.000611	7.80	0.003197	0.024949	0.000410
7.41	0.004487	0.033269	0.000605	7.81	0.003169	0.024764	0.000406
7.42	0.004448	0.033027	0.000599	7.82	0.003142	0.024580	0.000402
7.43	0.004410	0.032786	0.000593	7.83	0.003115	0.024397	0.000398
7.44	0.004372	0.032547	0.000587	7.84	0.003088	0.024216	0.000394
7.45	0.004334	0.032309	0.000582	7.85	0.003061	0.024036	0.000390
7.46	0.004297	0.032073	0.000576	7.86	0.003034	0.023858	0.000386
7.47	0.004260	0.031839	0.000570	7.87	0.003008	0.023680	0.000382
7.48	0.004223	0.031606	0.000564	7.88	0.002982	0.023504	0.000378
7.49	0.004187	0.031375	0.000559	7.89	0.002956	0.023329	0.000375
7.50	0.004150	0.031145	0.000553	7.90	0.002930	0.023155	0.000371
7.51	0.004115	0.030917	0.000548	7.91	0.002904	0.022983	0.000367
7.52	0.004079	0.030691	0.000542	7.92	0.002879	0.022811	0.000363
7.53	0.004044	0.030466	0.000537	7.93	0.002854	0.022641	0.000360
7.54	0.004009	0.030243	0.000531	7.94	0.002829	0.022473	0.000356
7.55	0.003974	0.030021	0.000526	7.95	0.002805	0.022305	0.000353
7.56	0.003940	0.029801	0.000521	7.96	0.002780	0.022138	0.000349
7.57	0.003906	0.029582	0.000516	7.97	0.002756	0.021973	0.000346
7.58	0.003872	0.029365	0.000511	7.98	0.002732	0.021809	0.000342
7.59	0.003839	0.029149	0.000506	7.99	0.002708	0.021646	0.000339
				8.00	0.002685	0.021484	0.000336

$x = \dfrac{h\nu_0}{kT}$	$\dfrac{x}{e^x-1}$	$\dfrac{x^2 e^x}{(e^x-1)^2}$	$-\ln(1-e^{-x})$	$x = \dfrac{h\nu_0}{kT}$	$\dfrac{x}{e^x-1}$	$\dfrac{x^2 e^x}{(e^x-1)^2}$	$-\ln(1-e^{-x})$
		x at Intervals of 0.05					
8.00	0.002685	0.021484	0.000336	10.00	0.000454	0.004540	0.000045
8.05	0.002570	0.020692	0.000320	10.05	0.000434	0.004362	0.000043
8.10	0.002459	0.019927	0.000304	10.10	0.000415	0.004191	0.000041
8.15	0.002354	0.019190	0.000289	10.15	0.000397	0.004026	0.000039
8.20	0.002253	0.018478	0.000275	10.20	0.000379	0.003867	0.000037
8.25	0.002156	0.017791	0.000262	10.25	0.000362	0.003715	0.000035
8.30	0.002063	0.017129	0.000249	10.30	0.000346	0.003568	0.000034
8.35	0.001974	0.016490	0.000237	10.35	0.000331	0.003427	0.000032
8.40	0.001889	0.015874	0.000225	10.40	0.000317	0.003292	0.000030
8.45	0.001808	0.015280	0.000214	10.45	0.000303	0.003161	0.000029
8.50	0.001730	0.014707	0.000204	10.50	0.000289	0.003036	0.000028
8.55	0.001655	0.014154	0.000194	10.55	0.000276	0.002916	0.000026
8.60	0.001584	0.013621	0.000184	10.60	0.000264	0.002800	0.000025
8.65	0.001515	0.013108	0.000175	10.65	0.000252	0.002688	0.000024
8.70	0.001450	0.012613	0.000167	10.70	0.000241	0.002581	0.000023
8.75	0.001387	0.012136	0.000159	10.75	0.000231	0.002478	0.000021
8.80	0.001327	0.011676	0.000151	10.80	0.000221	0.002379	0.000020
8.85	0.001269	0.011233	0.000143	10.85	0.000211	0.002284	0.000019
8.90	0.001214	0.010806	0.000136	10.90	0.000201	0.002193	0.000018
8.95	0.001161	0.010395	0.000130	10.95	0.000192	0.002105	0.000018
9.00	0.001111	0.009999	0.000123	11.00	0.000184	0.002021	0.000017
9.05	0.001063	0.009617	0.000117	11.05	0.000176	0.001940	0.000016
9.10	0.001016	0.009249	0.000112	11.10	0.000168	0.001862	0.000015
9.15	0.000972	0.008895	0.000106	11.15	0.000160	0.001787	0.000014
9.20	0.000930	0.008554	0.000101	11.20	0.000153	0.001715	0.000014
9.25	0.000889	0.008225	0.000096	11.25	0.000146	0.001646	0.000013
9.30	0.000850	0.007909	0.000091	11.30	0.000140	0.001580	0.000012
9.35	0.000813	0.007604	0.000087	11.35	0.000134	0.001516	0.000012
9.40	0.000778	0.007311	0.000083	11.40	0.000128	0.001455	0.000011
9.45	0.000744	0.007028	0.000079	11.45	0.000122	0.001396	0.000011
9.50	0.000711	0.006756	0.000075	11.50	0.000116	0.001340	0.000010
9.55	0.000680	0.006495	0.000071	11.55	0.000111	0.001285	0.000010
9.60	0.000650	0.006243	0.000068	11.60	0.000106	0.001233	0.000009
9.65	0.000622	0.006000	0.000064	11.65	0.000102	0.001183	0.000009
9.70	0.000594	0.005767	0.000061	11.70	0.000097	0.001135	0.000008
9.75	0.000568	0.005542	0.000058	11.75	0.000093	0.001089	0.000008
9.80	0.000543	0.005326	0.000055	11.80	0.000089	0.001045	0.000008
9.85	0.000520	0.005118	0.000053	11.85	0.000085	0.001002	0.000007
9.90	0.000497	0.004918	0.000050	11.90	0.000081	0.000962	0.000007
9.95	0.000475	0.004726	0.000048	11.95	0.000077	0.000922	0.000006

(Table continued)

TABLE 3 (*continued*)

$x=\dfrac{h\nu_0}{kT}$	$\dfrac{x}{e^x-1}$	$\dfrac{x^2 e^x}{(e^x-1)^2}$	$-\ln(1-e^{-x})$	$x=\dfrac{h\nu_0}{kT}$	$\dfrac{x}{e^x-1}$	$\dfrac{x^2 e^x}{(e^x-1)^2}$	$-\ln(1-e^{-x})$
12.00	0.000074	0.000885	0.000006	13.50	0.000019	0.000250	0.000001
12.05	0.000070	0.000849	0.000006	13.55	0.000018	0.000239	0.000001
12.10	0.000067	0.000814	0.000006	13.60	0.000017	0.000229	0.000001
12.15	0.000064	0.000781	0.000005	13.65	0.000016	0.000220	0.000001
12.20	0.000061	0.000749	0.000005	13.70	0.000015	0.000211	0.000001
12.25	0.000059	0.000718	0.000005	13.75	0.000015	0.000202	0.000001
12.30	0.000056	0.000689	0.000005	13.80	0.000014	0.000193	0.000001
12.35	0.000053	0.000660	0.000004	13.85	0.000013	0.000185	0.000001
12.40	0.000051	0.000633	0.000004	13.90	0.000013	0.000178	0.000001
12.45	0.000049	0.000607	0.000004	13.95	0.000012	0.000170	0.000001
12.50	0.000047	0.000582	0.000004	14.00	0.000012	0.000163	0.000001
12.55	0.000044	0.000558	0.000004	14.05	0.000011	0.000156	0.000001
12.60	0.000042	0.000535	0.000003	14.10	0.000011	0.000150	0.000001
12.65	0.000041	0.000513	0.000003	14.15	0.000010	0.000143	0.000001
12.70	0.000039	0.000492	0.000003	14.20	0.000010	0.000137	0.000001
12.75	0.000037	0.000472	0.000003	14.25	0.000009	0.000132	0.000001
12.80	0.000035	0.000452	0.000003	14.30	0.000009	0.000126	0.000001
12.85	0.000034	0.000434	0.000003	14.35	0.000008	0.000121	0.000001
12.90	0.000032	0.000416	0.000002	14.40	0.000008	0.000116	0.000001
12.95	0.000031	0.000398	0.000002	14.45	0.000008	0.000111	0.000001
13.00	0.000029	0.000382	0.000002	14.50	0.000007	0.000106	0.000001
13.05	0.000028	0.000366	0.000002	14.55	0.000007	0.000102	0.000000
13.10	0.000027	0.000351	0.000002	14.60	0.000007	0.000097	0.000000
13.15	0.000026	0.000336	0.000002	14.65	0.000006	0.000093	0.000000
13.20	0.000024	0.000322	0.000002	14.70	0.000006	0.000089	0.000000
13.25	0.000023	0.000309	0.000002	14.75	0.000006	0.000085	0.000000
13.30	0.000022	0.000296	0.000002	14.80	0.000006	0.000082	0.000000
13.35	0.000021	0.000284	0.000002	14.85	0.000005	0.000078	0.000000
13.40	0.000020	0.000272	0.000002	14.90	0.000005	0.000075	0.000000
13.45	0.000019	0.000261	0.000001	14.95	0.000005	0.000072	0.000000
				15.00	0.000005	0.000069	0.000000

TABLE 4

Restricted Internal Rotational Free Energy $(-G/T)$ Contributions[a]

(cal. per deg. mole)

V/RT	\multicolumn{12}{c}{$1/Q_f$}
	0.25	0.30	0.35	0.40	0.45	0.50	0.55	0.60	0.65	0.70	0.75	0.80
0.0	2.754	2.392	2.086	1.821	1.587	1.377	1.188					
0.2	2.710	2.359	2.061	1.803	1.574	1.368	1.182					
0.4	2.623	2.296	2.014	1.765	1.543	1.342	1.160					
0.6	2.518	2.208	1.944	1.708	1.498	1.309	1.134					
0.8	2.406	2.106	1.856	1.636	1.442	1.266	1.100					
1.0	2.296	2.004	1.764	1.559	1.379	1.214	1.059					
1.5	2.040	1.770	1.548	1.370	1.210	1.069	0.938					
2.0	1.819	1.563	1.360	1.193	1.052	0.927	0.817					
2.5	1.630	1.389	1.197	1.043	0.912	0.802	0.709					
3.0	1.473	1.240	1.059	0.914	0.793	0.695	0.612	0.53				
3.5	1.340	1.117	0.943	0.802	0.694	0.603	0.528	0.46				
4.0	1.225	1.013	0.847	0.713	0.613	0.527	0.456	0.40				
4.5	1.133	0.925	0.764	0.637	0.543	0.463	0.397	0.35	0.34			
5.0	1.053	0.849	0.696	0.577	0.483	0.408	0.347	0.30	0.29			
6	0.919	0.728	0.586	0.477	0.393	0.325	0.272	0.23	0.25	0.22		
7	0.819	0.636	0.503	0.402	0.325	0.267	0.218	0.183	0.19	0.16		
8	0.735	0.564	0.440	0.346	0.275	0.221	0.178	0.146	0.15	0.12	0.14	
9	0.667	0.504	0.388	0.300	0.235	0.186	0.149	0.119	0.119	0.10	0.10	
10	0.610	0.456	0.345	0.264	0.203	0.159	0.125	0.099	0.096	0.079	0.08	0.08
12	0.521	0.380	0.280	0.209	0.157	0.120	0.092	0.071	0.079	0.064	0.06	0.06
14	0.452	0.321	0.232	0.169	0.124	0.092	0.069	0.052	0.055	0.043	0.051	0.05
16	0.396	0.276	0.195	0.139	0.100	0.072	0.053	0.039	0.038	0.030	0.033	0.04
18	0.351	0.240	0.166	0.117	0.082	0.058	0.042	0.030	0.028	0.021	0.022	0.027
20	0.315	0.211	0.144	0.098	0.068	0.047	0.033	0.024	0.022	0.016	0.016	0.018
									0.017	0.012	0.012	0.012
											0.009	0.009
												0.006

[a] K. S. Pitzer and W. D. Gwinn, *J. Chem. Phys.*, **10**, 428 (1942).

TABLE 5

FREE ENERGY INCREASE FROM FREE ROTATION $(G - G_f)/T$[a]

$$-G_f/T = -G_f/T - (G - G_f)/T = R \ln Q_f - (G - G_f)/T$$

(cal. per deg. mole)

V/RT	0	0.05	0.10	0.15	0.20	$1/Q_f$ 0.25	0.30	0.35	0.40	0.45	0.50	0.55
0.0	0.0000	0.000	0.000	0.000	0.000	0.000	0.000	0.000	0.000	0.000	0.000	0.000
0.2	0.1937	0.154	0.117	0.085	0.061	0.044	0.033	0.025	0.018	0.013	0.009	0.005
0.4	0.3776	0.326	0.274	0.225	0.176	0.131	0.096	0.072	0.056	0.044	0.035	0.026
0.6	0.5516	0.489	0.424	0.361	0.298	0.236	0.184	0.142	0.113	0.089	0.068	0.054
0.8	0.7161	0.640	0.566	0.493	0.420	0.348	0.286	0.230	0.185	0.145	0.111	0.088
1.0	0.8711	0.784	0.699	0.617	0.537	0.461	0.389	0.322	0.262	0.208	0.163	0.129
1.2	1.2200	1.114	1.010	0.909	0.809	0.714	0.622	0.538	0.451	0.375	0.308	0.250
1.5	1.5182	1.395	1.276	1.159	1.045	0.935	0.829	0.726	0.628	0.535	0.450	0.371
2.0	1.7724	1.635	1.501	1.371	1.246	1.124	1.004	0.889	0.778	0.675	0.575	0.479
2.5	1.9893	1.839	1.693	1.552	1.415	1.282	1.152	1.027	0.907	0.794	0.682	0.576
3.0	2.1756	2.013	1.856	1.704	1.557	1.414	1.275	1.143	1.019	0.893	0.774	0.660
3.5	2.3366	2.163	1.996	1.833	1.676	1.525	1.379	1.239	1.108	0.974	0.850	0.732
4.0	2.4772	2.293	2.117	1.945	1.780	1.621	1.467	1.322	1.184	1.044	0.914	0.791
4.5	2.6012	2.408	2.221	2.042	1.868	1.703	1.543	1.392	1.244	1.104	0.969	0.841
5.0	2.8108	2.599	2.396	2.202	2.015	1.836	1.664	1.500	1.344	1.194	1.052	0.916
6.0	2.9833	2.755	2.537	2.328	2.129	1.936	1.757	1.583	1.418	1.262	1.111	0.971
7.0	3.1294	2.886	2.653	2.432	2.220	2.020	1.828	1.646	1.474	1.312	1.157	1.011
8.0	3.2563	2.998	2.753	2.520	2.298	2.087	1.888	1.698	1.520	1.351	1.192	1.039
9.0	3.3686	3.097	2.839	2.594	2.362	2.144	1.936	1.741	1.557	1.383	1.219	1.063
10.0	3.5602	3.263	2.982	2.718	2.468	2.233	2.013	1.806	1.612	1.429	1.258	1.096
12.0	3.7205	3.400	3.099	2.816	2.551	2.303	2.071	1.854	1.651	1.462	1.285	1.119
14.0	3.8584	3.517	3.197	2.897	2.618	2.358	2.116	1.891	1.682	1.486	1.305	1.135
16.0	3.9793	3.618	3.280	2.965	2.674	2.403	2.152	1.920	1.704	1.505	1.319	1.146
18.0	4.0872	3.707	3.353	3.024	2.720	2.440	2.181	1.942	1.722	1.519	1.331	1.155

[a] K. S. Pitzer and W. D. Gwinn, *J. Chem. Phys.*, **10** 428 (1942).

TABLE 6
Restricted Internal Rotational Entropy Contribution (S_r)[a]
(cal. per degree mole)

V/RT	0.25	0.30	0.35	0.40	0.45	$1/Q_f$ 0.50	0.55	0.60	0.65	0.70	0.75	0.80
0.0	3.748	3.386	3.079	2.814	2.580	2.371	2.181					
0.2	3.743	3.382	3.076	2.811	2.578	2.369	2.179					
0.4	3.730	3.370	3.065	2.801	2.568	2.359	2.171					
0.6	3.709	3.347	3.043	2.780	2.547	2.340	2.153					
0.8	3.679	3.318	3.013	2.750	2.519	2.315	2.128					
1.0	3.638	3.279	2.974	2.714	2.485	2.279	2.095					
1.5	3.512	3.156	2.854	2.600	2.376	2.173	1.992					
2.0	3.355	3.004	2.709	2.458	2.241	2.048	1.873					
2.5	3.180	2.836	2.548	2.303	2.091	1.907	1.741					
3.0	3.008	2.667	2.380	2.138	1.933	1.756	1.600	1.45				
3.5	2.838	2.500	2.218	1.978	1.782	1.610	1.459	1.34				
4.0	2.678	2.343	2.069	1.834	1.643	1.475	1.326	1.22				
4.5	2.528	2.199	1.926	1.698	1.511	1.348	1.204	1.10	1.10			
5.0	2.396	2.068	1.798	1.579	1.392	1.233	1.095	0.97	1.00			
6.0	2.166	1.844	1.585	1.370	1.192	1.040	0.913	0.79	0.88	0.81		
7.0	1.983	1.665	1.411	1.204	1.033	0.891	0.770	0.665	0.70	0.63		
8.0	1.830	1.519	1.272	1.071	0.906	0.770	0.656	0.564	0.57	0.50	0.56	
9.0	1.703	1.397	1.156	0.962	0.804	0.674	0.569	0.482	0.482	0.41	0.44	0.39
10.0	1.593	1.295	1.060	0.872	0.719	0.596	0.495	0.418	0.411	0.348	0.36	0.31
12.0	1.417	1.125	0.904	0.728	0.588	0.476	0.388	0.315	0.352	0.295	0.29	0.25
14.0	1.275	0.994	0.783	0.620	0.492	0.388	0.309	0.247	0.258	0.212	0.245	0.21
16.0	1.157	0.890	0.688	0.533	0.414	0.322	0.251	0.196	0.196	0.158	0.173	0.143
18.0	1.058	0.801	0.609	0.464	0.353	0.270	0.205	0.158	0.155	0.120	0.126	0.102
20.0	0.975	0.727	0.542	0.405	0.303	0.228	0.170	0.129	0.121	0.093	0.094	0.075
									0.097	0.073	0.072	0.056
											0.056	0.042

[a] K. S. Pitzer and W. D. Gwinn, *J. Chem. Phys.*, **10**, 428 (1942).

TABLE 7
Entropy Decrease from Free Rotation $(S_f - S_{r'})$[a]
$$S_r = S_f - (S_f - S_{r'}) = R(\tfrac{1}{2} + \ln Q_f) - (S_f - S)$$
(cal. per degree mole)

V/RT	0.0	0.05	0.10	0.15	0.20	1/Q_f 0.25	0.30	0.35	0.40	0.45	0.50	0.55
0.0	0.0000	0.000	0.000	0.000	0.000	0.000	0.000	0.000	0.000	0.000	0.000	0.000
0.2	0.0049	0.005	0.004	0.004	0.004	0.004	0.004	0.003	0.003	0.002	0.002	0.002
0.4	0.0198	0.020	0.018	0.018	0.018	0.018	0.016	0.014	0.013	0.012	0.012	0.010
0.6	0.0440	0.044	0.043	0.043	0.040	0.039	0.039	0.036	0.034	0.033	0.031	0.028
0.8	0.0771	0.077	0.077	0.075	0.072	0.069	0.068	0.066	0.064	0.061	0.056	0.053
1.0	0.1185	0.118	0.117	0.115	0.112	0.110	0.107	0.105	0.100	0.095	0.092	0.086
1.5	0.2527	0.252	0.250	0.248	0.242	0.236	0.230	0.225	0.214	0.204	0.198	0.189
2.0	0.4182	0.417	0.415	0.410	0.402	0.393	0.382	0.370	0.356	0.339	0.323	0.308
2.5	0.6001	0.599	0.594	0.585	0.577	0.568	0.550	0.531	0.511	0.489	0.464	0.440
3.0	0.7856	0.783	0.777	0.768	0.757	0.740	0.719	0.699	0.676	0.647	0.615	0.581
3.5	0.9660	0.964	0.957	0.944	0.929	0.910	0.886	0.861	0.836	0.798	0.761	0.722
4.0	1.1356	1.133	1.126	1.111	1.094	1.070	1.043	1.011	0.980	0.937	0.896	0.855
4.5	1.2918	1.289	1.280	1.265	1.244	1.220	1.187	1.153	1.116	1.069	1.023	0.977
5.0	1.4339	1.431	1.421	1.404	1.380	1.352	1.318	1.281	1.235	1.188	1.138	1.086
6.0	1.6781	1.674	1.662	1.643	1.616	1.582	1.542	1.494	1.444	1.388	1.331	1.268
7.0	1.8783	1.874	1.860	1.837	1.807	1.765	1.721	1.668	1.610	1.547	1.480	1.411
8.0	2.0447	2.040	2.024	1.998	1.962	1.918	1.867	1.807	1.743	1.674	1.601	1.525
9.0	2.1864	2.180	2.163	2.134	2.095	2.045	1.989	1.923	1.852	1.776	1.697	1.612
10.0	2.3095	2.303	2.284	2.252	2.208	2.155	2.091	2.019	1.942	1.861	1.775	1.686
12.0	2.5155	2.508	2.485	2.447	2.394	2.331	2.261	2.175	2.086	1.992	1.895	1.793
14.0	2.6847	2.676	2.650	2.607	2.547	2.473	2.392	2.296	2.194	2.088	1.983	1.872
16.0	2.8289	2.819	2.788	2.740	2.674	2.591	2.496	2.391	2.281	2.166	2.049	1.930
18.0	2.9545	2.943	2.910	2.855	2.781	2.690	2.585	2.470	2.350	2.227	2.101	1.976
20.0	3.0659	3.054	3.017	2.956	2.872	2.773	2.659	2.537	2.409	2.277	2.143	2.011

[a] K. S. Pitzer and W. D. Gwinn, *J. Chem. Phys.*, **10**, 28 (1942).

TABLE 8

RESTRICTED INTERNAL ROTATIONAL HEAT CONTENT CONTRIBUTION, $(H_T - H_0)/T$[a]

(cal per degree mole)

	$1/Q_f$																
V/RT	0.0	0.05	0.10	0.15	0.20	0.25	0.30	0.35	0.40	0.45	0.50	0.55	0.60	0.65	0.70	0.75	0.80
0.0	0.9934	0.993	0.993	0.993	0.993	0.993	0.993	0.993	0.993	0.993	0.993	0.993					
0.2	1.1822	1.142	1.106	1.074	1.050	1.032	1.022	1.015	1.008	1.004	1.000	0.997					
0.4	1.3513	1.300	1.249	1.200	1.151	1.106	1.073	1.051	1.036	1.025	1.015	1.009					
0.6	1.5011	1.437	1.374	1.311	1.251	1.190	1.138	1.099	1.072	1.049	1.030	1.019					
0.8	1.6324	1.556	1.482	1.411	1.340	1.272	1.211	1.157	1.114	1.077	1.048	1.028					
1.0	1.7460	1.660	1.576	1.495	1.418	1.344	1.275	1.211	1.155	1.106	1.065	1.037					
1.5	1.9607	1.856	1.753	1.654	1.561	1.472	1.385	1.306	1.230	1.164	1.103	1.054					
2.0	2.0934	1.971	1.854	1.742	1.636	1.536	1.440	1.350	1.265	1.190	1.120	1.056					
2.5	2.1657	2.031	1.900	1.779	1.662	1.550	1.448	1.351	1.260	1.179	1.104	1.032					
3.0	2.1971	2.049	1.909	1.777	1.651	1.535	1.426	1.321	1.224	1.140	1.060	0.988	0.92				
3.5	2.2030	2.043	1.893	1.753	1.621	1.497	1.382	1.275	1.176	1.088	1.006	0.931	0.88				
4.0	2.1944	2.024	1.864	1.715	1.577	1.448	1.329	1.221	1.121	1.030	0.947	0.870	0.82	0.77			
4.5	2.1788	1.998	1.829	1.673	1.529	1.394	1.273	1.162	1.061	0.968	0.884	0.807	0.75	0.71			
5.0	2.1607	1.971	1.794	1.631	1.481	1.344	1.218	1.104	1.002	0.909	0.824	0.748	0.67	0.63	0.59		
6.0	2.1261	1.918	1.727	1.552	1.392	1.247	1.115	0.999	0.893	0.799	0.714	0.641	0.56	0.51	0.47	0.42	
7.0	2.0984	1.875	1.670	1.484	1.315	1.164	1.029	0.908	0.802	0.708	0.624	0.553	0.482	0.42	0.38	0.34	0.31
8.0	2.0781	1.840	1.623	1.427	1.251	1.095	0.955	0.833	0.725	0.631	0.549	0.479	0.418	0.363	0.31	0.28	0.25
9.0	2.0634	1.811	1.583	1.379	1.196	1.035	0.892	0.768	0.661	0.569	0.488	0.420	0.363	0.315	0.269	0.23	0.20
10.0	2.0526	1.787	1.548	1.335	1.147	0.982	0.838	0.715	0.608	0.515	0.437	0.371	0.319	0.273	0.231	0.194	0.17
12.0	2.0382	1.749	1.492	1.264	1.067	0.896	0.745	0.624	0.519	0.431	0.356	0.296	0.244	0.203	0.169	0.140	0.116
14.0	2.0292	1.717	1.441	1.202	0.997	0.823	0.672	0.551	0.450	0.367	0.295	0.240	0.195	0.158	0.128	0.104	0.084
16.0	2.0229	1.690	1.401	1.150	0.937	0.760	0.613	0.493	0.394	0.314	0.249	0.198	0.157	0.127	0.099	0.078	0.063
18.0	2.0182	1.666	1.363	1.102	0.886	0.707	0.561	0.443	0.347	0.271	0.211	0.164	0.128	0.099	0.077	0.060	0.047
20.0	2.0147	1.646	1.329	1.061	0.841	0.660	0.515	0.399	0.307	0.236	0.181	0.138	0.105	0.080	0.061	0.047	0.036

[a] K. S. Pitzer and W. D. Gwinn, *J. Chem. Phys.*, **10**, 428 (1942).

TABLE 9
Restricted Internal Rotational Heat Capacity Contribution $(C_p°)$[a]
(cal. per degree mole)

V/RT	\\						$1/Q_f$								
	0.0	0.05	0.10	0.15	0.20	0.25	0.30	0.35	0.40	0.45	0.50	0.55	0.60	0.65	0.70	0.75	0.80
0.0	0.9934	0.993	0.993	0.993	0.993	0.993	0.993	0.993	0.993	0.993	0.99	0.99					
0.2	1.0033	1.003	1.003	1.002	1.001	1.000	0.999	0.998	0.998	0.998	1.00	1.00					
0.4	1.0326	1.033	1.032	1.030	1.028	1.025	1.024	1.021	1.019	1.017	1.02	1.01					
0.6	1.0799	1.080	1.079	1.076	1.073	1.068	1.065	1.060	1.056	1.051	1.05	1.04					
0.8	1.1433	1.143	1.141	1.138	1.133	1.128	1.121	1.114	1.106	1.099	1.09	1.08					
1.0	1.2201	1.219	1.217	1.212	1.206	1.199	1.190	1.180	1.169	1.157	1.14	1.13					
1.5	1.4506	1.449	1.444	1.435	1.423	1.408	1.391	1.370	1.348	1.324	1.30	1.27					
2.0	1.6975	1.695	1.687	1.673	1.655	1.632	1.606	1.574	1.541	1.505	1.469	1.43					
2.5	1.9211	1.917	1.908	1.888	1.866	1.840	1.801	1.756	1.717	1.670	1.623	1.58					
3.0	2.0986	2.095	2.082	2.062	2.033	1.996	1.952	1.900	1.846	1.794	1.738	1.68	1.7				
3.5	2.2223	2.218	2.204	2.180	2.146	2.106	2.054	1.995	1.934	1.869	1.803	1.74	1.7				
4.0	2.2986	2.294	2.276	2.249	2.213	2.168	2.110	2.048	1.980	1.907	1.834	1.76	1.69	1.6			
4.5	2.3354	2.330	2.312	2.280	2.238	2.190	2.129	2.062	1.990	1.911	1.832	1.75	1.67	1.6			
5.0	2.3443	2.338	2.318	2.285	2.241	2.186	2.120	2.056	1.972	1.890	1.808	1.719	1.62	1.55	1.4		
6.0	2.3155	2.307	2.283	2.245	2.192	2.130	2.059	1.979	1.893	1.803	1.711	1.616	1.51	1.43	1.33	1.2	
7.0	2.2647	2.256	2.228	2.185	2.126	2.055	1.973	1.883	1.787	1.688	1.588	1.491	1.394	1.30	1.21	1.12	
8.0	2.2157	2.205	2.174	2.125	2.058	1.979	1.888	1.788	1.684	1.576	1.468	1.362	1.260	1.159	1.07	0.99	1.0
9.0	2.1759	2.164	2.130	2.074	1.999	1.909	1.808	1.699	1.587	1.474	1.362	1.252	1.149	1.049	0.955	0.86	0.91
10.0	2.1454	2.133	2.094	2.033	1.951	1.854	1.745	1.630	1.507	1.382	1.262	1.151	1.047	0.949	0.853	0.762	0.79
12.0	2.1050	2.089	2.043	1.972	1.877	1.763	1.636	1.502	1.365	1.233	1.107	0.989	0.877	0.774	0.683	0.600	0.68
14.0	2.0810	2.063	2.009	1.923	1.814	1.686	1.546	1.400	1.254	1.112	0.978	0.855	0.744	0.644	0.555	0.476	0.519
16.0	2.0654	2.044	1.983	1.887	1.764	1.622	1.468	1.311	1.156	1.009	0.873	0.749	0.639	0.542	0.457	0.384	0.408
18.0	2.0544	2.031	1.961	1.853	1.717	1.562	1.397	1.232	1.070	0.919	0.780	0.657	0.549	0.456	0.378	0.312	0.321
20.0	2.0462	2.020	1.944	1.827	1.678	1.510	1.333	1.158	0.991	0.837	0.701	0.580	0.477	0.389	0.316	0.256	0.256

[a] K. S. Pitzer and W. D. Gwinn, *J. Chem. Phys.*, **10**, 428 (1942).

TABLE 10
Potential Barriers Hindering Internal Rotation

A	B	V_0 (kcal. per mole)	Compound
CH_3	$-CH_3$	2.88	Ethane[a]
CH_3	$-CH_2CH_3$	3.40	Propane[a]
CH_3	$-CH(CH_3)_2$	3.62	Isobutane[a]
CH_3	$-C(CH_3)_3$	4.30	Neopentane[a]
CH_3	$-CH=CH_2$	1.95	Propylene[a]
CH_3	$-CH_2-CH=CH_2$	3.60	Butene-1[b]
CH_3	$-OCH_3$	2.70	Dimethylether[c]
CH_3	$-SCH_3$	2.00	Dimethylsulphide[c]
CH_3	$-NHCH_3$	3.46	Dimethylamine[c]
CH_3	$-OH$	1.35	Methanol[c]
CH_3	$-SH$	1.46	Methyl mercaptan[c]
CH_3	$-NH_2$	1.9	Methylamine[d]
CH_3	$-CHO$	2.10	Acetaldehyde[e]
CH_3	$-CF_3$	3.45	Trifluoroethane[b]
CH_3	$-CCl_3$	2.70	Trichloroethane[f]
CH_3	$-COCH_3$	1.00	Acetone[c]
CH_3	$-CH_2CN$	5.2, 3.28	Ethylcyanide[g,o]
NH_2	$-NH_2$	2.80	Hydrazine[h]
CH_3	$-NHNH_2$	3.50	Methylhydrazine[g]
CH_3	$-N-NHCH_3$	3.0	Dimethylhydrazine[i]
CH_3	$-N(CH_3)NH_2$	3.7	*Unsym*-dimethylhydrazine[j]
CH_3-CH_2	$-CH_2CH_3$	3.3	n-Butane[k]
$ClCH_2$	$-CH_2Cl$	2.8	Dichloroethane[l]
CH_3	$-Si(CH_3)_3$	1.30	Tetramethylsilane[c]
CH_3	$-N(CH_3)_2$	4.27	Trimethylamine[c]
CF_3	$-CF_3$	4.35	Hexafluoroethane[m]
CH_3	$-C{\equiv}C(CH_3)$	0	Dimethylacetylene[a]
CH_3	$-C_6H_4(CH_3)$	0.5	*Meta-* and *para*-xylene[c]
CH_3	$-C_6H_4(CH_3)$	2.0	*Ortho*-xylene[c]
OH	$-NO_2$	7.0	Nitric acid[n]

[a] E. Blade and G. E. Kimball, *J. Chem. Phys.* **18**, 630 (1950).
[b] H. Russell, D. R. V. Goldbing, and D. M. Yost, *J. Am. Chem. Soc.* **66**, 16 (1944).
[c] K. S. Pitzer, *Discussions Faraday Soc.* **10**, 66 (1951).
[d] J. G. Aston and F. L. Gittler, *J. Chem. Phys.* **23**, 211 (1955).
[e] J. C. Morris, *J. Chem. Phys.* **11**, 230 (1943).
[f] T. R. Rubin, B. H. Levedahl, and D. M. Yost, *J. Am. Chem. Soc.* **66**, 279 (1944).
[g] N. E. Duncan and G. J. Janz, *J. Chem. Phys.* **23**, 434 (1955).
[h] D. W. Scott, G. D. Oliver, M. E. Gross, W. N. Hubbard, and H. M. Huffman, *J. Am. Chem. Soc.* **71**, 2293 (1949).
[i] J. G. Aston, H. L. Fink, G. J. Janz, and K. E. Russell, *J. Am. Chem. Soc.* **73**, 1943 (1951).
[j] J. G. Aston, J. L. Wood, and T. Zolki, *J. Am. Chem. Soc.* **75**, 6202 (1953).
[k] J. G. Aston and G. H. Messerly, *J. Am. Chem. Soc.* **62**, 1920 (1940).
[l] H. J. Bernstein, *J. Chem. Phys.* **17**, 262 (1949).
[m] E. L. Pace and J. G. Aston, *J. Am. Chem. Soc.* **70**, 566 (1948).
[n] W. R. Forsythe and W. F. Giauque, *J. Am. Chem. Soc.* **64**, 48 (1942).
[o] R. G. Lerner and, B. P. Dailey, *J. Chem. Phys.* **26**, 678 (1957).

TABLE 11

ADDITIVE INCREMENTS IN FREE ENERGY FUNCTION, $-(G° - H_0°)/T$, OF NORMAL PARAFFINS[a,b]

T, °K	$[G^0(T)]$ C_7 and above	[C—C$_{Str.}$]	[C—C$_{Bend.}$]	[I.Rot.]	[G$_{steric}$] A	B	[CH$_3$]	[CH$_2$]
298.16	49.524	0.707	1.252	2.951	4.320	1.037	0.882	0.032
300	49.573	0.707	1.258	2.960	4.335	1.040	0.898	0.033
400	51.860	0.749	1.554	3.444	4.997	1.195	1.399	0.118
500	53.634	0.813	1.823	3.849	5.425	1.295	1.932	0.263
600	55.083	0.890	2.065	4.189	5.724	1.365	2.487	0.460
700	56.308	0.976	2.284	4.482	5.944	1.416	3.055	0.698
800	57.370	1.067	2.482	4.736	6.113	1.456	3.630	0.965
900	58.306	1.158	2.663	4.954	6.246	1.487	4.203	1.254
1000	59.143	1.250	2.830	5.151	6.354	1.512	4.773	1.557
1100	59.901	1.340	2.984	5.326	6.443	1.533	5.335	1.870
1200	60.593	1.428	3.128	5.486	6.518	1.550	5.895	2.189
1300	61.229	1.514	3.262	5.625	6.582	1.565	6.440	2.511
1400	61.818	1.598	3.388	5.753	6.637	1.578	6.978	2.834
1500	62.366	1.679	3.506	5.868	6.685	1.589	7.503	3.156

[a] W. B. Person and G. C. Pimentel, *J. Am. Chem. Soc.*, **75**, 532 (1953).
[b] Units, cal./deg. mole.

TABLE 12

ADDITIVE INCREMENTS IN HEAT CONTENT, $(H° - H_0°)/T$, OF NORMAL PARAFFINS[a,b]

T, °K	$[G^0(T)]$ C_7 and above	[C—C$_{Str.}$]	[C—C$_{Bend.}$]	[I.Rot.]	[G$_{steric}$] A	B	[CH$_3$]	[CH$_2$]
298.16	7.949	0.085	0.921	1.572	2.524	0.592	1.432	0.159
300	7.949	0.087	0.926	1.577	2.515	0.589	1.443	0.164
400	7.949	0.215	1.132	1.766	2.083	0.487	2.075	0.462
500	7.949	0.359	1.275	1.864	1.762	0.411	2.735	0.865
600	7.949	0.496	1.378	1.895	1.522	0.355	3.382	1.315
700	7.949	0.620	1.455	1.896	1.337	0.312	4.004	1.778
800	7.949	0.730	1.515	1.886	1.191	0.278	4.594	2.234
900	7.949	0.825	1.563	1.865	1.074	0.250	5.147	2.672
1000	7.949	0.909	1.603	1.835	0.977	0.228	5.669	3.089
1100	7.949	0.982	1.635	1.804	0.896	0.209	6.160	3.482
1200	7.949	1.047	1.663	1.775	0.828	0.193	6.621	3.851
1300	7.949	1.104	1.686	1.743	0.769	0.179	7.053	4.196
1400	7.949	1.155	1.707	1.711	0.718	0.167	7.455	4.518
1500	7.949	1.201	1.724	1.681	0.673	0.157	7.825	4.819

[a] W. B. Person and G. C. Pimentel, *J. Am. Chem. Soc.*, **75**, 532 (1953).
[b] Units, cal./deg. mole.

TABLE 13

Additive Increments in Heat Capacity, $C_p°$, of Normal Paraffins[a,b]

T, °K	$[G^0(T)]$ C_7 and above	[C—C$_{Str.}$]	[C—C$_{Bend.}$]	[I.Rot.]	[G$_{steric}$] A	B	[CH$_3$]	[CH$_2$]
298.16	7.949	0.406	1.668	2.334	1.049	0.238	3.229	0.821
300	7.949	0.413	1.672	2.336	1.037	0.235	3.253	0.838
400	7.949	0.779	1.810	2.309	0.595	0.135	4.687	1.908
500	7.949	1.072	1.874	2.176	0.383	0.087	6.025	3.038
600	7.949	1.284	1.908	2.010	0.262	0.058	7.212	4.081
700	7.949	1.437	1.929	1.858	0.196	0.045	8.250	5.007
800	7.949	1.548	1.943	1.722	0.150	0.034	9.175	5.818
900	7.949	1.629	1.952	1.612	0.119	0.027	9.998	6.525
1000	7.949	1.691	1.959	1.521	0.096	0.022	10.726	7.140
1100	7.949	1.738	1.964	1.446	0.079	0.018	11.367	7.672
1200	7.949	1.776	1.967	1.385	0.067	0.015	11.931	8.133
1300	7.949	1.805	1.970	1.336	0.057	0.013	12.428	8.532
1400	7.949	1.829	1.973	1.292	0.049	0.011	12.864	8.878
1500	7.949	1.848	1.975	1.257	0.043	0.010	13.247	9.179

[a] W. B. Person and G. C. Pimentel, *J. Am. Chem. Soc.*, **75**, 532 (1953).
[b] Units, cal./deg. mole.

TABLE 14

Thermodynamic Properties of Normal Heptane[a,b]

T, °K	$-(G°-H_0°)/T$	$(H°-H_0°)/T$	$S°$	$C_p°$
298.16	76.70	25.54	102.24	39.67
300	76.86	25.63	102.50	39.86
400	84.91	30.51	115.42	50.42
500	92.25	35.49	127.74	60.07
600	99.13	40.26	139.39	68.33
700	105.68	44.77	150.45	75.38
800	111.95	49.00	160.95	81.43
900	117.93	52.91	170.84	86.68
1000	123.70	56.52	180.22	91.22
1100	129.25	59.86	189.11	95.16
1200	134.62	62.97	197.59	98.57
1300	139.76	65.83	205.59	101.55
1400	144.74	68.48	213.22	104.12
1500	149.54	70.92	220.46	106.37

[a] W. B. Person and G. C. Pimentel, *J. Am. Chem. Soc.*, **75**, 532 (1953).
[b] Units, cal./deg. mole.

TABLE 15

[CH$_2$] Increment for Normal Paraffins[a,b]

$N \geq 7$

T, °K	$-(G° - H_0°)/T$	$(H° - H_0°)/T$	$S°$	$C_p°$
298.16	5.979	3.330	9.309	5.466
300	5.999	3.344	9.343	5.494
400	7.061	4.063	11.123	6.941
500	8.043	4.774	12.817	8.246
600	8.970	5.439	14.409	9.342
700	9.856	6.062	15.918	10.276
800	10.705	6.643	17.348	11.065
900	11.516	7.176	18.692	11.746
1000	12.299	7.663	19.963	12.333
1100	13.053	8.112	21.165	12.838
1200	13.781	8.528	22.309	13.276
1300	14.477	8.908	23.386	13.657
1400	15.151	9.258	24.409	13.983
1500	15.798	9.582	25.381	14.269

[a] W. B. Person and G. C. Pimentel, *J. Am. Chem. Soc.*, **75**, 532 (1953).
[b] Units, cal./deg. mole.

TABLE 16

Additional Increments Extending the Infinite Chain Method to Branch Chain Paraffins[a,b]

T, °K	[CH]	Isobutane correction	Neopentane correction
		$-(G° - H_0°)/T$	
298.1	0.01	0.00	−1.39
400	0.05	−0.19	−1.80
500	0.13	−0.32	−1.95
600	0.23	−0.36	−2.05
800	0.50	−0.45	−2.20
1000	0.81	−0.60	−2.30
1500	1.63	−0.59	−2.16
		$(H° - H_0°)/T$	
298.1	0.07	−0.73	−1.39
400	0.23	−0.60	−1.16
500	0.46	−0.50	−0.90
600	0.72	−0.37	−0.55
800	1.18	−0.27	−0.11
1000	1.61	−0.22	0.11
1500	2.46	0.07	0.41

[a] K. S. Pitzer, *J. Chem. Phys.* **8**, 711 (1940); *Chem. Revs.* **27**, 39 (1940).
[b] Units, cal./deg. mole.

TABLE 17
Parent Group Properties[a,b]

Group	$\Delta H^\circ_{f\,298.1}$ (g) kcal./g. mole	$S^\circ_{298.1}$ (g) cal./g. mole(°K)	$C_p^\circ = a + bT + cT^2$ Ideal Gas at $T°$ K		
			a	$b\,(10^3)$	$c\,(10^6)$
Methane	−17.89	44.50	3.79	16.62	− 3.24
Cyclopentane	−18.46	70.00	− 9.02	109.28	−40.23
Cyclohexane	−29.43	71.28	−11.53	139.65	−52.02
Benzene	19.82	64.34	− 4.20	91.30	−36.63
Naphthalene	35.4	80.7	3.15	109.40	−34.79
Methylamine	− 7.1	57.7	4.02	30.72	− 8.70
Dimethylamine	− 7.8	65.2	3.92	48.31	−14.09
Trimethylamine	−10.9	—	3.93	65.85	−19.48
Dimethyl ether	−46.0	63.7	6.42	39.64	−11.45
Formamide	−49.5	—	6.51	25.18	− 7.47

[a] J. W. Andersen, G. H. Beyer, and K. M. Watson, *Natl. Petroleum News* **36**, R 476 (1944).

[b] J. M. Brown, Univ. Wisc. Dept. Chem. Eng. Spec. Probs. Proj. Rept. (June, 1953).

TABLE 18
Contributions of Primary CH₃ Substitution[a,b]

Base group	$\Delta(\Delta H^\circ_{f\,298.1})$ (g) kcal./g. mole	$\Delta S^\circ_{298.1}$ (g) cal./g. mole(°K)	Ideal gas at $T°$ K		
			Δa	$\Delta b\,(10^3)$	$\Delta c\,(10^6)$
1. Methane	− 2.50	10.35	−2.00	23.20	− 9.12
2. Cyclopentane					
(a) First primary substitution	− 7.04	11.24	1.87	17.55	− 6.68
(b) Second primary substitution					
To form 1,1	− 7.55	4.63	−0.67	24.29	−10.21
To form 1,2(*cis*)	− 5.46	6.27	−0.01	22.69	− 9.46
To form 1,2(*trans*)	− 7.17	6.43	0.28	21.97	− 9.18
To form 1,3(*cis*)	− 6.43	6.43	0.28	21.97	− 9.18
To form 1,3(*trans*)	− 6.97	6.43	0.28	21.97	− 9.18
(c) Additional substitutions, each	− 7.0	—	—	—	—

TABLE 18 *(continued)*

Base Group	$\Delta(\Delta H^°_{f\,298.1})$ (g) kcal./g. mole	$\Delta S^°_{298.1}$ (g) cal./g mole(°K)	Ideal gas at $T°$ K		
			Δa	$\Delta b\,(10^3)$	$\Delta c\,(10^6)$
3. Cyclohexane					
(a) Enlargement of ring over 6 C, per carbon atom added to ring	−10.97	1.28	−2.51	30.37	−11.79
(b) First primary substitution on ring	−7.56	10.78	2.13	18.66	−5.71
(c) Second primary substitution on ring					
To form 1,1	−6.27	5.18	−2.14	25.69	−10.09
To form 1,2(*cis*)	−4.16	7.45	−0.65	22.19	−8.84
To form 1,2(*trans*)	−6.03	6.59	−0.06	22.59	−2.56
To form 1,3(*cis*)	−7.18	6.48	−0.34	21.49	−7.95
To form 1,3(*trans*)	−5.21	7.86	0.29	19.29	−7.23
To form 1,4(*cis*)	−5.23	6.48	0.29	19.29	−7.23
To form 1,4(*trans*)	−7.13	5.13	−0.72	23.79	−9.91
(d) Additional substitutions on ring, each	−7.0	—	—	—	—
4. Benzene					
(a) First substitution	−7.87	12.08	0.78	16.68	−5.41
(b) Second substitution					
To form 1,2	−7.41	7.89	4.27	9.72	−1.87
To form 1,3	−7.83	9.07	0.77	17.46	−6.19
To form 1,4	−7.66	7.81	1.76	13.45	−3.41
(c) Third substitution					
To form 1,2,3	−6.83	9.19	1.41	12.78	−2.71
To form 1,2,4	−7.87	10.42	1.61	12.72	−2.77
To form 1,3,5	−7.96	6.66	2.41	11.30	−1.90
5. Naphthalene					
(a) First substitution	−4.5	12.0	0.36	17.65	−5.88
(b) Second substitution					
To form 1,2	−6.3	8.1	5.20	6.02	1.18
To form 1,3	−6.5	9.2	1.72	14.18	−3.74
To form 1,4	−8.0	7.8	1.28	14.57	−3.98
(*sym*)	—	8.0	0.57	16.51	−15.19
6. Methylamine	−5.7	—			
7. Dimethylamine	−6.3	—	−0.10	17.52	−5.35
8. Trimethylamine	−4.1	—			
9. Formamide Substitution on C atom	−9.0	—	6.11	−1.75	4.75

[a] J. W. Andersen, G. H. Beyer, and K. M. Watson, *Natl. Petroleum News* **36**, R 476 (1944).

[b] J. M. Brown: Univ. Wisc. Dept. Chem. Eng. Spec. Probs. Proj. Rept. (June, 1953).

TABLE 19
Secondary Methyl Substitutions[a,b]

A	B	$\Delta(\Delta H^\circ_{f298})$ kcal/g. mole	$\Delta S^\circ_{298.1}$ (g) cal./g. mole (°K)	Ideal gas at $T°$ K		
				Δa	Δb (10³)	Δc (10⁶)
1	1	−4.75	10.00	0.49	22.04	− 8.96
1	2	−4.92	9.18	1.09	17.79	− 6.47
1	3	−4.42	9.72	1.00	19.88	− 8.03
1	4	−5.0	11.0	1.39	17.12	− 5.88
1	5	−4.68	10.76	1.09	18.17	− 7.16
2	1	−6.31	5.57	−0.30	21.74	− 8.77
2	2	−6.33	7.15	−0.64	23.38	− 9.97
2	3	−5.25	6.53	0.80	19.27	− 7.70
2	4	−3.83	7.46	2.52	16.11	− 5.88
2	5	−6.18	6.72	0.37	19.25	− 7.72
3	1	−8.22	2.81	−0.28	24.21	−10.49
3	2	−7.0	3.87	−0.93	24.73	− 8.95
3	3	−5.19	3.99	−3.27	30.96	−14.06
3	4	−4.94	1.88	−0.14	27.57	−10.27
3	5	−9.2	1.3	0.42	16.20	− 4.68
1	—O— in ester or ether	−7.0	14.4	−0.01	17.58	− 5.33
Substitution of H of OH group to form ester		9.5	16.7	0.44	16.63	− 4.95

[a] J. W. Andersen, G. H. Beyer, and K. M. Watson, *Natl. Petroleum News* **36**, R 476 (1944).

[b] J. M. Brown: Univ. Wisc. Dept. Chem. Eng. Spec. Probs. Proj. Rept. (June, 1953)

TABLE 20
Multiple Bond Contributions and Additional Corrections[a,b]

Type of bond A B	$\Delta(\Delta H_f°)_{298.1}$ (g)kcal./g. mole	$\Delta S°_{298.1}$ (g) cal./g. mole (°K)	Ideal gas at $T°$ K Δa	Δb (10^3)	Δc (10^6)
1 = 1	32.88	−2.1	1.33	−12.69	4.77
1 = 2	30.00	0.8	1.56	−14.87	5.57
1 = 3	28.23	2.2	0.63	−23.65	13.10
2 = 2 cis	28.39	−0.6	0.40	−18.87	9.89
2 = 2 trans	27.40	−1.2	0.40	−18.87	9.89
2 = 3	26.72	1.6	0.63	−23.65	13.10
3 = 3	25.70	—	−4.63	−17.84	11.88
1 ≡ 1	74.58	−6.8	5.58	−31.19	11.19
1 ≡ 2	69.52	−7.8	6.42	−36.41	14.53
2 ≡ 2	65.50	−6.3	4.66	−36.10	15.28
Additional correction for length of each side chain on ring					
1. More than 2 C on cyclopentane side chain	−0.45	0.12	−0.48	1.5	1.15
2. More than 2 C on cyclohexane side chain	0.32	−0.39	0.76	2.10	1.30
3. More than 4 C on benzene side chain	−0.70	−0.62	0.22	−0.20	0.08
Additional correction for double-bond arrangement					
1. Adjacent double bonds	13.16	−3.74	2.24	1.16	−0.25
2. Alternate double bonds	−4.28	−5.12	−0.94	3.88	−3.49
3. Double bond adjacent to aromatic ring					
(a) Less than 5 C in side chain	−2.0	−2.65	1.01	−3.24	1.31
(b) Over 4 C in side chain	−1.16	−2.65	1.01	−3.24	1.31

[a] J. W. Andersen, G. H. Beyer, and K. M. Watson, *Natl. Petroleum News* **36**, R 476 (1944).

[b] J. M. Brown, Univ. Wisc. Dept. Chem. Eng. Spec. Probs. Proj. Rept. (June, 1953).

TABLE 21

Nonhydrocarbon Group Contributions Replacing [CH₃] Group[a,c]

Group	$\Delta(\Delta H_f°)_{298.1}$ (g) kcal./g. mole	$\Delta S°_{298.1}$ (g) cal./g. mole (°K)	Ideal gas at $T°$ K		
			Δa	Δb (10³)	Δc (10⁶)
—OH (aliphatic, meta, para)	−32.7	2.6	3.17	−14.86	5.59
—OH ortho	−47.7	—	—	—	—
—NO₂	1.2	2.0	6.3	−19.53	10.36
—CN	39.0	4.0	3.64	−13.92	4.53
—Cl	0 for first Cl on a carbon: 4.5 for each additional	0[b]	2.19	−18.85	6.26
—Br	10.0	3.0[b]	2.81	−19.41	6.33
—F	−35.0	−1.0[b]	2.24	−23.61	11.79
—I	24.8	5.0[b]	2.73	−17.37	4.09
=O aldehyde	−12.9	−12.3	3.61	−55.72	22.72
=O ketone	−13.2	−2.4	5.02	−66.08	30.21
—COOH	−87.0	15.4	8.50	−15.07	7.94
—SH	15.8	5.2	4.07	−24.96	12.37
—C₆H₆	32.3	21.7	−0.79	53.63	−19.21
—NH₂	12.3	−4.8	1.26	−7.32	2.23

[a] J. W. Andersen, G. H. Beyer, and K. M. Watson, *Natl. Petroleum News* **36**, R 476 (1944).

[b] Add 1.0 to the calculated entropy contributions of halides for methyl derivatives: for example, methyl chloride = 44.4 (base) + 10.4 (primary CH₃) − 0.0 (Cl substitution) +1.0.

[c] J. M. Brown, Univ. Wisc. Dept. Chem. Eng. Spec. Probs. Proj. Rept. (June, 1953).

TABLE 22. ($H° - H_0°$) Increments for Hydrocarbon Groups[a]
(kcal. per g. mole)

		0° K	298° K	400° K	500° K	600° K	800° K	1000° K	1200° K	1500° K
CH_3		0	1.43	2.15	3.00	4.01	6.38	9.14	12.22	17.25
CH_2		0	1.023	1.684	2.46	3.35	5.42	7.77	10.34	14.49
CH		0	−0.014	0.51	1.25	2.04	3.87	5.89	8.02	11.30
C		0	−0.68	−0.16	0.55	1.36	3.12	4.95	6.77	9.42
$H_2C=$		0	1.26	1.86	2.56	3.37	5.24	7.38	9.74	13.55
$H\!\!>\!\!C\!\!=\!\!CH_2$		0	1.81	2.84	4.08	5.48	8.77	12.55	16.66	23.32
$H\!\!>\!\!C\!\!=\!\!C\!\!<\!\!H$	(trans)	0	1.33	2.28	3.42	4.67	7.59	10.91	14.48	20.21
$H\!\!>\!\!C\!\!=\!\!C\!\!<\!\!H$	(cis)	0	1.12	1.84	2.84	3.99	6.75	9.95	13.44	19.12
$>\!\!C\!\!=\!\!CH_2$		0	1.22	2.20	3.41	4.73	7.73	11.09	14.71	20.50
$>\!\!C\!\!=\!\!C\!\!<\!\!H$		0	0.59	1.37	2.35	3.43	5.96	8.78	11.78	16.61
$>\!\!C\!\!=\!\!C\!\!<$		0	−0.02	0.61	1.45	2.36	4.48	6.84	(9.3)	(13.0)
$=C=$		0	0.50	0.90	1.35	1.83	2.85	3.92	5.00	6.71
$HC\equiv$		0	1.20	1.77	2.40	3.06	4.50	6.05	7.68	10.27
$-C\equiv$		0	0.47	0.81	1.18	1.59	2.49	3.48	4.52	6.15
⇉C−H		0	0.56	0.96	1.46	2.05	3.44	5.03	6.77	9.56
⇉C−		0	0.08	0.32	0.67	1.07	2.04	3.16	4.32	6.20
⇉C←→[b]		0	−0.74	−0.51	−0.20	0.19	1.36	1.95	2.34	—
←→CH_2		0	1.26	1.96	2.79	3.75	5.77	8.05	10.51	14.46
←→$C\!\!<\!\!H$		0	0.75	1.41	2.14	2.84	4.57	6.45	8.37	11.73
Cyclo C_6 ring		0	−1.55	−2.23	−2.77	−3.06	−3.22	−2.96	−2.51	−1.62
Cyclo C_5 ring		0	−1.04	−1.78	−2.44	−2.98	−3.72	−4.15	−4.50	−4.76
Cyclo C_4 ring[b]		0	−0.4	−0.8	−1.2	−1.8	−2.4	−3.2	−3.6	−4.4
Cyclo C_3 ring		0	−0.26	−0.47	−0.89	−1.06	−1.55	−1.94	−2.31	−2.87

Correction Factors for Paraffin Chains

Ethyl side chain	0	←	−0.7	→
3 adjacent CH groups	0	←	0.7	→
Adjacent quaternary C's	0	←	0.4	→
Quaternary C not adjacent to terminal methyl	0	←	−0.4	→

Correction Factor for Substituents on Aromatic Nucleus

1,2-Dimethyl- 1,2,3-trimethyl, and 1,2- or 1,3-methylethyl	0	←	0.5	→

[a] J. L. Franklin, *Ind. Eng. Chem.* **41**, 1070 (1949). [b] Estimated by extrapolation or interpolation.

TABLE 23. $\Delta H_f°$ Increments for Hydrocarbon Groups[a]
(kcal. per g. mole)

	0° K	298° K	400° K	500° K	600° K	800° K	1000° K	1200° K	1500° K
CH_3	−8.26	−10.12	−10.71	−11.22	−11.64	−12.27	−12.64	−12.82	−12.87
CH_2	−3.673	−4.926	−5.223	−5.465	−5.648	−5.871	−5.941	−5.886	−5.692
CH	0.18	−1.09	−1.17	−1.12	−1.05	−0.80	−0.50	−0.12	0.33
C	1.74	0.80	1.07	1.45	1.89	2.77	3.60	4.38	5.34
$H_2C=$	7.26	6.25	5.88	5.57	5.30	4.88	4.60	4.44	4.31
$H\!>\!\!C\!=\!CH_2$	16.73	15.00	14.47	14.01	13.62	13.03	12.67	12.50	12.39
$H\!>\!\!C\!=\!C\!<\!H$ (trans)	19.03	17.83	17.57	17.37	17.17	16.92	16.82	16.83	16.92
$H\!>\!\!C\!=\!C\!<\!H$ (cis)	20.31	18.88	18.42	18.07	17.77	17.37	17.14	17.09	17.12
$>\!C\!=\!CH_2$	18.20	16.89	16.68	16.53	16.40	16.23	16.16	16.24	16.39
$>\!C\!=\!C\!<\!H$	21.10	20.19	20.10	20.08	20.07	20.13	20.24	20.44	20.75
$>\!C\!=\!C\!<$	25.08	24.57	24.74	24.88	25.06	25.38	25.76	(26.2)	(26.9)
$=\!C\!=$	33.0[b]	33.42	33.59	33.71	33.82	33.95	34.03	34.06	34.06
$HC\!\equiv$	27.16	27.10	27.07	27.02	26.97	26.81	26.65	26.50	26.27
$-C\!\equiv$	27.12	27.34	27.42	27.48	27.50	27.53	27.49	27.51	27.46
↗CH	4.00	3.30	3.09	2.92	2.79	2.59	2.47	2.42	2.40
↗C−	5.76	5.57	5.59	5.63	5.62	5.72	5.85	5.99	6.14
↗C ↔[b]	5.29	4.28	4.33	4.28	4.27	4.56	4.16	3.51	—
↔CH_2	11.3[b]	10.08	9.81	9.64	9.48	9.25	9.12	9.06	9.05
↔$C\!<\!H$	12.65	12.04	12.11	12.16	12.17	12.27	12.43	12.59	12.66
C_6 cycloparaffin ring	1.10	−0.45	−1.13	−1.67	−1.96	−2.12	−1.86	−1.41	−0.52
C_5 cycloparaffin ring	6.72	5.68	4.94	4.28	3.74	3.00	2.57	2.22	1.96
C_4 cycloparaffin ring[b]	18.8	18.4	18.0	17.6	17.0	16.4	15.6	15.2	14.4
C_3 cycloparaffin ring	24.4	24.22	23.87	23.58	23.38	22.89	22.50	22.15	21.60
Correction Factors for Paraffin Chains									
Ethyl side chain	1.5	←			0.8				→
3 adjacent CH groups	1.6	←			2.3				→
Adjacent C and CH groups	2.5	←			2.5				→
Adjacent quaternary C's	5.0	←			5.4				→
Quaternary C not adjacent to terminal methyl	2.1	←			1.7				→
Correction Factors for Substituents on Aromatic Nucleus									
1,2-Dimethyl or 1,3-methylethyl	0.1	←			0.6				→
1,2-Methylethyl or 1,2,3-trimethyl	0.9	←			1.4				→

[a] J. L. Franklin, *Ind. Eng. Chem.* **41**, 1070 (1949).   [b] Estimated by extrapolation or interpolation.

TABLE 24. $(G° - H_0°)$ Increments for Hydrocarbon Groups[a]
(kcal. per g. mole)

	0° K	298° K	400° K	500° K	600° K	800° K	1000° K	1200° K	1500° K
CH_3	0	−6.96	−9.92	−13.04	−16.34	−23.47	−31.25	−39.60	−53.13
CH_2	0	−1.714	−2.747	−3.944	−5.306	−8.502	−12.257	−16.511	−23.651
CH	0	3.51	4.52	5.55	6.33	7.49	8.17	8.40	8.13
C	0	9.53	12.88	16.15	19.20	24.89	30.12	34.97	41.72
$H_2C=$	0	−6.62	−9.43	−12.36	−15.46	−22.12	−29.34	−37.07	−49.50
$H\!\!>\!\!C=CH_2$	0	−8.82	−12.64	−16.62	−20.89	−30.16	−40.32	−51.28	−69.02
$H\!\!>\!\!C=C\!\!<\!\!H$ (trans)	0	−3.42	−5.23	−7.21	−9.45	−14.95	−20.50	−27.12	−38.18
$H\!\!>\!\!C=C\!\!<\!\!H$ (cis)		−3.98	−5.86	−7.87	−10.11	−15.57	−21.05	−27.60	−38.50
$>\!\!C=CH_2$	0	−3.25	−5.07	−7.00	−9.19	−14.33	−20.17	−26.79	−37.80
$>\!\!C=C\!\!<\!\!H$	0	1.66	1.84	1.87	1.68	0.73	−0.87	−3.12	−7.41
$>\!\!C=C\!\!<$		1.89	9.06	11.13	13.00	16.22	18.94	(21.5)	(25.5)
$=C=$	0	−1.25	−1.90	−2.67	−3.51	−5.45	−7.65	−10.06	−14.03
$HC\equiv$	0	−6.17	−8.78	−11.47	−14.30	−20.32	−26.69	−33.40	−43.96
$-C\equiv$	0	−1.44	−2.14	−2.95	−3.80	−5.73	−7.92	−10.30	−14.20
⩚CH	0	−2.87	−4.11	−5.43	−6.86	−10.04	−13.59	−17.47	−23.85
⩚C−	0	2.42	3.24	3.90	4.51	5.52	6.29	6.77	7.18
⩚C ↔[b]	0	−0.33	0.17	0.63	1.07	1.08	1.16	1.54	
↔CH_2	0	−5.46	−7.85	−10.39	−13.12	−19.00	−25.46	−32.40	−43.56
↔$C\!\!<\!\!H$	0	−0.05	−0.45	−1.46	−1.64	3.45	−5.61	−8.26	−12.81
Cyclo C_6 ring	0	−7.45	−9.08	−10.71	−12.68	−15.60	−18.95	−21.70	−27.13
Cyclo C_5 ring	0	−9.38	−12.30	−14.78	−17.24	−21.97	−26.31	−30.74	−37.22
Cyclo C_4 ring[b]	0	−10.8	−14.8	−18.8	−22.8	−24.8	−31.6	−35.2	−44.0
Cyclo C_3 ring	0	−9.89	−13.07	−16.28	−19.34	−25.31	−31.22	−37.0	−45.76
Correction Factors for Paraffin Chains									
Ethyl side chain	0	←			−0.8				→
3 adjacent CH groups	0	←			0.5				→
Adjacent C and CH groups	0	←			1.1				→
Adjacent quaternary C's	0	0.8	0.9	1.1	1.3	1.5	2.0	—	—
Quaternary C not adjacent to terminal methyl	0	←			−1.1				→
Correction Factors for Substituents on Aromatic Nucleus									
2,2 substitution	0	←			0.5				→
1,2 substitution	0	←			0.8				→
Correction for symmetry	0	←			$RT \ln \sigma$				→

[a] J. L. Franklin, *Ind. Eng. Chem.* **41**, 1070 (1949).   [b] Estimated by extrapolation or interpolation.

II. NUMERICAL DATA

TABLE 25. ΔG_f° Increments for Hydrocarbon Groups[a]

(kcal. per g.mole)

	0° K	298° K	400° K	500° K	600° K	800° K	1000° K	1200° K	1500° K
CH₃	−8.26	−4.14	−2.00	0.24	2.57	7.40	12.38	17.40	24.96
CH₂	−3.673	2.048	4.479	6.931	9.428	14.484	19.579	24.668	32.331
CH	0.18	7.46	10.39	13.26	16.15	21.86	27.45	33.02	40.96
C	1.74	11.44	15.00	18.45	21.82	28.86	34.60	40.71	49.71
H₂C=	7.26	7.94	8.57	9.97	10.04	11.69	13.44	15.22	17.93
H\C=CH₂	16.73	19.13	20.61	22.21	23.89	27.41	31.05	34.75	40.32
H\C=C/H (trans)	19.03	23.19	25.03	26.93	28.86	32.80	36.79	40.79	46.82
H\C=C\H (cis)	20.32	23.92	25.68	27.56	29.49	33.48	37.52	41.60	47.79
\C=CH₂	18.20	22.45	24.36	26.31	28.28	32.28	36.28	40.30	46.36
\C=C/H	21.10	26.69	28.88	31.10	33.32	37.74	42.11	46.46	52.93
\C=C\	25.08	32.26	34.80	37.36	39.86	44.78	49.52	(54.2)	(61.2)
=C=	33.0[b]	32.09	31.61	31.08	30.56	29.44	28.25	27.16	25.43
HC≡	27.16	24.8	24.01	23.26	22.50	21.04	19.61	18.22	16.17
—C≡	27.12	25.65	26.28	24.72	24.19	23.07	21.96	20.84	19.19
↗CH	4.00	4.84	5.50	6.13	6.78	8.15	9.56	10.97	13.11
↗C—	5.76	8.76	9.34	10.23	11.18	13.08	14.76	16.59	19.23
↗C⟷[b]	5.29	5.43	5.80	6.49	7.28	8.05	9.18	10.95	12.29
⟷CH₂	11.3	13.17	14.17	15.26	16.41	18.75	21.14	23.56	27.19
⟷C\H	12.65	16.36	17.70	19.15	20.55	23.29	26.06	28.74	32.72
Cyclo C₆ ring	1.10	−6.35	−7.98	−9.61	−11.58	−14.50	−17.85	−20.60	−26.03
Cyclo C₅ ring	6.72	−2.66	−5.58	−8.06	−10.52	−15.25	−19.59	−24.02	−30.50
Cyclo C₄ ring[b]	18.8	8.0	4.0	0	−4.0	−6.0	−12.8	−16.4	−25.2
Cyclo C₃ ring	24.4[b]	14.51	11.33	8.12	5.06	−0.91	−6.82	−12.60	−21.36
Correction Factors for Paraffin chains									
Ethyl side chain	1.5	←			0.7			→ — →	
3 adjacent CH groups	1.6	←			2.1			→ — →	
Adjacent C and CH groups	2.5	←			3.6			→ — →	
Adjacent quaternary C's	5.0	5.8	5.9	6.1	6.3	6.5	7.0	—	—
Quaternary C not adjacent to terminal methyl	2.1	←			1.0			→ — →	
Correction Factors for Substituents on Aromatic Nucleus									
1,2 substitution		←			0.5			→	
1,3 substitution		←			−0.7			→	
1,2,3 substitution	0.9	←			1.0			→	
Correction for symmetry	0	←			$RT \ln \sigma$			→	

[a] J. L. Franklin, *Ind. Eng. Chem.*, **41**, 1070 (1949). [b] Estimated by extrapolation or interpolation.

TABLE 26. ΔG_f° AND ΔH_f° INCREMENTS FOR NONHYDROCARBON GROUPS[a]

(kcal. per g. mole)

	0° K		298° K		600° K		1000° K	
	ΔG_f°	ΔH_f°	ΔG_f°	ΔH_f°	ΔG_f°	ΔH_f°	ΔG_f°	ΔH_f°
—OH (primary)	−40.1	−40.1	−36.6	−41.9	−31.1	−42.2	−24.0	−40.6
—OH (sec.)	−43.1	−43.1	−42.1	−44.9	−37.4	−44.6	−33.6	−41.4
—OH (tert.)	−46.9	−46.9	−44.7	−49.2	−37.1	−48.9	−31.4	−44.4
—OH (phenol)	−44.0	−44.0	−39.9	−46.9	−31.1	−45.6	−23.2	−39.9
H\>C=O	−32.7	−32.7	−27.9	−33.9	−21.7	−34.0	−13.5	−33.9
\>C=O	−30.6	−30.6	−28.6	−31.6	−25.9	−31.2	−22.6	−30.5
—C(=O)OH	−93.1	−93.1	−87.1	−94.6	−79.6	−93.0	−70.9	−90.4
—C(=O)O— (ester)	—	—	−71.6	−79.8	—	—	—	—
—O— (ether)	—	—	−23.8	−27.2	—	—	—	—
—C(=O)O, —C(=O)O	—	—	−93.9	−102.6	—	—	—	—
—NH₂	—	—	—	2.8	—	—	—	—
—NH	—	—	—	12.0	—	—	—	—
—N—	—	—	—	−19.2	—	—	—	—
↔ NH₂	—	—	−6.4	−0.8	—	—	—	—
—NO₂	—	—	—	−8.5	—	—	—	—
—ONO	—	—	—	−10.9	—	—	—	—
—ONO₂	—	—	—	−18.4	—	—	—	—
—C≡N	—	—	—	29.5	—	—	—	—
—N=C	—	—	—	44.4	—	—	—	—
—SH	—	—	3.1	5.7	—	—	—	—
—S—	—	—	10.8	11.6	—	—	—	—
↔ S ↔	—	—	7.8	11.3	—	—	—	—

[a] J. L. Franklin, *Ind. Eng. Chem.* **41**, 1070 (1949).

TABLE 27. Vibrational Group Contributions to Heat Content[a,b] (zero pressure)

Temp., °F	Paraffinic				Cycloparaffinic		Aromatic		Acetylenic		Olefinic			
	CH₃—	—CH₂—	—CH	—C—	—CH₂—(5)	—CH₂—(6)	H—C=	—C=	HC≡	—C≡	H₂C=	HC=	—C=	=C=
−250	—	16.9	44.1	104.9	56.5	39.2	Liquid		0.8	60.3	—	14.4	85.5	54.9
−200	2.3	45.0	99.5	238.1	79.3	72.5	range		5.3	127.4	0.9	39.2	180.4	117.0
−100	14.9	143.3	322.9	688.0	152.2	181.1			57.4	342.0	13.3	146.0	501.5	324.5
0	66.8	317.7	678.6	1338	286.1	356.9	190.4	891.4	164.7	629.1	62.6	338.0	961.2	619.2
100	177.3	573.5	1146	2121	499.7	627.5	364.0	1363	329.4	964.8	168.7	604.4	1483	971.6
200	364.3	925.6	1706	3004	804.9	986.4	605.7	1883	540.5	1327	338.0	943.6	2083	1370
300	641.5	1375	2355	3970	1202	1452	916.2	2448	793.1	1732	579.6	1347	2725	1794
400	1012	1908	3082	5000	1685	2003	1287	3056	1078	2144	891.2	1821	3408	2241
500	1459	2522	3881	6085	2251	2636	1708	3699	1392	2577	1257	2341	4126	2705
600	1985	3208	4744	7213	2890	3347	2176	4372	1726	3024	1677	2909	4874	3184
700	2585	3962	5663	8377	3598	4122	2863	5073	2078	3484	2148	3522	5649	3674
800	3256	4776	6631	9571	4366	4959	3227	5800	2446	3955	2664	4174	6448	3172
900	3995	5655	7647	10790	5190	5858	3808	6545	2834	4436	3228	4863	7266	4681
1000	4795	6583	8702	12030	6065	6809	4421	7307	3239	4927	3830	5581	8101	5197
1100	5652	7555	9790	13287	6987	7807	5060	8085	3657	5425	4468	6327	8951	5721
1200	6560	8568	10909	14559	7950	8845	5724	8875	4090	5929	5141	7097	9813	6251
1300	7519	9622	12056	15843	8954	9922	6410	9676	4536	6440	5845	7892	10687	6787
1400	8519	10711	13227	17138	9995	11034	7117	10486	4998	6958	6586	8709	11571	7330
1500	9564	11832	14421	18442	11067	12179	7842	11302	5468	7478	7342	9544	12465	7874
1600	10647	12983	15635	19755	12171	13353	8583	12125	5954	8004	8134	10398	13369	8426
1700	11767	14161	16868	21074	13301	14555	9338	12953	6444	8530	8935	11268	14274	8978
1800	12923	15367	18119	22399	14459	15783	10113	13784	6954	9068	9774	12152	15193	9537
1900	14112	16593	19384	23729	15639	17032	10897	14619	7466	9605	10622	13051	16113	10095
2000	15327	17830	20663	25065	16844	18302	11696	15458	7992	10147	11499	14961	17040	10653
2200	17836	20388	23260	27748	19311	20899	13327	17148	9065	11239	13290	15814	18903	11789
2400	20435	23012	25901	30445	21850	23561	14999	18848	10169	12339	15145	17704	20780	12929
2600	23114	25695	28579	33152	24447	26276	16708	20561	11300	13446	17051	19627	22669	14074
2800	25862	28429	31288	35871	27099	29040	18436	22284	12456	14563	19010	21580	24564	15225
3000	28669	31204	34024	38595	29796	31846	20190	24017	13632	15687	21008	23554	26470	16380
					Cyclopentane type	Cyclohexane type								Allene type

[a] Heat content = 0 at 0° R, units, B.t.u./lb. mole.
[b] M. Souders, C. S. Matthews, and C. O. Hurd, *Ind. Eng. Chem.* **41**, 1037, 1048 (1949).

TABLE 28. Characteristic Internal Rotational Contributions to Heat Content[a,b] (zero pressure)

Temp., °F	I $-CH_2-CH_2-$	II CH_3-CH_3-	III $R-CH$	IV $R-\overset{\|}{C}-$	Va $R-CH=$	Vb $R-CH=$	VI $=CH-CH=$	VII $\equiv C-C\equiv$	VIII $R-\overset{\|}{C}=$	IX CH_3-CH_3	X $\frac{1}{2}RT$	$4 RT$
-250	173.9	71.1	78.5	59.9	114.3	240.1	71.8	59.0	95.9	55.0	208.3	1666.2
-200	261.0	128.3	139.5	112.7	193.0	328.0	126.3	90.5	164.2	104.0	258.0	2063.5
-100	491.6	282.5	298.7	254.4	381.6	483.1	298.3	165.4	337.7	247.5	357.3	2858.1
0	779.0	468.9	488.3	427.5	588.4	627.5	538.3	252.9	539.5	429.7	456.6	3652.7
100	1095	674.8	695.8	622.0	792.2	755.6	827.3	344.1	753.0	628.3	555.9	4447.3
200	1399	888.6	912.1	829.8	986.8	876.6	1146	430.7	968.2	832.0	655.2	5241.9
300	1676	1105	1131	1046	1171	992.6	1471	508.1	1179	1033	754.6	6036.5
400	1932	1318	1347	1267	1344	1105	1783	575.7	1382	1228	853.9	6831.0
500	2171	1526	1558	1489	1509	1215	2078	634.6	1576	1416	953.2	7625.6
600	2393	1727	1763	1708	1665	1323	2356	686.5	1761	1597	1052.5	8420.2
700	2603	1921	1960	1924	1814	1430	2619	732.8	1937	1769	1151.9	9214.8
800	2798	2105	2149	2136	1958	1535	2869	774.7	2105	1935	1251.2	10009.3
900	2984	2283	2330	2340	2096	1640	3108	811.2	2267	2094	1350.5	10804.0
1000	3160	2454	2505	2538	2228	1744	3336	845.9	2424	2247	1448.7	11598.5
1100	3328	2620	2674	2731	2357	1847	3554	880.6	2575	2395	1549.1	12393
1200	3490	2779	2837	2916	2485	1950	3764	915.9	2720	2539	1648.5	13188
1300	3645	2937	2995	3096	2608	2053	3967	947.4	2861	2678	1747.8	13982
1400	3793	3088	3147	3271	2728	2155	4163	977.4	2999	2814	1847.1	14777
1500	3939	3234	3296	3443	2848	2257	4353	1007	3135	2948	1946.4	15571
1600	4080	3376	3440	3608	2965	2359	4539	1034	3266	3079	2045.7	16366
1700	4219	3517	3583	3771	3081	2460	4719	1061	3396	3209	2145.1	17161
1800	4352	3653	3721	3927	3194	2561	4895	1086	3523	3334	2244.3	17955
1900	4484	3788	3858	4081	3307	2662	5067	1111	3649	3458	2343.8	18750
2000	4612	3919	3991	4231	3419	2763	5235	1135	3772	3580	2443.0	19544
2200	4861	4177	4253	4524	3640	2965	5557	1181	4013	3822	2641.8	21134
2400	5100	4426	4507	4808	3859	3167	5867	1225	4248	4055	2849.4	22723
2600	5333	4668	4752	5078	4075	3368	6166	1265	4477	4285	3039.0	24312
2800	5565	4908	4995	5346	4291	3568	6455	1305	4706	4513	3237.6	25901
3000	5794	5141	5231	5603	4505	3768	6733	1341	4930	4737	3436.3	27490
	n-Butane	Propane	Isobutane	Neopentane	Propylene	cis-2-Butene	1,3-Butadiene	Diacetylene conj. fact.	Isobutylene	Ethane	Free rotation	$trans. +$ $ext.rot.+$ $C_p°-C_v°$

[a] Heat content = 0 at 0° R.: units, B.t.u./lb. mole.
[b] M. Souders, C. S. Matthews, and C. O. Hurd, *Ind. Eng. Chem.* **41**, 1037, 1048 (1949).

TABLE 29. VIBRATIONAL GROUP CONTRIBUTIONS TO HEAT CAPACITY[a],[b] (zero pressure)

Temp., °F	Paraffinic				Cycloparaffinic		Aromatic		Acetylenic		Olefinic			
	CH_3-	$-CH_2-$	$\overset{\mid}{-CH-}$	$\overset{\mid}{-\underset{\mid}{C}-}$	$-CH_2-$ (5)	$-CH_2-$ (6)	$\overset{H}{\underset{}{-C=}}$	$\overset{\mid}{-C=}$	$HC\equiv$	$-C\equiv$	$H_2C=$	$HC=$	$\overset{\mid}{-C=}$	$=C=$
-250	0.02	0.33	0.86	2.12	0.39	0.57	Liquid		0.09	1.10	—	0.33	1.56	1.01
-200	0.06	0.66	1.51	3.42	0.53	0.80	range		0.24	1.68	0.04	0.69	2.46	1.61
-100	0.29	1.35	2.97	5.58	1.00	1.40			0.76	2.58	0.27	1.50	3.94	2.58
0	0.77	2.14	4.10	7.19	1.71	2.18	1.41	4.37	1.37	3.13	0.74	2.30	4.98	3.27
100	1.47	3.04	5.10	8.36	2.58	3.11	2.10	4.93	1.90	3.53	1.38	3.03	5.68	3.77
200	2.33	3.96	6.05	9.28	3.50	4.09	2.76	5.43	2.33	3.83	2.06	3.74	6.20	4.12
300	3.20	4.87	6.89	10.01	4.41	5.05	3.38	5.88	2.67	4.07	2.74	4.36	6.64	4.37
400	4.06	5.76	7.65	10.61	5.25	5.93	3.96	6.26	2.96	4.26	3.37	4.94	7.02	4.56
500	4.89	6.54	8.34	11.09	6.02	6.74	4.46	6.60	3.20	4.42	3.94	5.46	7.34	4.72
600	5.66	7.22	8.94	11.48	6.74	7.46	4.90	6.90	3.42	4.55	4.47	5.92	7.63	4.85
700	6.39	7.87	9.47	11.81	7.38	8.11	5.28	7.15	3.61	4.67	4.96	6.34	7.88	4.95
800	7.06	8.47	9.93	12.08	7.97	8.70	5.64	7.36	3.79	4.77	5.41	6.71	8.09	5.04
900	7.69	9.02	10.34	12.30	8.50	9.24	5.96	7.55	3.95	4.86	5.82	7.03	8.27	5.12
1000	8.28	9.50	10.71	12.49	8.99	9.74	6.26	7.71	4.11	4.94	6.20	7.32	8.43	5.19
1100	8.83	9.93	11.04	12.64	9.43	10.18	6.52	7.84	4.26	5.01	6.56	7.58	8.56	5.27
1200	9.34	10.33	11.34	12.78	9.84	10.58	6.75	7.96	4.40	5.07	6.88	7.83	8.68	5.33
1300	9.82	10.71	11.60	12.90	10.22	10.95	6.97	8.06	4.53	5.13	7.19	8.06	8.79	5.39
1400	10.25	11.06	11.84	13.00	10.57	11.30	7.16	8.14	4.66	5.19	7.48	8.27	8.89	5.44
1500	10.65	11.37	12.04	13.09	10.89	11.61	7.34	8.20	4.78	5.23	7.75	8.46	8.98	5.48
1600	11.03	11.66	12.24	13.16	11.18	11.89	7.50	8.26	4.89	5.27	7.99	8.63	9.05	5.52
1700	11.39	11.93	12.42	13.22	11.45	12.15	7.65	8.30	5.00	5.32	8.21	8.78	9.12	5.56
1800	11.72	12.18	12.58	13.28	11.80	12.38	7.79	8.34	5.11	5.36	8.42	8.91	9.18	5.59
1900	12.01	12.40	12.73	13.33	11.93	12.60	7.92	8.38	5.20	5.40	8.62	9.04	9.24	5.61
2000	12.29	12.61	12.86	13.38	12.14	12.80	8.04	8.41	5.29	5.43	8.81	9.16	9.28	5.63
2200	12.78	12.97	13.10	13.45	12.52	13.15	8.26	8.47	5.45	5.48	9.14	9.37	9.36	5.68
2400	13.21	13.28	13.30	13.51	12.85	13.46	8.45	8.53	5.60	5.52	9.43	9.55	9.41	5.72
2600	13.58	13.55	13.47	13.57	13.14	13.71	8.59	8.59	5.72	5.56	9.67	9.69	9.46	5.74
2800	13.90	13.78	13.62	13.61	13.38	13.93	8.71	8.65	5.83	5.60	9.89	9.81	9.50	5.76
3000	14.17	13.96	13.74	13.64	13.59	14.11	8.82	8.69	5.93	5.64	10.08	9.92	9.55	5.78
					Cyclopentane type	Cyclohexane type								Allene type

[a] M. Souders, C. S. Matthews, and C. O. Hurd, *Ind. Eng. Chem.* **41**, 1037, 1048 (1949). [b] Units, B.t.u./lb. mole.

TABLE 30. Characteristic Internal Rotational Contributions to Heat Capacity[a],[b] (zero pressure)

Temp., °F	I —CH$_2$—CH$_2$—	II CH$_3$—CH$_2$—	III R—CH	IV R—C—	Va R—CH=	Vb R—CH=	VI =CH—CH=	VII ≡C—C≡	VIII R—C=	IX CH$_3$—CH$_3$	X ½ R
−250	1.60	1.03	1.12	0.93	1.44	1.80	0.84	0.60	1.24	0.87	0.99
−200	1.99	1.33	1.39	1.21	1.73	1.69	1.34	0.67	1.54	1.20	
−100	2.64	1.74	1.76	1.60	2.02	1.48	2.09	0.81	1.91	1.66	
0	3.08	1.98	2.00	1.85	2.08	1.34	2.70	0.91	2.10	1.93	
100	3.17	2.11	2.13	2.02	2.00	1.24	3.07	0.91	2.16	2.03	
200	2.90	2.17	2.19	2.13	1.89	1.18	3.27	0.83	2.14	2.04	Same at all temperatures
300	2.65	2.16	2.18	2.20	1.79	1.14	3.21	0.72	2.07	1.99	
400	2.46	2.10	2.14	2.22	1.69	1.11	3.03	0.63	1.98	1.92	
500	2.30	2.05	2.08	2.22	1.60	1.09	2.86	0.55	1.89	1.84	
600	2.15	1.98	2.01	2.18	1.52	1.07	2.70	0.49	1.81	1.76	
700	2.02	1.90	1.93	2.13	1.45	1.06	2.56	0.44	1.72	1.69	
800	1.90	1.82	1.85	2.08	1.40	1.05	2.44	0.39	1.65	1.62	
900	1.80	1.74	1.76	2.02	1.36	1.04	2.33	0.36	1.58	1.56	
1000	1.72	1.68	1.71	1.95	1.32	1.04	2.23	0.34	1.53	1.50	
1100	1.64	1.63	1.66	1.89	1.28	1.03	2.14	0.33	1.48	1.45	
1200	1.58	1.57	1.60	1.83	1.25	1.03	2.06	0.32	1.44	1.41	
1300	1.53	1.53	1.56	1.78	1.22	1.02	1.99	0.31	1.40	1.38	
1400	1.48	1.48	1.51	1.73	1.20	1.02	1.93	0.30	1.36	1.34	
1500	1.43	1.43	1.47	1.68	1.18	1.02	1.88	0.28	1.33	1.32	
1600	1.40	1.40	1.43	1.64	1.16	1.02	1.83	0.27	1.31	1.29	
1700	1.38	1.38	1.40	1.60	1.15	1.01	1.78	0.26	1.28	1.27	
1800	1.36	1.36	1.38	1.56	1.14	1.01	1.74	0.25	1.26	1.25	
1900	1.33	1.33	1.35	1.52	1.12	1.01	1.70	0.24	1.23	1.23	
2000	1.30	1.30	1.33	1.49	1.11	1.01	1.66	0.23	1.21	1.22	
2200	1.27	1.27	1.29	1.44	1.10	1.01	1.58	0.22	1.18	1.19	
2400	1.22	1.24	1.25	1.39	1.09	1.00	1.52	0.21	1.16	1.16	
2600	1.18	1.21	1.22	1.35	1.08	1.00	1.47	0.20	1.14	1.14	
2800	1.16	1.18	1.20	1.31	1.08	1.00	1.42	0.19	1.12	1.12	
3000	1.15	1.16	1.18	1.28	1.07	1.00	1.36	0.18	1.10	1.11	
	1.14										
	n-Butane	Propane	Isobutane	Neopentane	Propylene	cis-2-Butene	1,3-Butadiene	Diacetylene conj. fact.	Isobutylene	Ethane	Free rotation

[a] M. Souders, C. S. Matthews, and C. O. Hurd, *Ind. Eng. Chem.* **41**, 1037, 1048 (1949). [b] Units, B.t.u./lb. mole.

TABLE 31

Group Contributions to $\Delta S_f^°$ and $\Delta H_f^{°a}$

(Type I groups)

Structural group	$\Delta S_f^°$, cal.g./mole °K				$\Delta H_f^°$, kcal.g./mole			
	Aliphatic hydro-carbon	Six carbon naphthenic ring	Five carbon naphthenic ring	Aromatic ring	Aliphatic hydro-carbon	Six carbon naphthenic ring	Five carbon naphthenic ring	Aromatic ring
—CH$_3$	−19.66	—	—	—	−10.05	—	—	—
—CH$_2$—	−23.37	−20.09	−17.68	—	−4.95	−4.91	−3.68	—
—CH (2nd carbon)[b]	−30.19	−26.08	−24.71	—	−1.57	−1.53	−1.63	—
(3rd or higher)[b]	−30.19	—	—	—	−0.88	—	—	—
—C— (2nd carbon)[b]	−37.17	(−32.82)	(−30.80)	—	0.85	(0.85)	(0.85)	—
(3rd or higher)[b]	−37.17	—	—	—	2.45	—	—	—
H$_2$C=	−5.05	—	—	—	5.80	—	—	—
HC=[c]	−8.61	−3.93	(−1.80)	−5.43	9.28	9.20	9.57	3.33
HC= (trans)	−9.28	—	—	—	8.70	—	—	—
—C=	−14.55	(−9.00)	(−6.50)	−9.49	10.84	(10.75)	(11.10)	5.48
=C=	4.51	—	—	—	34.09	—	—	—
HC≡	7.64	—	—	—	27.04	—	—	—
—C≡	4.68	—	—	—	27.65	—	—	—

[a] M. Souders, C. S. Matthews, and C. O. Hurd, *Ind. Eng. Chem.* **41**, 1037, 1048 (1949).

[b] Indicates position of group in the longest chain of an aliphatic hydrocarbon (measured from shortest end).

[c] To be used when groups are in the adjacent (cis) position or when there is no cis-trans effect.

TABLE 32

Conjugation and Adjacency Contributions to ΔS_f° and $\Delta H_f^{\circ\, a,b}$

(Type I groups)

Group	ΔS_f° cal./deg.mole	Group	ΔH_f° kcal./mole
$=\!\!\underset{}{\overset{R\ \ H}{C}}\!\!-\!\!\underset{}{\overset{}{C}}\!\!=$	−3.73	$=\!\!\underset{}{\overset{H\ \ H}{C}}\!\!-\!\!\underset{}{\overset{}{C}}\!\!=$ (aliphatics)	−3.38
$\equiv\!C\!-\!C\!\equiv$	−8.83	$=\!\!\underset{}{\overset{\mid\ \ H}{C}}\!\!-\!\!\underset{}{\overset{}{C}}\!\!=$ (aliphatics)	−4.45
⌬-$\underset{C=CH}{\overset{H\ R}{\mid}}$	−2.05	$=\!\!\underset{}{\overset{\mid\ \ \mid}{C}}\!\!-\!\!\underset{}{\overset{}{C}}\!\!=$ (aliphatics)	−2.10
Each pair ortho groups in aromatics	−0.8	$=\!\!\underset{}{\overset{H\ \ H}{C}}\!\!-\!\!\underset{}{\overset{}{C}}\!\!=$ (5-member naphthenic ring)	−2.88
		$=\!\!\underset{}{\overset{H\ \ H}{C}}\!\!-\!\!\underset{}{\overset{}{C}}\!\!=$ (6-member naphthenic ring)	−1.76
Symmetry Contributions		⌬-$\underset{C=CH}{\overset{H\ R}{\mid}}$	−2.01
σ	$-R \ln \sigma$	Each pair ortho groups in aromatics	0.69
		Ethyl side chain (aliphatics)	−0.88
2	−1.38	$-\!\!\underset{H\ \ H}{\overset{\mid\ \ \mid}{C\!-\!C}}\!-$	0.75
3	−2.18		
4	−2.76	$-\!\!\underset{H\ \ \mid}{\overset{\mid\ \ \mid}{C\!-\!C}}\!-$	2.39
5	−3.20	$-\!\!\underset{H\ H\ H}{\overset{\mid\ \ \mid\ \ \mid}{C\!-\!C\!-\!C}}\!-$	3.30
6	−3.56		
10	−4.57	$-\!\!\underset{\mid\ \ \mid}{\overset{\mid\ \ \mid}{C\!-\!C}}\!-$	4.61
12	−4.94	$-\!\!\underset{\mid\ \ \mid}{\overset{\mid\ \ \mid}{C\!=\!C}}\!-$	2.61

[a] To be added to the group contributions whenever these molecular groups appear. The symbol — indicates a C—C bond; R indicates either H or C.

[b] M. Souders, C. S. Matthews, and C. O. Hurd, *Ind. Eng. Chem.* **41**, 1037, 1048 (1949).

II. NUMERICAL DATA

TABLE 33. VIBRATIONAL GROUP CONTRIBUTIONS TO $(S°_{fT} - S°_{f\,298})$[a,b] (Type II groups)

Temp., °K	Paraffinic				Cycloparaffinic		Aromatic		Acetylenic		Olefinic			
	—CH₃	—CH₂—	>CH—	>C<	—CH₂— (5)	—CH₂— (6)	HC=	=C<	HC≡	—C≡	H₂C=	HC=	=C<	=C=
300	−0.068	−0.040	−0.004	+0.037	−0.041	−0.037	−0.022	+0.017	−0.023	+0.008	−0.047	−0.016	+0.021	+0.010
350	−1.764	−0.955	−0.065	1.011	−1.026	−0.939	−0.557	0.446	−0.598	0.214	−1.228	−0.406	0.566	0.253
400	−3.185	−1.693	−0.062	1.905	−1.826	−1.657	−0.998	0.823	−1.101	0.372	−2.225	−0.717	1.046	0.447
450	−4.392	−2.291	−0.013	2.731	−2.480	−2.236	−1.364	1.160	−1.550	0.496	−3.078	−0.967	1.474	0.604
500	−5.425	−2.778	+0.070	3.493	−3.020	−2.704	−1.670	1.466	−1.960	0.593	−3.815	−1.169	1.859	0.733
550	−6.315	−3.176	0.177	4.198	−3.467	−3.084	−1.930	1.746	−2.336	0.667	−4.459	−1.334	2.210	0.836
600	−7.087	−3.507	0.301	4.851	−3.841	−3.395	−2.154	2.003	−2.688	0.724	−5.026	−1.469	2.530	0.919
650	−7.766	−3.785	0.435	5.460	−4.158	−3.654	−2.351	2.240	−3.009	0.766	−5.532	−1.581	2.826	0.984
700	−8.365	−4.017	0.576	6.026	−4.427	−3.869	−2.525	2.459	−3.314	0.795	−5.986	−1.676	3.097	1.035
750	−8.896	−4.211	0.721	6.556	−4.657	−4.048	−2.678	2.662	−3.601	0.817	−6.396	−1.755	3.351	1.075
800	−9.370	−4.373	0.868	7.054	−4.852	−4.197	−2.815	2.852	−3.870	0.857	−6.768	−1.822	3.588	1.106
850	−9.795	−4.511	1.016	7.522	−5.022	−4.322	−2.937	3.030	−4.125	0.867	−7.107	−1.880	3.810	1.131
900	−10.177	−4.628	1.165	7.966	−5.167	−4.425	−3.047	3.199	−4.364	0.874	−7.418	−1.928	4.020	1.152
950	−10.522	−4.726	1.313	8.385	−5.292	−4.510	−3.146	3.358	−4.590	0.876	−7.702	−1.969	4.217	1.169
1000	−10.837	−4.809	1.460	8.784	−5.400	−4.582	−3.236	3.508	−4.805	0.878	−7.966	−2.003	4.406	1.183
1100	−11.381	−4.934	1.752	9.527	−5.572	−4.684	−3.391	3.787	−5.200	0.876	−8.433	−2.052	4.753	1.205
1200	−11.842	−5.022	2.034	10.207	−5.702	−4.752	−3.522	4.041	−5.558	0.871	−8.841	−2.086	5.076	1.221
1300	−12.240	−5.083	2.303	10.833	−5.803	−4.798	−3.635	4.272	−5.886	0.862	−9.203	−2.109	5.373	1.232
1400	−12.585	−5.122	2.562	11.412	−5.878	−4.825	−3.733	4.484	−6.187	0.852	−9.525	−2.124	5.650	1.239
1500	−12.891	−5.149	2.809	11.948	−5.939	−4.841	−3.820	4.678	−6.467	0.839	−9.817	−2.135	5.906	1.240
1600	−13.164	−5.167	3.045	12.450	−5.985	−4.848	−3.897	4.858	−6.727	0.826	−10.083	−2.142	6.144	1.240
1800	−13.648	−5.193	3.480	13.359	−6.059	−4.856	−4.039	5.183	−7.208	0.795	−10.561	−2.156	6.572	1.228
2000	−14.069	−5.210	3.869	14.166	−6.120	−4.861	−4.172	5.468	−7.645	0.758	−10.992	−2.173	6.948	1.206
					Cyclopentane type	Cyclohexane type								Allene type

[a] M. Souders, C. S. Matthews, and C. O. Hurd, *Ind. Eng. Chem.* **41**, 1037, 1048 (1949). [b] Units, cal./mole, °K.

TABLE 34. CHARACTERISTIC INTERNAL ROTATIONAL CONTRIBUTIONS TO $(S^\circ_{fT} - S^\circ_{f298})^{a,b}$

Temp., °K	I —CH$_2$—CH$_2$—	II CH$_3$—CH$_2$—	III R—CH	IV R—C	Va R—CH=	Vb R—CH=	VI =CH—CH=	VII ≡C—C≡	VIII R—C=	IX CH$_3$—CH$_3$	X ½ R ln T/298.16	XI 4 R ln T/298.16
300	0.020	0.013	0.013	0.012	0.018	0.008	0.018	0.006	0.013	0.012	0.006	0.049
350	0.502	0.341	0.343	0.328	0.320	0.197	0.501	0.144	0.346	0.326	0.159	1.274
400	0.885	0.631	0.635	0.614	0.572	0.355	0.936	0.252	0.631	0.596	0.292	2.335
450	1.197	0.885	0.892	0.873	0.782	0.489	1.313	0.336	0.875	0.830	0.409	3.271
500	1.459	1.107	1.118	1.107	0.960	0.606	1.635	0.402	1.085	1.032	0.513	4.108
550	1.680	1.302	1.342	1.318	1.113	0.710	1.910	0.455	1.266	1.208	0.608	4.865
600	1.869	1.475	1.519	1.509	1.247	0.804	2.148	0.499	1.425	1.363	0.695	5.557
650	2.034	1.629	1.675	1.681	1.365	0.889	2.357	0.535	1.566	1.500	0.774	6.193
700	2.179	1.765	1.815	1.837	1.470	0.967	2.542	0.566	1.691	1.622	0.848	6.781
750	2.307	1.888	1.940	1.979	1.563	1.039	2.706	0.591	1.803	1.732	0.916	7.330
800	2.422	1.999	2.052	2.108	1.650	1.106	2.854	0.614	1.904	1.831	0.980	7.842
850	2.525	2.100	2.155	2.225	1.729	1.169	2.988	0.634	1.996	1.922	1.041	8.324
900	2.617	2.192	2.248	2.333	1.802	1.228	3.110	0.653	2.080	2.003	1.097	8.778
950	2.702	2.277	2.335	2.433	1.869	1.283	3.221	0.670	2.158	2.079	1.151	9.208
1000	2.781	2.355	2.415	2.525	1.932	1.335	3.324	0.686	2.229	2.150	1.202	9.615
1100	2.921	2.495	2.558	2.688	2.046	1.432	3.506	0.713	2.358	2.278	1.297	10.373
1200	3.042	2.617	2.683	2.830	2.147	1.520	3.665	0.737	2.471	2.391	1.383	11.064
1300	3.148	2.725	2.793	2.955	2.238	1.600	3.804	0.757	2.572	2.491	1.463	11.700
1400	3.242	2.892	2.892	3.066	2.321	1.675	3.927	0.774	2.663	2.581	1.536	12.289
1500	3.327	2.909	2.981	3.161	2.397	1.744	4.037	0.790	2.746	2.663	1.605	12.837
1600	3.404	2.989	3.062	3.251	2.467	1.809	4.137	0.803	2.822	2.739	1.669	13.350
1800	3.541	3.131	3.205	3.409	2.593	1.927	4.309	0.827	2.958	2.872	1.786	14.286
2000	3.659	3.255	3.331	3.523	2.705	2.031	4.455	0.846	3.076	2.991	1.890	15.123
	n-Butane	Propane	Isobutane	Neopentane	Propylene	cis-2-Butene	1,3-Butadiene	Diacetylene	Isobutylene	Ethane	Free rotation	trans.+ ext.rot.+ $C_p^\circ - C_V^\circ$
								conjugation				

[a] Souders, C. S. Matthews, and C. O. Hurd, *Ind. Eng. Chem.* **41**, 1037, 1048 (1949). [b] Units, cal./mole, °K

TABLE 35. VIBRATIONAL GROUP CONTRIBUTIONS TO $(H°_{fT} - H°_{f298})^{a,b}$ (Type II groups)

Temp., °K	Paraffinic —CH₃	Paraffinic —CH₂—	Paraffinic >CH—	Paraffinic >C<	Cycloparaffinic —CH₂— (5)	Cycloparaffinic —CH₂— (6)	Aromatic HC=	Aromatic —C=	Acetylenic HC≡	Acetylenic —C≡	Olefinic H₂C=	Olefinic HC=	Olefinic —C=	Olefinic =C=
300	0.016	−0.008	+0.001	+0.012	−0.008	−0.008	−0.004	+0.007	−0.004	+0.004	−0.011	−0.003	+0.007	+0.003
350	−0.566	−0.305	−0.019	0.328	−0.328	−0.300	−0.178	0.146	−0.190	0.070	−0.393	−0.129	0.183	0.082
400	−1.099	−0.582	−0.017	0.664	−0.627	−0.570	−0.343	0.287	−0.379	0.129	−0.767	−0.245	0.363	0.156
450	−1.611	−0.835	+0.004	1.015	−0.904	−0.815	−0.498	0.431	−0.570	0.182	−1.129	−0.351	0.544	0.224
500	−2.101	−1.066	0.043	1.377	−1.160	−1.036	−0.643	0.577	−0.764	0.227	−1.478	−0.447	0.727	0.285
550	−2.578	−1.275	0.100	1.747	−1.395	−1.236	−0.780	0.723	−0.956	0.266	−1.815	−0.533	0.911	0.339
600	−3.008	−1.463	0.174	2.126	−1.607	−1.412	−0.905	0.873	−1.154	0.300	−2.139	−0.608	1.098	0.389
650	−3.434	−1.636	0.257	2.506	−1.804	−1.574	−1.029	1.021	−1.356	0.326	−2.454	−0.679	1.283	0.430
700	−3.839	−1.793	0.353	2.888	−1.985	−1.719	−1.146	1.169	−1.563	0.346	−2.762	−0.742	1.468	0.465
750	−4.227	−1.934	0.457	3.272	−2.152	−1.850	−1.259	1.316	−1.772	0.361	−3.060	−0.801	1.651	0.493
800	−4.594	−2.060	0.572	3.657	−2.304	−1.965	−1.364	1.463	−1.980	0.373	−3.348	−0.853	1.835	0.517
850	−4.944	−2.174	0.695	4.045	−2.443	−2.066	−1.465	1.611	−2.189	0.382	−3.628	−0.901	2.018	0.539
900	−5.278	−2.276	0.825	4.432	−2.570	−2.156	−1.561	1.758	−2.399	0.387	−3.898	−0.944	2.202	0.557
950	−5.598	−2.367	0.963	4.821	−2.686	−2.235	−1.653	1.905	−2.609	0.390	−4.162	−0.982	2.386	0.573
1000	−5.905	−2.448	1.107	5.210	−2.791	−2.303	−1.740	2.052	−2.819	0.392	−4.419	−1.014	2.569	0.587
1100	−6.480	−2.580	1.412	5.991	−2.974	−2.413	−1.904	2.345	−3.234	0.390	−4.911	−1.067	2.939	0.610
1200	−7.010	−2.682	1.736	6.774	−3.123	−2.491	−2.054	2.637	−3.646	0.384	−5.380	−1.105	3.311	0.629
1300	−7.505	−2.756	2.075	7.557	−3.245	−2.544	−2.194	2.926	−4.054	0.376	−5.830	−1.132	3.685	0.643
1400	−7.973	−2.821	2.425	8.339	−3.348	−2.580	−2.325	3.212	−4.460	0.364	−6.265	−1.152	4.058	0.652
1500	−8.419	−2.864	2.782	9.119	−3.435	−2.603	−2.451	3.494	−4.866	0.347	−6.689	−1.167	4.428	0.655
1600	−8.851	−2.896	3.146	9.897	−3.510	−2.616	−2.572	3.774	−5.271	0.327	−7.105	−1.179	4.796	0.654
1800	−9.674	−2.933	3.884	11.445	−3.629	−2.622	−2.809	4.327	−6.082	0.274	−7.918	−1.197	5.523	0.630
2000	−10.527	−2.963	4.623	12.978	−3.735	−2.622	−3.054	4.870	−6.905	0.207	−8.725	−1.221	6.236	0.607
					Cyclopentane type	Cyclohexane type								Allene type

^{a} Units, kcal./mole. ^{b} M. Souders, C. S. Matthews, and C. O. Hurd, *Ind. Eng. Chem.* **41**, 1037, 1048 (1949).

TABLE 36. CHARACTERISTIC INTERNAL ROTATIONAL CONTRIBUTIONS TO $(H°_{fT} - H°_{f298})$[a,b]

Temp, °K	I —CH_2—CH_2—	II CH_3—CH_3—	III R—CH | CH_3	IV R—C— |	Va R—CH=	Vb R—CH=	VI =CH—CH=	VII ≡C—C≡	VIII R—C=	IX CH_3—CH_3	X ½R (T − 298.16)	XI 4R (T − 298.16)
300	0.005	0.004	0.003	0.004	0.003	0.002	0.005	0.001	0.003	0.004	0.002	0.015
350	0.161	0.110	0.111	0.106	0.102	0.064	0.162	0.045	0.111	0.105	0.052	0.412
400	0.304	0.218	0.221	0.214	0.196	0.123	0.325	0.086	0.218	0.207	0.101	0.809
450	0.436	0.326	0.330	0.324	0.285	0.179	0.485	0.122	0.321	0.306	0.151	1.207
500	0.560	0.432	0.437	0.435	0.370	0.235	0.637	0.154	0.421	0.402	0.201	1.604
550	0.676	0.534	0.541	0.545	0.450	0.290	0.781	0.182	0.516	0.495	0.250	2.001
600	0.785	0.634	0.642	0.655	0.527	0.343	0.918	0.207	0.608	0.584	0.300	2.398
650	0.889	0.730	0.740	0.762	0.601	0.397	1.048	0.229	0.695	0.670	0.350	2.796
700	0.986	0.822	0.834	0.868	0.673	0.449	1.173	0.250	0.779	0.752	0.399	3.193
750	1.079	0.911	0.925	0.970	0.742	0.502	1.293	0.268	0.860	0.832	0.449	3.590
800	1.168	0.997	1.013	1.070	0.809	0.554	1.408	0.286	0.939	0.909	0.498	3.988
850	1.253	1.081	1.098	1.167	0.873	0.606	1.518	0.303	1.015	0.984	0.548	4.385
900	1.335	1.161	1.181	1.261	0.938	0.656	1.625	0.321	1.089	1.056	0.598	4.782
950	1.414	1.240	1.261	1.352	1.000	0.708	1.728	0.337	1.160	1.127	0.647	5.179
1000	1.488	1.319	1.339	1.441	1.061	0.760	1.828	0.353	1.230	1.196	0.697	5.577
1100	1.636	1.465	1.488	1.613	1.181	0.862	2.019	0.382	1.366	1.330	0.796	6.371
1200	1.775	1.606	1.631	1.776	1.297	0.963	2.201	0.409	1.496	1.460	0.896	7.166
1300	1.908	1.742	1.769	1.932	1.410	1.064	2.375	0.434	1.623	1.585	0.995	7.961
1400	2.036	1.873	1.902	2.082	1.522	1.165	2.542	0.458	1.746	1.707	1.094	8.755
1500	2.159	2.001	2.032	2.227	1.632	1.266	2.702	0.480	1.865	1.827	1.194	9.550
1600	2.278	2.125	2.158	2.368	1.741	1.367	2.856	0.502	1.982	1.943	1.293	10.344
1800	2.510	2.366	2.402	2.637	1.957	1.568	3.150	0.542	2.211	2.172	1.492	11.933
2000	2.739	2.599	2.638	2.894	2.171	1.769	3.424	0.579	2.435	2.396	1.690	13.523
	n-Butane	Propane	Isobutane	Neopentane	Propylene	cis-2-Butene	1,3-Butadiene	Diacetylene conjugation	Isobutylene	Ethane	Free rotation	trans. + ext. rot. + $C_P° − C_V°$

[a] M. Souders, C. S. Matthews, and C. O. Hurd, *Ind. Eng. Chem.* **41**, 1037, 1048 (1949). [b] Units, kcal./mole.

TABLE 37

$\Delta G_f°$ Increments for Hydrocarbon Groups[a, b]

Group	300 — 600° K		600 — 1500° K	
	A	B	A	B
CH_4	−18.948	2.225	−21.250	2.596
—CH_3	−10.943	2.215	−12.310	2.436
—CH_2—	−5.193	2.430	−5.830	2.544
—CH<	−0.705	2.910	−0.705	2.910
—C<	1.958	3.735	4.385	3.350
$H_2C=CH_2$	11.552	1.545	9.450	1.888
$H_2C=CH$—	13.737	1.655	12.465	1.762
$H_2C=C<$	16.467	1.915	16.255	1.966
$^HC=C^H$ (cis)	17.663	1.965	16.180	2.116
$^HC=C_H$ (trans)	17.187	1.915	15.815	2.062
$^HC=C<$	20.217	2.295	19.584	2.354
$>C=C<$	25.135	2.573	25.135	2.573
$H_2C=C=CH_2$	45.250	1.027	43.634	1.311
$H_2C=C=CH$—	49.377	1.035	48.170	1.208
$H_2C=C=C<$	51.084	1.474	51.084	1.474
$^HC=C=C^H$	52.460	1.483	52.460	1.483
$H_2C\leftrightarrow$	5.437	0.675	4.500	0.832
$CH\leftrightarrow$	7.407	1.035	6.980	1.088
$>C\leftrightarrow$	9.152	1.505	10.370	1.308
$HC\equiv$	27.048	−0.765	26.700	−0.704
—$C\equiv$	26.938	−0.525	26.555	−0.550
$HC\diagup\diagdown$	3.047	0.615	2.505	0.706
—$C\diagup\diagdown$	4.675	1.150	5.010	0.988
$\leftrightarrow C\diagup\diagdown$	6.608	0.514	6.260	0.583

[a] D. W. Van Krevelen and H. A. G. Chermin, *Chem. Eng. Sci.* **1**, 66 (1951).
[b] Units, kcal./mole.

TABLE 38. $\Delta G_f°$ CORRECTIONS FOR RING FORMATION AND BRANCHING EFFECTS[a],[b]

Group	300—600° K		600—1500° K		Group	300—600° K		600—1500° K	
	A	B	A	B		A	B	A	B
Ring Formation					**Branching in 5 ring**				
3 ring	23.458	−3.045	22.915	−2.966	Single branching	−1.04	0	−1.69	0
4 ring	10.73	−2.65	10.60	−2.50	Double branching				
5 ring	4.275	−2.350	2.665	−2.182	1,1 position	−1.85	0	−1.190	−0.160
6 ring	−1.128	−1.635	−1.930	−1.504	cis 1,2 position	−0.38	0	−0.38	0
Pentene ring	−3.657	−2.395	−3.915	−2.250	trans 1,2 position	−2.55	0	−0.945	−0.266
Hexene ring	−9.102	−2.045	−8.810	−2.071	cis 1,3 position	−1.20	0	−0.370	−0.166
					trans 1,3 position	−2.35	0	−0.800	−0.264
Branching in paraffin chains					**Branching in 6 ring**				
					Single branching	−0.93	0	0.230	−0.192
					Double branching				
	12.86	−0.63	12.86	−0.63	1,1 position	0.835	−0.367	1.745	−0.556
					cis 1,2 position	−0.19	0	1.470	−0.276
	−5.82	0.25	−3.53	−0.16	trans 1,2 position	−2.41	0	0.045	−0.398
					cis 1,3 position	−2.70	0	−1.647	−0.185
					trans 1,3 position	−1.60	0	0.260	−0.290
					cis 1,4 position	−1.11	0	−1.11	0
					trans 1,4 position	−2.80	0	−0.995	−0.245
Side chain with 2 or more C-atoms	1.31	0	1.31	0	**Branching in aromatics**				
					Double branching				
3 Adjacent —CH— groups	2.12	0	2.12	0	1,2 position	1.02	0	1.02	0
					1,3 position	−0.31	0	−0.31	0
Adjacent —CH and —C— groups	1.80	0	1.80	0	1,4 position	0.93	0	0.93	0
					Triple branching				
2 Adjacent —C— groups	2.58	0	2.58	0	1,2,3 position	1.91	0	2.10	0
					1,2,4 position	1.10	0	1.10	0
					1,3,5 position	0	0	0	0

[a] D. W. Van Krevelen and H. A. G. Chermin, *Chem. Eng. Sci.* **1**, 66 (1951). [b] Units, kcal./mole.

TABLE 39
ΔG_f° Increments for Nonhydrocarbon Groups[a,b]

Group	300 — 600° K		600 — 1500° K	
	A	B	A	B
	Oxygen containing groups			
H_2O	−58.076	1.154	−59.138	1.316
—OH	−41.56	1.28	−41.56	1.28
—O—	−15.79	−0.85	—	—
O (ring)	−18.37	0.80	−16.07	0.40
H_2CO	−29.118	0.653	−30.327	0.854
—CHO	−29.28	0.77	−30.15	0.83
>C=O	−28.08	0.91	−28.08	0.91
HC(=O)—OH	−87.660	2.473	−90.569	2.958
—C(=O)—OH	−98.39	2.86	98.83	2.93
—C(=O)—O—	−92.62	2.61	−92.62	2.61
$H_2C=C=O$	−14.515	0.295	−14.515	0.295
CH=C=O	−12.86	0.46	−12.86	0.46
>C=C=O	−9.62	0.72	−9.38	0.73
	Nitrogen containing groups			
HCN	31.179	−0.826	30.874	−0.775
—C≡N	30.75	−0.72	30.75	−0.72
—N≡C	46.32	−0.89	46.32	−0.89
NH_3	−11.606	2.556	−12.972	2.784
—NH_2	2.82	2.71	−6.78	3.98
>NH	12.93	3.16	12.93	3.16
>N—	19.46	3.82	19.46	3.82
≡N	11.32	1.11	12.26	0.96
—NO_2	−9.0	3.70	−14.19	4.38

[a] D. W. Van Krevelen and H. A. G. Chermin, *Chem. Eng. Sci.* **1**, 66 (1951).
[b] Units, kcal./mole.

(Table continued)

TABLE 39 (*continued*)

Group	300 — 600° K		600 — 1500° K	
	A	B	A	B
	Halogen containing groups			
HF	− 64.476	− 0.145	− 64.884	− 0.081
—F	− 45.10	− 0.20	—	—
HCl	− 22.100	− 0.215	− 22.460	− 0.156
—Cl	− 8.25	0	− 8.25	0
HBr	− 12.533	− 0.234	13.010	− 0.158
—Br	− 1.62	− 0.26	− 1.62	− 0.26
HI	− 1.330	− 0.225	− 1.718	− 0.176
—I	7.80	0	+ 7.80	0
	Sulphur containing groups			
H_2S	− 20.552	1.026	− 21.366	1.167
—SH	− 10.68	1.07	− 10.68	1.07
—S—	− 3.32	1.42	− 3.32	1.44
$\searrow\!\!\!\nearrow\!\!S$	− 0.97	0.51	− 0.65	0.44
>SO	− 30.19	3.39	− 30.19	3.39
>SO_2	− 82.58	5.58	− 80.69	5.26

TABLE 40

$\Delta G_f°$ Parameters for Some Simple Organic and Inorganic Compounds[a]

Compound	Formula	Free energy of formation, $A + \dfrac{B}{100} \cdot T$ (kcal./mole)			
		300 — 600° K		600 — 1500° K	
		A	B	A	B
Ethane	C_2H_6	− 21.539	4.583	− 24.588	5.093
Carbon monoxide	CO	− 26.582	− 2.122	− 26.582	− 2.122
Carbon dioxide	CO_2	− 94.076	− 0.061	− 94.234	− 0.035
Methanol	CH_3OH	− 49.440	3.480	− 51.330	3.795
Phosgene	$COCl_2$	− 52.01	0.94	− 52.01	0.94
Cyanogen chloride	CNCl	36.751	− 0.430	36.751	− 0.430
Cyanogen iodide	CNI	46.768	− 0.596	46.768	− 0.596
Cyanogen	$(CN)_2$	71.889	− 1.059	71.889	− 1.059
Sulphur dioxide	SO_2	− 86.657	1.735	− 86.657	1.735
Sulphur trioxide	SO_3	− 109.813	3.943	− 109.813	3.943
Nitric oxide	NO	21.543	− 0.301	21.543	− 0.301
Nitrogen dioxide	NO_2	7.933	1.486	7.933	1.486
Nitrogen tetroxide	N_2O_4	2.212	7.112	—	—
Carbon oxi-sulphide	COS	− 49.414	− 0.245	− 49.269	− 0.265
Carbon disulphide	CS_2	− 3.301	− 0.150	− 3.104	− 0.177

[a] D. W. Van Krevelen and H. A. G. Chermin, *Chem. Eng. Sci.* **1**, 66 (1951).

TABLE 41

$\Delta H_f°_{298}$ Group Increments for Radicals[a]

Group	ΔH_f kcal./mole-group	calculated from
—$\dot{C}H_2$	34	$n\dot{C}_3H_7$
>$\dot{C}H$	34	$2\dot{C}_3H_7$
>$\dot{C}$<	33	$t\dot{C}_4H_9$
—$\dot{O}$	7.5	$C\dot{C}\dot{C}O$
—O—O·	7.5	$\begin{array}{c}C\\CCO\text{—}O\\C\end{array}$
—S·	43	$CH_3\dot{S}$

[a] J. L. Franklin, *J. Chem. Phys.* **21**, 2029 (1953).

TABLE 42

$\Delta H_f°_{298}$ Group Increments for Ions[a]

Group	ΔH_f kcal./mole-group	calculated from
—$\overset{+}{C}H_2$	212	nC_4H_{10}
>$\overset{+}{C}H$	209	iC_4H_{10}
>$\overset{+}{C}$<	195	tC_4H_9Cl
—$\overset{+}{C}$=CH_2	240	butene-1
$\overset{H}{\underset{}{-}C}$=$\overset{+}{C}H$	245	butene-2
—S+	232	CH_3SH

[a] J. L. Franklin, *J. Chem. Phys.* **21**, 2029 (1953).

TABLE 43
Atomic Contributions to $C_p°$ and $S°$ (25° C, 1 atmos.)[a,b]

Atom	$C_p°$	$S°$[c]
H	0.85	21.0
D	1.20	21.7
C	3.75	-32.6; $(-13.5)_3$; $(+5.3)_2$; $(22.0)_1$
N	3.40	-12.1_3; $(5.8)_2$; $(22.9)_1$
O	3.40	8.8_2; $(25.5)_1$
F	2.40	25.5
Cl	3.70	28.4_1; $(10.5)_2$
Br	4.20	31.3
I	4.60	33.3
Si	5.90	-29.3_4
P	—	-9.5_3
S	4.70	12.8_2; $(-11.0)_3$; $(-33.5)_4$; $(27.0)_1$

[a] S. W. Benson and J. H. Buss, *J. Chem. Phys.* **29**, 546 (1958).

[b] Note: Units throughout are kcal./mole for H and G and cal./mole° K for C_p and S.

[c] Subscript designates ligancy of atom. The entropy contribution is uncorrected for symmetry so that from the entropy sum obtained for any particular molecule must be subtracted $R \ln \sigma$ where σ is the total symmetry number of the molecule (e.g., 12 for CH_4, 18 for the other linear paraffins).

TABLE 44
Bond Contributions to $C_p°$, $S°$, and $\Delta H_f°$ at 25° C, 1 atmos.[a]

Bond	$C_p°$	$S°$	$\Delta H_f°$	Bond	$C_p°$	$S°$	$\Delta H_f°$
C—H	1.74	12.90	-3.83	S—S	5.4	11.6	—
C—D	2.06	13.60	-4.73	C_{vi}—C[b]	2.6	-14.3	6.7
C—C	1.98	-16.40	2.73	C_{vi}—H	2.6	13.8	3.2
C—F	3.34	16.90	—	C_{vi}—F	4.6	18.6	—
C—Cl	4.64	19.70	-7.4	C_{vi}—Cl	5.7	21.2	-0.7
C—Br	5.14	22.65	2.2	C_{vi}—Br	6.3	24.1	9.7
C—I	5.54	24.65	15.0	C_{vi}—I	6.7	26.1	—
C—O	2.7	-4.0	-02.0	>CO—H[c]	4.2	26.8	-13.9
O—H	2.7	24.0	-27.0	>CO—C	3.7	-0.6	-14.4
O—D	3.1	24.8	-27.9	>CO—O	2.2	9.8	-50.5
O—O	4.9	9.1	21.5	>CO—F	5.7	31.6	—
O—Cl	5.5	32.5	9.1	>CO—Cl	7.2	35.2	-27.0
C—N	2.1	-12.8	9.3	ϕ—H[d]	3.0	11.7	3.25
N—H	2.3	17.7	-2.6	ϕ—C[d]	4.5	-17.4	7.25
C—S	3.4	-1.5	-6.7	(NO_2)—O[d]	—	43.1	-3.0
S—H	3.2	27.0	-0.8	(NO)—O[d]	—	35.5	$+9.0$

All substances in ideal gas state.

[a] S. W. Benson and J. H. Buss, *J. Chem. Phys.* **29**, 546 (1958).

[b] C_{vi} represents the vinyl group carbon atom. The vinyl group is here considered a tetravalent unit.

TABLE 45

Group Contributions to $C_p°$, $S°$, and $\Delta H_f°$ for Ideal Gases at 25°C, 1 atmos. Hydrocarbons[a]

Group	$C_p°$	Contribution to $S°$	$\Delta H_f°$
C—(H)$_3$(C)	6.20	30.41	−10.08
C—(H)$_2$(C)$_2$	5.45	9.42	− 4.95
C—(H)(C)$_3$	4.47	−12.07	− 1.48(−1.90)[c]
C—(C)$_4$	4.35	−35.10	1.95(0.50)[c]
Correction for each *gauche*[c] configuration[b] of large groups	—	—	0.70[c]
C$_d$—(C$_d$)(H)$_2$	5.20	27.6	6.25
[C$_d$—(C$_d$)(C)(H)] + [C—(C$_d$)(H)$_3$]$_{Av}$	10.00	38.5	− 1.20
Correction for *cis/trans* isomers[d]	∓ 1.00[d]	± 0.6[d]	± 0.50[d]
[C$_d$—(C$_d$)$_2$] + 2[C—(C$_d$)(H)$_3$]	16.10	48.7	− 9.85
[C—(C$_d$)(C)(H)$_2$] − [C—(C$_d$)(H)$_3$]	− 0.8	−20.9	− 5.2
[C—(C$_d$)(C)$_2$(H)] − [C—(C$_d$)(H)$_3$]	1.0	−42.9	8.2
[C—(C$_d$)(C)$_3$] − [C—(C$_d$)(H)$_3$]	− 3.5	−66.5	10.9
C$_t$—(C$_t$)(H)	5.3	24.7	27.1
[C$_t$—(C$_t$)(C)] + [C—(C$_t$)(H)$_3$]	9.3	37.6	17.5
[C—(C$_t$)(C)(H)$_2$] − [C—(C$_t$)(H)$_3$]	− 1.1	−20.2	5.2
[C—(C$_t$)(C)$_2$(H)] − [C—(C$_t$)(H)$_3$]	− 2.0	−41.6	8.4
C$_B$—(C$_B$)$_2$(H)	3.25	11.55	3.30
[C$_B$—(C$_B$)$_2$(C)] + [C—(C$_B$)(H)$_3$]	9.25(8.75)[e]	22.0(22.5)[e]	− 4.20(−4.50)[e]
Correction for each *ortho*-configuration of large groups[e]	1.2	− 1.3	0.3
[C—(C$_B$)(C)(H)$_2$] − [C—(C$_B$)(H)$_3$]	− 0.8(−0.4)	−20.4(−20.8)	4.9(5.2)
[C—(C$_B$)(C)$_2$(H)] − (C—(C$_B$)(H)$_3$]	− 0.8(−0.4)	−42.2	9.0
[C—(C$_B$)$_2$(C$_d$)] + [C$_d$—(C$_B$)(C$_d$)(H)]	7.7	− 1.5	12.5

[a] S. W. Benson and J. H. Buss, *J. Chem. Phys.* **29**, 546 (1958).

[b] *Note:* C$_d$ represents a double-bonded C atom; C$_t$, a triple-bonded atom, and C$_B$ is a C atom in a benzene ring.

[c] Values in parentheses are to be used when corrections for *gauche*-configurations are made. The correction applies to each such configuration. Thus 2 methylbutane has 1 *gauche*-correction; 2,3 dimethylbutane has 2 and 2,2,3 trimethylpentane has 5.

[d] The upper sign represents correction for *cis*-isomer while lower sign is that for *trans*-isomer. In molecules such as 2 methylbutene-2 where two methyl groups are cis to each other a *cis*-correction should be applied. *Cis*-corrections should be applied for every pair of large groups that are on the same side of a double bond (e.g., 2 corrections in tetramethylethylene).

[e] Correction is applied to each independent pair of large groups that are ortho to each other in which case the values in parentheses should be used as well (e.g., 2 corrections in 1,2,3 trimethylbenzene and 6 corrections in hexamethylbenzene).

⚜⚜⚜⚜⚜⚜⚜⚜⚜⚜⚜⚜⚜⚜⚜⚜⚜⚜⚜⚜⚜⚜⚜⚜⚜⚜⚜⚜⚜⚜⚜⚜

[c] >CO— represents the bond to carbonyl carbon, the latter being considered a bivalent unit.

[d] NO and NO$_2$ are here considered as univalent, terminal groups, while the phenyl group, C$_6$H$_5$(φ) is considered as a hexavalent unit.

TABLE 46

Group Contributions to $C_p°$, $S°$, and $\Delta H_f°$ for Ideal Gases at 25° C, 1 atmos. Halogen Compounds[a]

Group	Contribution to		
	$C_p°$	$S°$	$\Delta H_f°$
C—(Cl)(H)$_2$(C)	8.8	37.8	15.7
C—(Br)(H)$_2$(C)	9.1	40.5	− 5.2
C—(I)(H)$_2$(C)	—	—	8.8
C—(Cl)(H)(C)$_2$	8.5	17.7	−14.4
C—(Br)(H)(C)$_2$	—	(20.3)[b]	− 2.3
C—(Cl)(C)$_3$	8.7	− 7.6	−12.7
C—(Br)(C)$_3$	9.3	− 4.2	− 0.4
C—(Cl)$_2$(H)(C)	12.2	44.2	—
C—(Cl)$_3$(C)	15.8	50.1	—
C—(F)$_3$(C)	12.5	42.0	—
C—(Br)(Cl)(H)(C)	12.5	47.0	—
C—(F)$_2$(Cl)(C)	13.6	43.0	—
C—(F)(Cl)$_2$(C)	15.0	45.6	—
C$_d$—(C$_d$)(Cl)(H) Av[c]	7.9	35.3	2.7
C$_d$—(C$_d$)(Cl)$_2$	11.5	42.3	—
C$_d$—(C$_d$)(Br)(H) Av[c]	8.1	38.1	12.4
C$_d$—(C$_d$)(I)(H) Av[c]	8.6	40.5	—
C$_d$—(C$_d$)(F)$_2$	9.6	37.2	—
C$_d$—(C$_d$)(Br)$_2$	11.6	46.9	—
C$_d$—(C$_d$)(F)(Cl)	10.4	39.8	—

[a] S. W. Benson and J. H. Buss, *J. Chem. Phys.* **29**, 546 (1958).
[b] Estimate by the authors.
[c] These are averages for *cis-trans* isomers.

TABLE 47

GROUP CONTRIBUTIONS TO $C_p°$, $S°$, AND $\Delta H_f°$ FOR IDEAL GASES AT 25°C, 1 atmos.
MISCELLANEOUS[a]

Group	Contribution to		
	$C_p°$	$S°$	$\Delta H_f°$
[**O**—(C)(H)] + [(**C**—(O)(H)₃]	10.5	59.5	− 48.1
[**O**—(C)(C)] + [2(**C**—(O)(H)₃]	15.7	69.4	− 45.3
[**C**—(O)(C)(H)₂] − [**C**—(O)(H)₃]	−1.1	−20.4	2.0
[**C**—(O)(C)₂(H)] − [**C**—(O)(H)₃]	(−1.2)[b]	−44.3	3.0
[**C**—(O)(C)₂] − [**C**—(O)(H)₃]	—	—	3.3
O—(O)(H)	5.2	28.5	− 16.3
[**O**—(O)(C)] + [**C**—(O)(H)₃]	(9.3)[b]	(40.0)[b]	− 15.8
[**CO**—(C)(H)] + [**C**—(CO)(H)₃]	13.1	65.3	− 39.6
[**CO**—(C)₂] + 2[**C**—(CO)(H)₃]	17.9	76.2	− 51.7
[**C**—(CO)(C)(H)₂] − [**C**—(CO)(H)₃]	(−0.8)[b]	(−20.5)[b]	3.3
[**CO**—(O)(H)] + [(**O**—(CO)(H)]	9.1	60.1	− 89.6
[**CO**—(O)(C)] + [**O**—(CO)(H)] + [(**C**—(CO)(H)₃]	—	72.3	−104.0
[**O**—(CO)(C)] − [**O**—(CO)(H)] + [**C**—(O)(H)₃]	—	12.2	5.5
[**N**—(C)(H)₂] + (**C**—(N)(H)₃]	12.9	59.9	− 6.7
[(**N**—(C)₂(H)] + 2[**C**—(N)(H)₃]	16.6	69.7	− 6.6
[**C**—(C)(N)(H)₂] − [**C**—(N)(H)₃]	−2.4[c]	(−20.7)[b]	5.2
[**N**—(C)₃] + 3[**C**—(N)(H)₃]	—	77.7	− 6.6
[**S**—(C)(H)] + [**C**—(S)(H)₃]	12.12	63.1	− 5.5
[**S**—(C)₂] + 2[**C**—(S)(H)₃]	17.7	74.0	− 9.0
[**S**—(S)(C)] + [**C**—(S)(H)₂]	11.0	43.1	—
[**C**—(S)(C)(H)₂] − [**C**—(S)(H)₃]	−1.1	−20.5	4.6
[**C**—(S)(C)₂(H)] − [**C**—(S)(H)₃]	−1.6	−42.2	7.7
[**C**—(S)(C)₃] − [**C**—(S)(H)₃]	−1.8	−64.8	9.3
C—(C)(H)₂(NO₃)	—	58.3	− 26.7
C—(C)₂(H)(NO₃)	—	(36.5)[b]	− 25.4
C(C)(H)₂(CN)	11.0	39.6	—

[a] S. W. Benson and J. H. Buss, *J. Chem. Phys.* **29**, 546 (1958).
[b] Estimates by the authors.
[c] Although obtained from the data on C₂H₃NH₂ this value seems too low by about 1.5 units.

TABLE 48

VALUES OF GROUP CONTRIBUTIONS TO THE HEAT OF FORMATION USED IN BRYANT CALCULATIONS[a]

Nature of contribution				$\Delta H_{298.15}$ kcal./g.mole	Table (Part II)
Base group	Primary CH_3 substitution	Secondary CH_3 substitutions[b]			
		A	B		
Methane	—	—	—	−17.9	17
	Methane	—	—	− 2.4*	18
		1	1	− 4.5	
		1	2	− 5.0*	
		1	3	− 5.5	
		1	4	− 5.0	
		2	1	− 6.6	
		2	2	− 6.5*	
		2	3	− 6.5*	19
		2	4	− 5.1	
		3	1	− 8.1	
		3	2	− 8.0	
		3	3	− 6.5*	
		3	4	− 5.0*	

[a] W. M. D. Bryant, *J. Polymer. Sci.* **56**, 277 (1962).
[b] As in Table 19, Pt. II, A indicates the type of carbon atoms on which substitution is made: Type 1, CH_3; Type 2, CH_2; Type 3, CH; Type 4, C. B indicates the highest type number of an adjacent carbon atom.

TABLE 49

VALUES OF SUBSTITUTION GROUP CONTRIBUTIONS TO THE HEAT OF FORMATION INVOLVING THE HALOGENS[a]

Nature of contribution	$\Delta H_{298.15}(g)$ kcal./g.mole	Table (Part II)
F in CF_4	−44.5*	—
F in CF_3	−43.7*	—
F in CF_2	−39.0*	—
F in CF	−34.5*	
Cl (monosubstitution)	0.0	
Cl (higher degree of substitution)	4.5	21
Br (mono and higher substitution, provisional)	11.6*	
I (mono and higher substitution, provisional)	24.8	

[a] W. M. D. Bryant, *J. Polymer Sci.* **56**, 277 (1962).

TABLE 50
Log K_{fG} for Paraffinic Groups[a]

Temperature (°K)	—CH$_3$	—CH$_2$—	$-\overset{\mid}{\underset{\mid}{C}}-H$	$-\overset{\mid}{\underset{\mid}{C}}-$	$CH_3-\overset{\overset{\displaystyle CH_3}{\mid}}{\underset{\mid}{C}}-CH_3$	$-\overset{\overset{\displaystyle CH_3}{\mid}}{\underset{\underset{\displaystyle CH_3}{\mid}}{C}}-\overset{CH_3}{}$
300	2.986	−1.496	−4.962	−8.373	+ 0.571	− 2.401
400	1.093	−2.421	−5.299	−8.205	− 4.926	− 6.019
500	−0.105	−3.008	−5.498	−8.069	− 8.384	− 8.279
600	−0.937	−3.417	−5.628	−7.951	−10.762	− 9.823
700	−1.551	−3.718	−5.716	−7.847	−12.500	−10.949
800	−2.024	−3.948	−5.777	−7.743	−13.815	−11.791
900	−2.400	−4.130	−5.822	−7.652	−14.852	−12.452
1000	−2.705	−4.277	−5.852	−7.570	−15.685	−12.797

[a] R. Ciola, *Ind. Eng. Chem.* **49**, 1789 (1957).

TABLE 51
Log K_{fG} for Olefinic Groups[a]

Temperature (°K)	$H_2C=\overset{\mid}{C}-$	$\overset{H}{\underset{H}{\diagdown}}C=C\overset{\diagup}{\diagdown}$ *cis*	$\overset{H}{\underset{H}{\diagdown}}C=C\overset{H}{\underset{\diagdown}{\diagup}}$ *trans*	$C-\overset{\mid}{H}-\overset{\mid}{C}=$	$-\overset{\mid}{C}=\overset{\mid}{C}-$
300	−13.872	−17.144	−16.695	−19.508	−23.344
400	−11.259	−13.800	−13.543	−15.809	−18.847
500	− 9.701	−11.857	−11.656	−13.616	−16.152
600	− 8.689	−10.568	−10.405	−12.149	−14.340
700	− 8.159	− 9.658	− 9.517	−11.097	−13.030
800	− 7.446	− 8.984	− 8.872	−10.310	−12.044
900	− 7.040	− 8.459	− 8.348	− 9.691	−11.271
1000	− 6.714	− 8.039	− 7.896	− 9.196	−10.650

Temperature (°K)	$H_2C=\overset{\mid}{C}-$	$H_2C=C=\overset{\mid}{\underset{H}{C}}-$	$H_2C=C=\overset{\mid}{C}-$	$HC=C=CH$
300	−15.863	−37.575	−40.328	−42.897
400	−12.927	−28.706	−30.999	−32.203
500	−11.194	−23.419	−25.412	−26.383
600	−10.045	−19.916	−21.668	−22.527
700	− 9.232	−17.427	−19.004	−19.773
800	− 8.625	−15.569	−17.022	−17.708
900	− 8.153	−14.129	−15.467	−16.110
1000	− 7.778	−12.980	−14.231	−14.823

[a] R. Ciola, *Ind. Eng. Chem.* **49**, 1789 (1957).

TABLE 52
Log K_{fG} for Acetylenic Groups[a]

Temp. (°K)	H—C≡C—	—C≡C—
300	−36.000	−37.770
400	−26.684	−27.459
500	−20.725	−21.400
600	−16.761	−17.360
700	−13.929	−14.472
800	−11.809	−12.303
900	−10.170	−10.619
1000	−8.913	−9.272

[a] R. Ciola, *Ind. Eng. Chem.* **49**, 1789 (1957).

TABLE 53
Log K_{fG} of Conjugate Olefin Groups[a]

Temp. (°K)	H$_2$C↔	—C↔	—C↔
300	−5.436	−7.658	−9.956
400	−4.446	−6.309	−8.290
500	−3.852	−5.499	−7.289
600	−3.456	−4.960	−6.623
700	−3.223	−4.557	−6.096
800	−3.047	−4.283	−5.690
900	−2.911	−4.073	−5.377
1000	−2.801	−3.903	−5.125

[a] R. Ciola, *Ind. Eng. Chem.* **49**, 1789 (1957).

TABLE 54
Log K_{fG} for Aromatic Groups[a]

Temp. (°K)	H—C↔	—C↔	↔C↔
300	−3.576	−6.175	−5.937
400	−3.008	−5.088	−4.734
500	−2.678	−4.477	−4.012
600	−2.470	−4.069	−3.530
700	−2.328	−3.775	−3.228
800	−2.226	−3.547	−2.983
900	−2.148	−3.376	−2.794
1000	−2.088	−3.233	−2.642

[a] R. Ciola, *Ind. Eng. Chem.* **49**, 1789 (1957).

TABLE 55
Log K_{fG} for Oxygen-Containing Groups[a]

Temp. (°K)	—CH$_2$OH	—OH	—CHO	$\underset{\mathrm{O}}{\overset{}{-\!\!\mathrm{C}\!\!-}}$	—CO$_2$H	$\underset{\mathrm{O}}{\overset{}{-\!\!\mathrm{C}\!\!-\!\mathrm{O}\!-}}$	—O—	↔O↔	$\underset{\mathrm{H}}{\overset{-\mathrm{C}=\mathrm{O}=\mathrm{C}-}{}}$	$-\mathrm{C}\!\!=\!\!\mathrm{C}\!\!=\!\!\mathrm{O}$
300	26.089	27.585	20.130	18.467	51.819	61.770	9.648	11.637	8.363	5.430
400	17.656	20.077	14.726	13.353	34.473	44.901	6.768	8.287	6.021	3.638
500	12.552	15.550	11.457	10.285	24.018	34.778	5.043	6.280	4.616	2.631
600	9.128	12.545	9.268	8.239	17.111	28.033	4.203	4.942	3.679	1.931
700	6.669	10.387	7.699	6.778	12.206	23.212		4.142	3.010	1.333
800	4.820	8.768	6.516	5.681	8.529	19.592		3.515	2.507	0.967
900	3.380	7.510	5.591	4.830	5.690			3.027	2.117	0.682
1000	2.226	6.503	4.846	4.148	3.432			2.637		

[a] R. Ciola, *Ind. Eng. Chem.* **49**, 1789 (1957).

TABLE 56
Log K_{fG} for Nitrogen-Containing Groups[a]

Temp. (°K)	—C≡N	—N=C	—NH$_2$	$\underset{}{-\mathrm{N}-}$	$-\underset{\underset{\swarrow}{\nwarrow}}{\mathrm{N}}-$	—N—H	—NO$_2$
300	−20.821	−31.799	−7.977	−22.525	−16.673	−16.326	−1.846
400	−15.227	−23.363	−7.463	−18.981	−8.611	−13.971	−3.293
500	−11.867	−18.301	−7.155	−16.854	−7.374	−12.558	−4.164
600	−9.627	−14.927	−6.950	−15.804	−6.549	−11.616	−4.793
700	−8.026	−12.506	−6.581	−14.424	−4.926	−10.942	−5.179
800	−6.691		−6.844	−13.663	−5.446	−10.436	−5.496
900	−5.894		−7.052	−13.074	−5.075	−10.046	−5.740
1000	−5.116		−7.216	−12.601	−4.777	−9.731	−5.935

[a] R. Ciola, *Ind. Eng. Chem.* **49**, 1789 (1957).

TABLE 57
Log K_{fG} for Sulfur-Containing Groups[a]

Temp. (°K)	$\overset{O}{\underset{\|}{-S-}}$	$\overset{-S-}{\underset{O}{\overset{\|}{\overset{\|}{\underset{O}{\|}}}}}\hspace{-0.5em}$	—S—	↔S↔	—SH
300	14.584	47.964	−0.685	−0.408	5.442
400	9.086	32.924	−1.289	−0.584	3.497
500	5.787	23.900	−1.652	−0.691	2.330
600	3.588	17.885	−1.894	−0.761	1.552
700	2.168	13.696	−2.110	−0.758	0.996
800	0.838	10.545	−2.240	−0.783	0.579
900	−0.078	8.098	−2.341	−0.804	0.254
1000	−0.811	6.139	−2.421	−0.820	−0.044

[a] R. Ciola, *Ind. Chem. Eng.* **49**, 1789 (1957).

TABLE 58
Log K_{fG} of Halogen-Containing Groups[a]

Temp. (°K)	—F	—Cl	—Br	—I
300	33.292	6.100	1.748	−5.680
400	25.078	4.508	1.453	−4.260
500	20.149	3.606	1.276	−3.409
600	16.864	3.005	1.158	−2.841
700		2.576	1.073	−2.435
800		2.252	1.010	−2.130
900		2.003	0.959	−1.894
1000		1.802	0.922	−1.705

[a] R. Ciola, *Ind. Eng. Chem.* **49**, 1789 (1957).

TABLE 59
Log K_{fG} for Some Organic Compounds[a]

Temp. (°K)	$COCl_2$	CH_3OH	HCO_2H	$H_2C=C=O$	$H_2C=O$	CH_4	C_2H_4	C_6H_{12}*	C_2H_2
300	35.835	28.411	58.534	9.929	19.785	8.818	−11.878	−5.693	−36.406
400	26.362	19.407	42.568	7.286	14.482	5.490	−9.657	−11.286	−26.541
500	20.678	14.004	32.989	5.700	11.300	3.427	−8.412	−14.893	−20.629
600	16.890	9.310	26.604	4.642	9.179	2.000	−7.619	−17.432	−16.695
700	14.183	7.732	21.811	3.887	7.602	0.953	−7.080	−19.310	−13.893
800	12.150	5.727	18.272	3.320	6.690	0.150	−6.690	−20.750	−11.798
900	10.575	4.171	15.528	2.880	5.498	−0.488	−6.400	−21.885	−10.170
1000	9.312	2.924	13.328	—	4.760	−1.008	−6.174	−22.794	−8.874

* Cyclohexane
[a] R. Ciola, *Ind. Eng. Chem.* **49**, 1789 (1957).

TABLE 60
CORRECTION FACTORS FOR CYCLIZATION[a]

Temp. (°K)	(3-ring structure)	(5-ring with O structure)	Cyclo-pentene	Cyclo-hexene	Number of C atoms in ring			
					3	4	5	6
300	−9.231	3.694	3.187	7.078	−10.528	−8.714	1.580	4.507
400	−6.889	2.634	2.521	5.420	− 6.190	−4.371	2.640	4.377
500	−5.484	1.998	2.121	4.425	− 3.549	−1.748	3.210	4.248
600	−4.547	1.574	1.855	3.762	− 1.843	− 0.00	3.558	4.143
700	−4.012	1.451	1.714	3.203	− 0.671	+1.236	3.784	4.063
800	−3.376	1.335	1.561	2.859	+ 0.248	+2.168	3.944	4.004
900	−2.986	1.206	1.443	2.591	+ 0.945	+2.889	4.121	3.961
1000	−2.673	1.120	1.347	2.378	+ 1.490	+2.797	4.147	3.929

[a] R. Ciola, *Ind. Eng. Chem.* **49**, 1789 (1957).

TABLE 61
Correction Factors Necessitated by Introducing Lateral Chains[a]

Temp. (°K)	cis-Cyclopentane		Cyclohexane					Branching in aromatics				
								Double		Triple		
	1.2	1.3	1.1	1.2, cis	1.2, trans	1.3, trans	1.4, cis	1.2	1.3	1.2.4	1.2.4	1.3.5
300	1.281	0.544	−0.989	−1.979	−0.798	−1.120	−1.407	−0.353	0.182	−0.904	−0.178	0.582
400	0.990	0.425	−0.798	−1.470	−0.615	−0.791	−1.112	−0.375	0.103	−0.849	−0.310	0.350
500	0.828	0.372	−0.637	−1.173	−0.513	−0.600	−0.897	−0.302	0.111	−0.669	−0.241	0.348
600	0.714	0.331	−0.592	−0.997	−0.428	−0.497	−0.774	−0.422	0.124	−0.550	−0.194	0.346
700	0.629	0.293	−0.551	−0.825	−0.358	−0.413	−0.691	−0.200	0.130	−0.468	−0.162	0.332
800	0.557	0.246	−0.518	−0.715	−0.333	−0.345	−0.604	−0.172	0.130	−0.418	−0.150	0.331
900	0.559	0.287	−0.501	−0.639	−0.295	−0.286	−0.583	−0.140	0.137	−0.363	−0.125	0.336
1000	0.459	0.212	−0.475	−0.570	−0.273	−0.262	−0.563	−0.122	0.137	−0.333	−0.118	0.317

[a] R. Ciola, *Ind. Eng. Chem.* **49**, 1789 (1957).

TABLE 62

Thermodynamic Properties for n-Propyl Halides[a]

Temp. (°K)	$\left[\dfrac{-(G° - H_o°)}{T}\right]$ Substituent Halogens				$\left[\dfrac{H_0 - H_o}{T}\right]$ Substituent Halogens			
	F	Cl	Br	I	F	Cl	Br	I
298.16	58.65	62.28	65.06	65.16	13.17	13.60	14.23	14.27
400	62.81	66.55	69.57	69.67	15.29	15.81	16.36	16.53
500	66.46	70.31	73.48	73.60	17.52	18.09	18.60	18.83
600	69.86	73.79	77.08	77.24	19.75	20.35	20.82	21.07
700	73.06	77.08	80.45	80.65	21.89	22.50	23.15	23.19
800	76.11	80.21	83.65	83.87	23.90	24.51	24.89	25.17

Temp. (°K)	$[S°]$ Substituent Halogens				$[C_p°]_x$ Substituent Halogens			
	F	Cl	Br	I	F	Cl	Br	I
298.16	71.81	75.90	79.29	79.43	19.02	19.69	20.11	20.60
400	78.10	82.36	85.94	86.20	24.06	24.84	25.21	25.68
500	83.98	88.43	92.08	92.43	28.77	29.54	29.84	30.27
600	89.61	94.14	97.89	98.31	32.89	33.60	33.82	34.19
700	95.95	99.58	103.37	103.84	36.42	37.07	37.21	37.53
800	100.01	104.72	108.54	109.04	39.44	40.03	40.11	40.38

[a] J. P. Morgan and J. Lielmezs, *Ind. Eng. Chem. Fundamentals* **4**, 383 (1965).

TABLE 63

Aliphatic Hydrocarbon Groups[a]

Group	300–850° K		850–1500° K		Highest Temperature (°K)
	A	$B \times 10^2$	A	$B \times 10^2$	
—CH_3	−8.948	−0.436	−12.800	0.000	1500
—CH_2—	−4.240	−0.235	− 6.720	0.090	1500
—CH—	−1.570	0.095	− 2.200	0.172	1500
—C—[b]	−0.650	0.425	0.211	0.347	1500
=CH_2	7.070	−0.295	4.599	−0.0114	1500

TABLE 63 (*continued*)

Group	300–850° K		850–1500° K		Highest Temperature (°K)
	A	$B \times 10^2$	A	$B \times 10^2$	
—C≡	27.276	0.036	27.600	−0.010	1500
≡CH[c]	27.242	−0.046	27.426	−0.077	1500
=C=	33.920	−0.563	33.920	−0.563	1500
H\C=CH₂	16.323	−0.437	12.369	0.128	1500
\C=CH₂	16.725	−0.150	15.837	0.038	1500
\C=C/	29.225	0.415	30.129	0.299	1500
\C=C/H	20.800	−0.100	19.360	0.080	1500
\C=C/H, H	20.100	0.000	19.212	0.102	1500
H\C=C/H (*cis*)	19.088	−0.378	17.100	0.000	1500
H\C=C/H (*trans*)	18.463	−0.211	16.850	0.000	1500
\C=C=CH₂	51.450	−0.050	50.200	0.100	1500
\C=C=CH₂, H	50.163	−0.233	48.000	0.000	1500
\C=C=C/H	54.964	0.027	53.967	0.133	1500

[a] K. K. Verma and L. K. Doriaswamy, *Ind. Eng. Chem. Fundamentals* **4**, 389 (1965).
[b] Temperature ranges: 300–1100° K and 1100–1500° K.
[c] Temperature ranges: 300–600° K and 600–1500° K.

TABLE 64
Aromatic Hydrocarbon Groups[a]

Group	300–750° K		850–1500° K		Highest Temperature (° K)
	A	$B \times 10^2$	A	$B \times 10^2$	
HC⟨	3.768	−0.167	2.616	−0.016	1500
—C⟨	5.437	0.037	5.279	0.058	1500
↔C⟨	4.208	0.092	4.050	0.100	1500

[a] K. K. Verma and L. K. Doriaswamy, *Ind. Eng. Chem. Fundamentals* **4**, 389 (1965).

TABLE 65
Branching in Paraffin Chains[a]

Group	300–750° K		850–1500° K		Highest Temperature (°K)
	A	$B \times 10^2$	A	$B \times 10^2$	
Side chain with 2 or more C—atoms	0.800	0.000	0.800	0.000	1500
3 adjacent —CH groups	−1.200	0.000	—	—	1500
Adjacent —CH and —C— groups	0.600	0.000	0.600	0.000	1500

[a] K. K. Verma and L. K. Doriaswamy, *Ind. Eng. Chem. Fundamentals* **4**, 389 (1965).

TABLE 66
Branching in Cycloparaffins[a]

Group	300–850° K		850–1000° K		Highest Temperature (°K)
	A	$B \times 10^2$	A	$B \times 10^2$	
Branching in six-membered rings					
Single branching	0.00	0.00	2.85	−0.40	1500
Double branching					
1,1 position	1.10	0.45	−0.40	0.00	1500
cis-1,2 position	3.05	−1.09	1.46	−0.13	1500
trans-1,2 position	−0.90	−0.60	−1.50	0.00	1500
cis-1,3	0.00	−1.00	−2.60	0.00	1500
trans-1,3	0.00	−0.16	2.80	−0.32	1500
cis-1,4	0.00	−0.16	2.80	−0.32	1500
trans-1,4	0.00	−1.00	−2.60	0.00	1500
Branching in five-membered rings					
Single branching	0.00	0.00	1.40	−0.20	1500
Double branching					
1,1 position	0.30	0.00	1.90	−0.25	1500
cis-1,2	0.70	0.00	0.00	0.00	1500
trans-1,2	−1.10	0.00	−1.60	0.00	1500
cis-1,3	−0.30	0.00	0.15	0.00	1500
trans-1,3	−0.90	0.00	−1.40	0.00	1500

[a] K. K. Verma and L. K. Doriaswamy, *Ind. Eng. Chem. Fundamentals* **4**, 389 (1965).

TABLE 67
Branching in Aromatics[a]

Group	300–850° K		850–1500° K		Highest Temperature (°K)
	A	$B \times 10^2$	A	$B \times 10^2$	
Double branching					
1,2 position	0.85	0.03	0.85	0.03	1500
1,3 position	0.56	−0.06	0.56	−0.06	1500
1,4 position	1.00	−0.14	1.40	−0.12	1500
Triple branching					
1,2,3 position	2.01	−0.07	1.50	0.00	1500
1,2,4 position	1.18	−0.25	1.50	−0.10	1500
1,3,5 position	1.18	−0.25	1.80	−0.08	1500

[a] K. K. Verma and L. K. Doriaswamy, *Ind. Eng. Chem. Fundamentals* **4**, 389 (1965).

TABLE 68
Ring Correction[a]

Group	300–750° K		750°–1500° K		Highest Temperature (°K)
	A	$B \times 10^2$	A	$B \times 10^2$	
C_2-cycloparaffin ring	24.850	−0.240	24.255	−0.174	1500
C_4-cycloparaffin ring	19.760	−0.440	17.950	−0.231	1500
C_5-cycloparaffin ring	7.084	−0.552	4.020	−0.140	1500
C_6-cycloparaffin ring	0.378	−0.382	−4.120	0.240	1500

[a] K. K. Verma and L. K. Doriaswamy, *Ind. Eng. Chem. Fundamentals* **4**, 389 (1965).

TABLE 69
Oxygen-Containing Groups[a]

Group	300–850° K		850–1500° K		Highest Temperature (°K)
	A	$B \times 10^2$	A	$B \times 10^2$	
$\diagup$C=O	−31.505	0.007	−32.113	0.073	1500
[—O— (—C—O—C—)]	−24.200	0.000	−24.200	0.000	1000
[—O— (H_3C—C—O—C—CH_3)]	−30.500	0.000	−30.500	0.000	1000
O (epoxide)	−21.705	0.030	−21.600	0.020	1500
—CHO[b]	−29.167	−0.183	−30.500	0.000	1000
—C(=O[b])OH	−94.488	−0.063	−94.880	0.000	1500
Contribution of —OH group					
HO---CH_3	−37.207	−0.259	−37.993	−0.136	1000
HO---CH_2—	−40.415	−0.267	−41.265	−0.116	1000
HO---CH	−43.200	−0.200	−43.330	−0.143	1000
HO---C—	−46.850	−0.250	−47.440	−0.146	1000
HO---C	−44.725	−0.125	−45.220	−0.021	1000

[a] K. K. Verma and L. K. Doriaswamy, *Ind. Eng. Chem. Fundamentals* **4**, 389 (1965).
[b] Temperature ranges: 300–700° K and 750–1500° K.

TABLE 70

Nitrogen- and Sulfur-Containing Groups[a]

Group	300–750° K		750–1500° K		Highest Temperature (°K)
	A	$B \times 10^2$	A	$B \times 10^2$	
—NO_2 (aliphatic)	− 7.813	−0.043	− 9.250	0.143	1500
—C≡N	36.580	0.080	37.170	0.000	1000
—NH_2 (aliphatic)[b]	3.832	−0.208	2.125	0.002	1500
=NH (aliphatic)	13.666	−0.067	12.267	0.133	1500
≡N (aliphatic)	18.050	0.300	18.050	0.300	1500
—NH_2 (aromatic)[b]	− 0.713	−0.188	− 1.725	0.000	1000
=NH (aromatic)	9.240	−0.250	8.460	−0.140	1000
≡N (aromatic)	18.890	0.110	16.200	0.250	1000
SH[c]	−10.580	−0.080	—	—	1000
—S—[c]	− 4.725	0.160	—	—	1000

[a] K. K. Verma and L. K. Doriaswamy, *Ind. Eng. Chem. Fundamentals* **4**, 389 (1965).
[b] Temperature ranges: 300–600° K.
[c] Temperature ranges: 300–1000° K.

Standard state of sulfur is S_2 (gas) at 298° K. For conversion to S (crystal, rhombic) at 298° K, the factor is 15.42 kcal./mole.

TABLE 71

Halogen-Containing Groups[a]

Group	300–750° K		750–1500° K		Highest Temperature (°K)
	A	$B \times 10^2$	A	$B \times 10^2$	
H—CH$_2$—Cl	− 9.322	−0.045	− 9.475	−0.025	1000
H_2C—CH$_2$—Cl	−10.007	−0.033	−10.438	0.029	1000
H_3C—CHCl—Cl	−14.780	−0.040	−14.780	−0.040	1500

TABLE 71 (continued)

Group	300–750 °K		750–1500° K		Highest Temperature, (°K)
	A	$B \times 10^2$	A	$B \times 10^2$	
H—CHCl—Cl (CH$_2$Cl$_2$) H,Cl,Cl	−13.222	−0.029	−13.222	−0.029	1500
H—CCl$_2$—Cl (CHCl$_3$)	− 6.684	−0.033	− 6.684	−0.033	1500
Cl—CCl$_2$—Cl (CCl$_4$)	− 6.400	−0.050	− 6.400	−0.050	1500
H$_2$C=CHCl	− 7.622	0.029	− 7.390	0.000	1500
HClC=CHCl (1,1)	− 6.171	−0.029	− 6.171	−0.029	1500
HClC=CHCl (cis-1,2)	− 5.916	0.071	− 5.386	−0.007	1500
HClC=CHCl (trans-1,2)	− 6.532	0.233	− 5.480	0.106	1500
Cl$_2$C=CCl$_2$	− 6.047	0.236	− 6.047	0.236	1500

[a] K. K. Verma and L. K. Doriaswamy, *Ind. Eng. Chem. Fundamentals* **4**, 389 (1965).

TABLE 72

Bond Contributions for Heats of Atomization, Formation, and Combustion of Gases and Liquids[a,b]

Bond	Symbol	Heat of atomization		Heat of formation		Heat of combustion	
		Gas	Liquid	Gas	Liquid	Gas	Liquid
C—C	c_1	85.4	85.4	−0.45	−0.45	47.48	47.48
C=C	c_2	145.3	(145.3)	30.2	(30.2)	32.5	(32.5)
C≡C	c_3	202.3	(202.3)	−55.3	(−55.3)	196.3	(196.3)
C—H	p	98.47	98.96	3.45	3.95	54.22	53.73
C—H	s	97.65	98.23	2.64	3.21	55.04	54.46
C—H	t	96.80	97.32	1.78	2.30	55.89	55.37
C—H	p'	98.90	(99.40)	3.89	(4.38)	53.79	(53.29)
C—H	s'	98.00	(97.58)	2.99	(3.57)	54.69	(54.11)
C—H	t'	97.48	(98.00)	2.47	(2.99)	55.20	(54.69)
C—H	s_2	99.26	(99.84)	−10.07	(−9.49)	75.58	(75.00)
C—H	t_2	98.01	(98.53)	−25.62	(−24.11)	98.97	(98.45)
C—H	t_3	96.30	(96.82)	1.29	(1.80)	56.39	(55.87)
C—OH	a_{lc}	198.1	204.9	49.93	50.69	13.75	6.99
C—O—C	e	172.0	173.3	27.0	28.3	20.0	18.7
C—CHO	a_{ld}	355.4	360.6	29.5	34.7	122.2	117.0
C—CO—C	k	350.1	356.9	33.4	40.2	107.7	100.9
C—COOH	a_c	480.1	498.6	95.1	113.9	56.7	37.9
C—NH$_2$	a_m	256.6	—	−3.5	—	95.3	—
C—NO$_2$	n_i	285.2	292.2	11.0	18.0	12.6	5.5
C—ONO$_2$	n_a	354.5	361.0	21.1	27.5	2.4	−4.0
C—NHNO$_2$	n_{m1}	—	440.2	—	0.9	—	56.8
C\ NNO$_2$ /C	n_{m2}	—	421.3	—	−8.9	—	55.9

[a] K. J. Laidler, *Canad. J. Chem.* **34**, 628 (1956).

[b] The values in brackets are based on the assumption that the contributions for the heats of vaporization of C—H are the same irrespective of the proximity of double and triple bonds.

c_1, c_2, c_3 are the ordinary C—C, C=C, and C≡C bond types; p, s, t, the primary, secondary, and tertiary type C—H bonds; the prime (') and subscript 2 designate C—H bonds at sites next but one and adjacent to a double bond, respectively, and the subscript 3, the C—H bond adjacent to a triple bond. The remaining symbols stand for specific groups' contributions (alcohols, ethers, ketenes, acids, amines, nitro and nitrates, and primary and secondary nitramines).

TABLE 73. EMPIRICAL EQUATIONS FOR HEAT OF COMBUSTION[a,b]

$$-\Delta H_{c\ 18°\ C\ (liq.)} = 26.05\ N + \text{(Correction factors)}$$

Type	Corr. Factor	Type	Corr. Factor
sat. aliph. hydrocarbons:		*acids* (continued)	
open chain, cyclic	0	hydroxy polybasic	$+ 6.5_k + 6.5_v$
arom. hydrocarbons:		olef. and acetyl.	
mono-polycyclic,		polybasic	$+ 13_y + 16.5_x$
fused rings	$-3.5_a - 6.5_b$	arom. mono- and poly	
olefins:		basic, phenyl aliph.	$-3.5_a - 6.5_b$
aliph. mono-	$+ 13_c$	hydroxy	$-3.5_a - 6.5_b + 3.5_m$
aromatic	$+ 13_c - 6.5_d$	hydroarom. and poly-	
cyclic	$+ 6.5_e$	methylene	$+ 13_z + 13_{aa}$
terpenes	$+ 13_c + 6.5_e$	*acid anhydrides*	$+ 10_{bb}$
acetylenes	$+ 33.1_h + 46.1_g$	*lactones*	$+ 13_{cc}$
alcohols:		*esters:*	
1[ary] (all types)	$+ 13_j$	aliph. mono- and	
2[ary]	$+ 6.5_k$	polybasic acid	$+ 16.5_{dd}$
3[ary]	$+ 3.5_e$	arom. mono- and poly-	
polyhydroxy		monobasic acid	$+ 16.5_{dd} - 3.5_a - 6.5_d$
(all types)	$+ 13_j + 6.5_k + 3.5_l$	phenol	$+ 16.5_{dd} - 6.5_b - 3.5_a$
ethers: aliph., arom.	$+ 19.5_{o,\ p}$		$+ 3.5_m$
aldehydes:		*amines:*	
aliphatic	$+ 13.0_q$	aliph. 1[ary]	$+ 13_{ee}$
aromatic	$+ 13.0_q - 3.5_a + 3.5_m$	aliph. 2[ary]	$+ 19.5_{ff}$
phenols	$+ 3.5_n - 3.5_a - 6.5_b$	aliph. 3[ary]	$+ 26_{gg}$
aliph. acetals	$+ 19.5_o$	arom. 1[ary]	$+ 6.5_{jj} - 3.5_{hh} - 6.5_b - 3.5_a$
ketones:		arom. 2[ary]	$+ 13_{kk} - 3.5_{hh}$
aliphatic	$+ 6.5_s$	arom. 3[ary]	$+ 19.5_{ll} - 3.5_{hh}$
aromatic	$+ 6.5_t - 3.5_a - 6.5_b$	*acid amides:*	
quinones	$+ 33.1 - 3.5_a - 6.5_b$	aliphatic	0
hydroarom. and poly-		aromatic	$- 3.5_{hh}$
methylene	$+6.5$	*imides*	$- 3.5_{hh}$
carbohydrates:		*nitriles:*	
monosaccharides	$+ 13_q + 13_j + 6.5_k + 3.5_l$	aliphatic	$+ 16.5_{nn}$
disaccharides	$+ 19.5_o + 13_j + 6.5_k +$	aromatic	$+ 16.5_{nn} - 6.5_{pp}$
	$3.5_l + 13_q$	*carbylamines:*	
acids:		isocyanates	$+ 33.1_{qq}$
sat. aliph. mono- and		*nitro compounds:*	
poly-basic	0	aliph. arom.	$+ 13_{rr,\ ss}$
hydroxy- and keto	$6.5_s + 13_j + 6.5_k + 3.5_l$	subst. arom.	$+ 13_{ss} - 3.5_a$
olef. monobasic	$+ 13$	phenols, phenetoles,	
acetylenic monobasic	$+ 33.1_h$	anilines,	$+ 13_{ss} + 3.5_m + 6.5_{jj} +$
		acetanilides	$13_{kk} + 19.5_p$
		arom. aldehydes	$+ 13_{ss} + 13_r - 3.5_a$
		arom. acids	$+ 13_{ss} - 3.5_a + 13_c$

[a] M. S. Kharasch, *J. Research Natl. Bur. Standards* **2**, 359 (1929). [b] Units, kcal./mole.

TABLE 74

Structural Correlations for Calculation of Heat of Combustion[a, b]

Symbol	Pair of electrons held between—	For each such grouping add or substract as indicated
a	Aromatic radical ⦂ Aliphatic radical	− 3.5
b	Aromatic radical ⦂ Aromatic radical	− 6.5
c	Ethylene bond —C=C— ⬡—⦂ C=C—	+ 13.0
d	Aromatic radical carbon ⦂ Vinyl radical	− 6.5
e	Ethylene bond in ring system, as in cyclohexene	+ 6.5

Acetylene bond

g	One or more replaceable hydrogens —C≡C—	+ 46.1
h	No replaceable hydrogens R—C≡C—R	+ 33.1
i	Aromatic radical carbon ⦂ Acetylene radical ⬡—⦂ C≡C—	− 6.5

Alcohols

j	Aliphatic radical ⦂ Hydroxyl group (Primary alcohol)	+ 13.0
k	Secondary radical ⦂ Hydroxyl group (Secondary alcohol)	+ 6.5
l	Tertiary radical ⦂ Hydroxyl group (Tertiary alcohol)	+ 3.5
m	Aromatic radical ⦂ Hydroxyl group ⬡—⦂ OH	+ 3.5
o	Acetal linkage R.$\overset{H}{C}$ ⦂ -(O—R_1)$_2$ where R and R_1 are aliphatic radicals	+ 19.5
p	Aromatic ethers ⬡—⦂ O—⦂ CH_3	+ 19.5
q	Aliphatic aldehydes R—$\overset{H}{C}$=O	+ 13.0

(Table continued)

TABLE 74 (*continued*)

Symbol	Pair of electrons held between–	For each such grouping add or substract as indicated
r	Aromatic aldehydes R.C(H)=O	+ 13.0
s	Aliphatic ketones R—C(=O)—R	+ 6.5
t	Aromatic ketones R—C(=O)—C$_6$H$_5$	+ 6.5
u	If R—C=O radical is next to COOH radical, as in R—C(=O)······ ⁚ ······—C(=O)OH	+ 13
v	If R—C(R)(OH) radical is next to COOH, as in R—C(R)(OH)— ⁚ —C(=O)OH	+ 6.5
w	If R—C=O is next to CO, as in R.C(=O)····· ⁚ —C(=O)·····R	+ 6.5
x	For —C=C— bond in *cis* compounds	+ 16.5
y	For —C=C— bond in *trans* compounds	+ 13.0
z	For trimethylene ring in carboxylic acids, as in	+ 13.0
aa	For cyclobutane ring in carboxylic acids	+ 13.0
bb	Acid anhydride R.C(=O)—O—C(=O)—R	+ 10.0

(*Table continued*)

TABLE 74 (continued)

Symbol	Pair of electrons held between–	For each such grouping add or substract as indicated
cc	Lactone $\text{HC}-\text{C}-\text{C}-\text{C}=\text{O}$ with H H H H H and O bridge	+ 13.0
dd	Esters (Aliphatic) $\text{R}-\text{C}(=\text{O})-\text{O}-\text{R}$	+ 16.5
ee	Aliphatic radical $\overset{\circ}{_\circ}\text{NH}_2$ (Primary aliphatic amine)	+ 13.0
ff	Aliphatic radical $\overset{\circ}{_\circ}\overset{H}{N}\overset{\circ}{_\circ}$—Aliphatic radical (Secondary aliphatic amine)	+ 19.5
gg	Aliphatic radical $\overset{\circ}{_\circ}N\overset{\circ}{_\circ}$(Aliphatic radical)$_2$ (Tertiary aliphatic amines)	+ 26.0
hh	Aromatic radical carbon $\overset{\circ}{_\circ}$—N (Ammonia type of nitrogen)	− 3.5
jj	Aromatic radical $\overset{\circ}{_\circ}$—N—N—H with H on first N (Primary aromatic amine)	+ 6.5
kk	Aromatic radical $\overset{\circ}{_\circ}\overset{H}{N}\overset{\circ}{_\circ}$—Aromatic radical (Secondary aromatic amine)	+ 13.0
ll	Aromatic radical $\overset{\circ}{_\circ}N\overset{\circ}{_\circ}$(Aromatic radical)$_2$	+ 19.5
mm	For substituted amides, as in O=C and $\overset{R}{N}\overset{\circ}{_\circ}-\overset{H}{N}-\text{CH}_3$	+ 6.5
nn	For Nitrile radical aliphatic or aromatic RC≡N	+ 16.5
pp	Aromatic radical carbon $\overset{\circ}{_\circ}$C≡N	− 6.5
qq	For Carbylamine radical, aliphatic R—N=C	+ 33.1
rr	Aliphatic radical $\overset{\circ}{_\circ}$N with =O and O	+ 13.0
ss	Aromatic radical $\overset{\circ}{_\circ}$N with =O and O	+ 13.0

[a] M. S. Kharasch, *J. Research Natl. Bur. Standards* **2**, 359 (1929).   [b] Units, kcal./mole

TABLE 75

Atomic and Structural Parachor Contributions

Carbon	4.8	Double bond	23.2
Hydrogen	17.1	Triple bond	46.6
Nitrogen	12.5	3-membered ring	16.7
Oxygen	20.0	4-membered ring	11.6
Chlorine	54.3	5-membered ring	8.5
Bromine	68.0	6-membered ring	6.1
Iodine	91.0	Naphthalene ring	12.2
Fluorine	25.7	Semipolar bond	−1.6
Sulfur	48.2	Esters	−3.2
Phosphorus	37.7		

TABLE 76

Generalized Bonding Frequencies and Constants to Evaluate Einstein Functions, $(C_p°)_i = a_i + b_iT + c_iT^2$.[a,b]

Bonding	ν Frequency, cm.$^{-1}$	a_i	$b_i \times 10^3$	$c_i \times 10^6$	δ Frequency, cm.$^{-1}$	a_i	$b_i \times 10^3$	$c_i \times 10^6$
C—I, S—S	500	0.181	4.664	−3.338	260	1.461	1.730	−1.272
C—Br	560	−0.073	5.158	−3.591	280	1.242	2.046	−1.501
C—Cl, C—S	650	−0.562	6.385	−4.495	330	1.023	2.590	−1.874
C—C, C—N, N—N	990	−1.090	6.000	−3.441	390	0.730	3.414	−2.577
C—O, N—O	1030	−1.173	6.132	−3.555	205	1.461	1.633	−1.414
C—F, C=S	1050	−1.128	5.845	−3.253	530	0.011	5.119	−3.699
C=C, C=N	1620	−0.432	1.233	0.935	845	−1.140	7.254	−4.936
C=O, N=O	1700	−0.324	0.724	1.308	390	0.730	3.414	−2.577
S—H	2570	0.129	−1.333	2.263	1050	−1.128	5.845	−3.253
C—H, al. N—H	2920	0.229	−1.224	1.658	1320	−0.938	3.900	−1.342
C—H arom.	3050	0.171	−0.934	1.284	1320	−0.938	3.900	−1.342
O—H	3420	0.150	−0.810	1.055	1150	−1.135	5.363	−2.740

[a] C. J. Dobratz, *Ind. Eng. Chem.* **33**, 759 (1941).
[b] Units, cal/deg. mole.

TABLE 77

Contribution of Generalized Bond Vibrational Frequencies to Heat Capacity[a,b,c]

$T°$ K	C—H (Al) $\nu=2914$	C—C (Al) $\nu=989$	C=C (Al)(Sy) $\nu=1618$	C=C (Al)(Un) $\nu=1664$	C≡C (Al) $\nu=1115$	C—H (Ar) $\nu=3045$	C—C (Ar) $\nu=989$	C=C (Ar) $\nu=1618$	C≡C $\nu=2215$
250	0.0000	0.2214	0.0161	0.0129	0.1260	0.0000	0.2214	0.0161	0.0009
300	0.0004	0.3991	0.0520	0.0442	0.2765	0.0002	0.3991	0.0520	0.0055
350	0.0018	0.5866	0.1156	0.1009	0.4388	0.0012	0.5866	0.1156	0.0166
400	0.0063	0.7634	0.2040	0.1827	0.6040	0.0043	0.7634	0.2040	0.0445
450	0.0159	0.9206	0.3076	0.2810	0.7606	0.0111	0.9206	0.3076	0.0840
500	0.0324	1.0569	0.4210	0.3899	0.9010	0.0244	1.0569	0.4210	0.1398
600	0.0907	1.2721	0.6489	0.6126	1.1338	0.0728	1.2721	0.6489	0.2821
700	0.1812	1.4268	0.8548	0.8169	1.3084	0.1510	1.4268	0.8548	0.4466
800	0.2940	1.5383	1.0290	0.9942	1.4390	0.2544	1.5383	1.0290	0.6743
900	0.4187	1.6218	1.1732	1.1415	1.5365	0.3716	1.6218	1.1732	0.7696
1000	0.5474	1.6848	1.2923	1.2620	1.6120	0.4927	1.6848	1.2923	0.9104
1100	0.6713	1.7327	1.3894	1.3623	1.6707	0.6136	1.7327	1.3894	1.0345
1200	0.7868	1.7710	1.4680	1.4447	1.7170	0.7292	1.7710	1.4680	1.1423
1300	0.8932	1.8007	1.5335	1.5120	1.7542	0.8360	1.8007	1.5335	1.2356
1400	0.9907	1.8249	1.5881	1.5683	1.7841	0.9346	1.8249	1.5881	1.3156
1500	1.0785	1.8441	1.6339	1.6162	1.8084	1.0238	1.8441	1.6339	1.3852
	$\delta=1247$	$\delta=390$	$\delta=599$	$\delta=421$	$\delta=470$	$\delta=1318$	$\delta=390$	$\delta=844$	$\delta=333$
250	0.0793	1.3296	0.8044	1.2491	1.1190	0.0591	1.3296	0.3735	1.4782
300	0.1826	1.4980	1.0443	1.4336	1.3250	0.1450	1.4980	0.5922	1.6152
350	0.3160	1.6120	1.2298	1.5593	1.4702	0.2640	1.6120	0.7966	1.7052
400	0.4638	1.6920	1.3710	1.6490	1.5750	0.3996	1.6920	0.9740	1.7665
450	0.8109	1.7496	1.4786	1.7140	1.6530	0.5420	1.7496	1.1235	1.8103
500	0.7514	1.7916	1.5614	1.7624	1.7107	0.6790	1.7916	1.2468	1.8423
600	0.9937	1.8481	1.6793	1.8272	1.7900	0.9220	1.8481	1.4305	1.8853
700	1.1833	1.8844	1.7553	1.8686	1.8388	1.1190	1.8844	1.5571	1.9114
800	1.3305	1.9080	1.8060	1.8957	1.8738	1.2735	1.9080	1.6474	1.9290
900	1.4438	1.9238	1.8415	1.9162	1.8967	1.3940	1.9238	1.7126	1.9409
1000	1.5313	1.9354	1.8691	1.9272	1.9134	1.4870	1.9354	1.7613	1.9497
1100	1.6001	1.9446	1.8914	1.9375	1.9257	1.5623	1.9446	1.7977	1.9560
1200	1.6566	1.9510	1.9045	1.9456	1.9350	1.6235	1.9510	1.8263	1.9609
1300	1.7011	1.9565	1.9161	1.9514	1.9428	1.6711	1.9565	1.8483	1.9648
1400	1.7371	1.9610	1.9253	1.9564	1.9487	1.7107	1.9610	1.8678	1.9679
1500	1.7674	1.9650	1.9331	1.9606	1.9536	1.7438	1.9650	1.8825	1.9703

[a] D. R. Stull and F. D. Mayfield, *Ind. Eng. Chem.* **35,** 639 (1943); D. R. Stull, *ibid.* **35,** 1301 (1943).

[b] Naphthenic ring bondings were treated as aromatic.

[c] Units, cal/deg. mole.

TABLE 78
Heat Capacity Solutions to Einstein Functions[a,b]
(one degree of freedom)

x^c	C_{vib}	x	C_{vib}	x	C_{vib}	x	C_{vib}
0.000	1.98700	0.200	1.98039	0.400	1.96072	0.600	1.92845
0.005	1.98699	0.205	1.98006	0.405	1.96006	0.605	1.92748
0.010	1.98698	0.210	1.97971	0.410	1.95940	0.610	1.92652
0.015	1.98696	0.215	1.97936	0.415	1.95873	0.615	1.92554
0.020	1.98693	0.220	1.97900	0.420	1.95805	0.620	1.92455
0.025	1.98690	0.225	1.97864	0.425	1.95736	0.625	1.92356
0.030	1.98685	0.230	1.97826	0.430	1.95666	0.630	1.92256
0.035	1.98680	0.235	1.97788	0.435	1.95596	0.635	1.92156
0.040	1.98674	0.240	1.97749	0.440	1.95525	0.640	1.92054
0.045	1.98666	0.245	1.97709	0.445	1.95453	0.645	1.91952
0.050	1.98659	0.250	1.97668	0.450	1.95381	0.650	1.91849
0.055	1.98650	0.255	1.97627	0.455	1.95307	0.655	1.91746
0.060	1.98640	0.260	1.97584	0.460	1.95233	0.660	1.91642
0.065	1.98630	0.265	1.97541	0.465	1.95158	0.665	1.91537
0.070	1.98619	0.270	1.97497	0.470	1.95082	0.670	1.91431
0.075	1.98607	0.275	1.97453	0.475	1.95006	0.675	1.91324
0.080	1.98594	0.280	1.97407	0.480	1.94928	0.680	1.91217
0.085	1.98580	0.285	1.97361	0.485	1.94850	0.685	1.91109
0.090	1.98566	0.290	1.97313	0.490	1.94772	0.690	1.91001
0.095	1.98551	0.295	1.97265	0.495	1.94692	0.695	1.90891
0.100	1.98534	0.300	1.97216	0.500	1.94612	0.700	1.90781
0.105	1.98518	0.305	1.97167	0.505	1.94530	0.705	1.90671
0.110	1.98500	0.310	1.97116	0.510	1.94449	0.710	1.90559
0.115	1.98481	0.315	1.97065	0.515	1.94366	0.715	1.90447
0.120	1.98462	0.320	1.97013	0.520	1.94283	0.720	1.90334
0.125	1.98441	0.325	1.96960	0.525	1.94198	0.725	1.90221
0.130	1.98420	0.330	1.96906	0.530	1.94113	0.730	1.90106
0.135	1.98399	0.335	1.96852	0.535	1.94028	0.735	1.89991
0.140	1.98376	0.340	1.96797	0.540	1.93941	0.740	1.89875
0.145	1.98352	0.345	1.96741	0.545	1.93854	0.745	1.89759
0.150	1.98328	0.350	1.96684	0.550	1.93766	0.750	1.89642
0.155	1.98303	0.355	1.96626	0.555	1.93677	0.755	1.89524
0.160	1.98277	0.360	1.96568	0.560	1.93588	0.760	1.89406
0.165	1.98250	0.365	1.96509	0.565	1.93497	0.765	1.89287
0.170	1.98222	0.370	1.96449	0.570	1.93406	0.770	1.89167
0.175	1.98194	0.375	1.96388	0.575	1.93315	0.775	1.89065
0.180	1.98164	0.380	1.96326	0.580	1.93222	0.780	1.88924
0.185	1.98134	0.385	1.96264	0.585	1.93129	0.785	1.88803
0.190	1.98103	0.390	1.96201	0.590	1.93035	0.790	1.88680
0.195	1.98071	0.395	1.96137	0.595	1.92940	0.795	1.88557

[a] D. R. Stull and F. D. Mayfield, *Ind. Eng. Chem.* **35**, 639 (1943).

[b] Units, cal./deg.mole.

[c] $x = h\nu_0/kT$, ν_0 is in sec^{-1}.

TABLE 78 (continued)

x	C_{vib}	x	C_{vib}	x	C_{vib}	x	C_{vib}
0.800	1.88433	1.025	1.82180	1.250	1.74730	1.475	1.66280
0.805	1.88309	1.030	1.82027	1.255	1.74552	1.480	1.66082
0.810	1.88183	1.035	1.81873	1.260	1.74374	1.485	1.65884
0.815	1.88057	1.040	1.81719	1.265	1.74195	1.490	1.65685
0.820	1.87931	1.045	1.81564	1.270	1.74016	1.495	1.65487
0.825	1.87803	1.050	1.81408	1.275	1.73837	1.500	1.65288
0.830	1.87675	1.055	1.81252	1.280	1.73657	1.505	1.65089
0.835	1.87547	1.060	1.81095	1.285	1.73476	1.510	1.64889
0.840	1.87417	1.065	1.80938	1.290	1.73295	1.515	1.64689
0.845	1.87287	1.070	1.80780	1.295	1.73113	1.520	1.64488
0.850	1.87157	1.075	1.80622	1.300	1.72929	1.525	1.64287
0.855	1.87025	1.080	1.80463	1.305	1.72749	1.530	1.64086
0.860	1.86893	1.085	1.80303	1.310	9.72566	1.535	1.63884
0.865	1.86761	1.090	1.80143	1.315	1.72382	1.540	1.63682
0.870	1.86627	1.095	1.79982	1.320	1.72199	1.545	1.63480
0.875	1.86493	1.100	1.79817	1.325	1.72014	1.550	1.63277
0.880	1.86359	1.105	1.79659	1.330	1.71829	1.555	1.63074
0.885	1.86224	1.110	1.79496	1.335	1.71644	1.560	1.62871
0.890	1.86088	1.115	1.79333	1.340	1.71458	1.565	1.62667
0.895	1.85951	1.120	1.79170	1.345	1.71272	1.570	1.62463
0.900	1.85814	1.125	1.79006	1.350	1.71085	1.575	1.62258
0.905	1.85676	1.130	1.78841	1.355	1.70898	1.580	1.62053
0.910	1.85538	1.135	1.78676	1.360	1.70711	1.585	1.61848
0.915	1.85399	1.140	1.78510	1.365	1.70523	1.590	1.61643
0.920	1.85259	1.145	1.78344	1.370	1.70334	1.595	1.61437
0.925	1.85118	1.150	1.78177	1.375	1.70146	1.600	1.61231
0.930	1.84977	1.155	1.78010	1.380	1.69956	1.605	1.61024
0.935	1.84836	1.160	1.77842	1.385	1.69763	1.610	1.60818
0.940	1.84693	1.165	1.77673	1.390	1.69576	1.615	1.60610
0.945	1.84551	1.170	1.77504	1.395	1.69386	1.620	1.60403
0.950	1.84407	1.175	1.77335	1.400	1.69195	1.625	1.60195
0.955	1.84263	1.180	1.77165	1.405	1.69003	1.630	1.59987
0.960	1.84118	1.185	1.76994	1.410	1.68811	1.635	1.59779
0.965	1.83973	1.190	1.76823	1.415	1.68619	1.640	1.59570
0.970	1.83827	1.195	1.76651	1.420	1.68426	1.645	1.59361
0.975	1.83680	1.200	1.76479	1.425	1.68233	1.650	1.59151
0.980	1.83533	1.205	1.76307	1.430	1.68040	1.655	1.58942
0.985	1.83385	1.210	1.76134	1.435	1.67846	1.660	1.58732
0.990	1.83233	1.215	1.75960	1.440	1.67651	1.665	1.58522
0.995	1.83087	1.220	1.75786	1.445	1.67456	1.670	1.58311
1.000	1.82938	1.225	1.75611	1.450	1.67262	1.675	1.58100
1.005	1.82787	1.230	1.75436	1.455	1.67066	1.680	1.57889
1.010	1.82640	1.235	1.75260	1.460	1.66870	1.685	1.57678
1.015	1.82485	1.240	1.75084	1.465	1.66674	1.690	1.57466
1.020	1.82333	1.245	1.74907	1.470	1.66477	1.695	1.57254

(Table continued)

TABLE 78 (*continued*)

x	C_{vib}	x	C_{vib}	x	C_{vib}	x	C_{vib}
1.700	1.57042	1.925	1.47231	2.150	1.37061	2.375	1.26729
1.705	1.56829	1.930	1.47009	2.155	1.36832	2.380	1.26499
1.710	1.56616	1.935	1.46785	2.160	1.36603	2.385	1.26269
1.715	1.56403	1.940	1.46562	2.165	1.36375	2.390	1.26039
1.720	1.56189	1.945	1.46339	2.170	1.36146	2.395	1.25809
1.725	1.55976	1.950	1.46115	2.175	1.35917	2.400	1.25579
1.730	1.55762	1.955	1.45892	2.180	1.35688	2.405	1.25349
1.735	1.55548	1.960	1.45668	2.185	1.35459	2.410	1.25119
1.740	1.55333	1.965	1.45444	2.190	1.35230	2.415	1.24889
1.745	1.55118	1.970	1.45219	2.195	1.35001	2.420	1.24659
1.750	1.54903	1.975	1.44995	2.200	1.34772	2.425	1.24429
1.755	1.54688	1.980	1.44771	2.205	1.34543	2.430	1.24200
1.760	1.54472	1.985	1.44546	2.210	1.34313	2.435	1.23970
1.765	1.54257	1.990	1.44321	2.215	1.34084	2.440	1.23740
1.770	1.54041	1.995	1.44096	2.220	1.33855	2.445	1.23510
1.775	1.53824	2.000	1.43871	2.225	1.33625	2.450	1.23281
1.780	1.53608	2.005	1.43646	2.230	1.33396	2.455	1.23051
1.785	1.53391	2.010	1.43420	2.235	1.33166	2.460	1.22821
1.790	1.53174	2.015	1.43195	2.240	1.32937	2.465	1.22592
1.795	1.52957	2.020	1.42969	2.245	1.32707	2.470	1.22362
1.800	1.52739	2.025	1.42743	2.250	1.32477	2.475	1.22133
1.805	1.52521	2.030	1.42517	2.255	1.32248	2.480	1.21904
1.810	1.52303	2.035	1.42291	2.260	1.32018	2.485	1.21674
1.815	1.52085	2.040	1.42065	2.265	1.31788	2.490	1.21445
1.820	1.51867	2.045	1.41839	2.270	1.31558	2.495	1.21216
1.825	1.51648	2.050	1.41612	2.275	1.31329	2.500	1.20986
1.830	1.51429	2.055	1.41386	2.280	1.31099	2.505	1.20757
1.835	1.51210	2.060	1.41159	2.285	1.30869	2.510	1.20528
1.840	1.50991	2.065	1.40932	2.290	1.30639	2.515	1.20299
1.845	1.50772	2.070	1.40705	2.295	1.30409	2.520	1.20070
1.850	1.50552	2.075	1.40478	2.300	1.30179	2.525	1.19841
1.855	1.50332	2.080	1.40251	2.305	1.29949	2.530	1.19612
1.860	1.50112	2.085	1.40024	2.310	1.29719	2.535	1.19384
1.865	1.49891	2.090	1.39797	2.315	1.29489	2.540	1.19155
1.870	1.49671	2.095	1.39569	2.320	1.29259	2.545	1.18926
1.875	1.49450	2.100	1.39341	2.325	1.29029	2.550	1.18698
1.880	1.49229	2.105	1.39114	2.330	1.28799	2.555	1.18469
1.885	1.49008	2.110	1.38886	2.335	1.28569	2.560	1.18241
1.890	1.48786	2.115	1.38658	2.340	1.28339	2.565	1.18012
1.895	1.48565	2.120	1.38430	2.345	1.28109	2.570	1.17784
1.900	1.48343	2.125	1.38202	2.350	1.27879	2.575	1.17556
1.905	1.48121	2.130	1.37974	2.355	1.27649	2.580	1.17328
1.910	1.47899	2.135	1.37746	2.360	1.27419	2.585	1.17100
1.915	1.47677	2.140	1.37517	2.365	1.27189	2.590	1.16872
1.920	1.47454	2.145	1.37289	2.370	1.26959	2.595	1.16644

II. NUMERICAL DATA

x	$c_{vib.}$	x	$c_{vib.}$	x	$c_{vib.}$	x	$c_{vib.}$
2.600	1.16416	2.825	1.06282	3.10	0.94329	3.55	0.76247
2.605	1.16189	2.830	1.06059	3.11	0.93906	3.56	0.75870
2.610	1.15961	2.835	1.05837	3.12	0.93484	3.57	0.75493
2.615	1.15733	2.840	1.05616	3.13	0.93062	3.58	0.75118
2.620	1.15506	2.845	1.05394	3.14	0.92641	3.59	0.74744
2.625	1.15279	2.850	1.05173	3.15	0.92222	3.60	0.74371
2.630	1.15051	2.855	1.04951	3.16	0.91803	3.61	0.74000
2.635	1.14824	2.860	1.04730	3.17	0.91385	3.62	0.73629
2.640	1.14597	2.865	1.04509	3.18	0.90968	3.63	0.73260
2.645	1.14370	2.870	1.04288	3.19	0.90552	3.64	0.72891
2.650	1.14144	2,875	1.04068	3.20	0.90137	3.65	0.72524
2.655	1.13917	2.880	1.03847	3.21	0.89723	3.66	0.72158
2.660	1.13690	2.885	1.03627	3.22	0.89310	3.67	0.71793
2.655	1.13464	2.890	1.03407	3.23	0.88897	3.68	0.71430
2.670	1.13237	2.895	1.03187	3.24	0.88486	3.69	0.71067
2.675	1.13011	2.900	1.02967	3.25	0.88076	3.70	0.70706
2.680	1.12785	2.905	1.02747	3.26	0.87666	3.71	0.70346
2.685	1.12559	2.910	1.02528	3.27	0.87258	3.72	0.69987
2.690	1.12333	2.915	1.02308	3.28	0.86851	3.73	0.69629
2.695	1.12107	2.920	1.02089	3.29	0.86444	3.74	0.69273
2.700	1.11882	2.925	1.01870	3.30	0.86039	3.75	0.68917
2.705	1.11656	2.930	1.01652	3.31	0.85635	3.76	0.68563
2.710	1.11431	2.935	1.01433	3.32	0.85231	3.77	0.68210
2.715	1.11205	2.940	1.01215	3.33	0.84829	3.78	0.67858
2.720	1.10980	2.945	1.00996	3.34	0.84428	3.79	0.67507
2.725	1.10755	2.950	1.00778	3.35	0.84028	3.80	0.67158
2.730	1.10530	2.955	1.00561	3.36	0.83628	3.81	0.66810
2.735	1.10305	2.960	1.00343	3.37	0.83230	3.82	0.66463
2.740	1.10080	2.965	1.00125	3.38	0.82833	3.83	0.66117
2.745	1.09856	2.970	0.99908	3.39	0.82437	3.84	0.65772
2.750	1.09631	2.975	0.99691	3.40	0.82042	3.85	0.65429
2.755	1.09407	2.980	0.99474	3.41	0.81648	3.86	0.65086
2.760	1.09183	2.985	0.99258	3.42	0.81255	3.87	0.64745
2.765	1.08959	3.990	0.99041	3.43	0.80863	3.88	0.64406
2.770	1.08735	2.995	0.98825	3.44	0.80473	3.89	0.64067
2.775	1.08511	3.00	0.98609	3.45	0.80083	3.90	0.63730
2.780	1.08287	3.01	0.98177	3.46	0.79695	3.91	0.63393
2.785	1.08064	3.02	0.97746	3.47	0.79307	3.92	0.63058
2.790	1.07835	3.03	0.97316	3.48	0.78921	3.93	0.62725
2.795	1.07618	3.04	0.96887	3.49	0.78536	3.94	0.62392
2.800	1.07395	3.05	0.96458	3.50	0.78151	3.95	0.62061
2.805	1.07172	3.06	0.96031	3.51	0.77768	3.96	0.61730
2.810	1.06949	3.07	0.95604	3.52	0.77386	3.97	0.61402
2.815	1.06726	3.08	0.95178	3.53	0.77006	3.98	0.61074
2.820	1.06504	3.09	0.94753	3.54	0.76626	3.99	0.60747

(Table continued)

TABLE 78 (*continued*)

x	$C_{vib.}$	x	$C_{vib.}$	x	$C_{vib.}$	x	$C_{vib.}$
4.00	0.60422	4.45	0.47045	4.90	0.36061	5.35	0.27262
4.01	0.60098	4.46	0.46775	4.91	0.35843	5.36	0.27090
4.02	0.59775	4.47	0.46507	4.92	0.35626	5.37	0.26918
4.03	0.59454	4.48	0.46240	4.93	0.35409	5.38	0.26747
4.04	0.59133	4.49	0.45974	4.94	0.35194	5.39	0.26576
4.05	0.58814	4.50	0.45709	4.95	0.34980	5.40	0.26407
4.06	0.58496	4.51	0.45445	4.96	0.34767	5.41	0.26239
4.07	0.58179	4.52	0.45183	4.97	0.34556	5.42	0.26072
4.08	0.57864	4.53	0.44921	4.98	0.34345	5.43	0.25906
4.09	0.57550	4.54	0.44661	4.99	0.34135	5.44	0.25740
4.10	0.57236	4.55	0.44402	5.00	0.33926	5.45	0.25576
4.11	0.56925	4.56	0.44145	5.01	0.33719	5.46	0.25412
4.12	0.56614	4.57	0.43888	5.02	0.33512	5.47	0.25249
4.13	0.56305	4.58	0.43633	5.03	0.33307	5.48	0.25087
4.14	0.55996	4.59	0.43379	5.04	0.33102	5.49	0.24926
4.15	0.55690	4.60	0.43125	5.05	0.32899	5.50	0.24766
4.16	0.55384	4.61	0.42873	5.06	0.32696	5.51	0.24607
4.17	0.55079	4.62	0.42623	5.07	0.32495	5.52	0.24449
4.18	0.54776	4.63	0.42373	5.08	0.32294	5.53	0.24291
4.19	0.54474	4.64	0.42125	5.09	0.32095	5.54	0.24135
4.20	0.54173	4.65	0.41877	5.10	0.31897	5.55	0.23979
4.21	0.53873	4.66	0.41631	5.11	0.31700	5.56	0.23824
4.22	0.53575	4.67	0.41386	5.12	0.31503	5.57	0.23670
4.23	0.53276	4.68	0.41142	5.13	0.31308	5.58	0.23517
4.24	0.52982	4.69	0.40900	5.14	0.31114	5.59	0.23365
4.25	0.52687	4.70	0.40658	5.15	0.30920	5.60	0.23213
4.26	0.52393	4.71	0.40418	5.16	0.30728	5.61	0.23063
4.27	0.52101	4.72	0.40178	5.17	0.30537	5.62	0.22913
4.28	0.51810	4.73	0.39940	5.18	0.30347	5.63	0.22764
4.29	0.51520	4.74	0.39703	5.19	0.30158	5.64	0.22617
4.30	0.51231	4.75	0.39467	5.20	0.29970	5.65	0.22469
4.31	0.50944	4.76	0.39232	5.21	0.29782	5.66	0.22323
4.32	0.50658	4.77	0.38998	5.22	0.29596	5.67	0.22177
4.33	0.50372	4.78	0.38766	5.23	0.29411	5.68	0.22033
4.34	0.50088	4.79	0.38534	5.24	0.29226	5.69	0.21889
4.35	0.49806	4.80	0.38304	5.25	0.29043	5.70	0.21746
4.36	0.49524	4.81	0.38075	5.26	0.28861	5.71	0.21604
4.37	0.49244	4.82	0.37847	5.27	0.28679	5.72	0.21462
4.38	0.48965	4.83	0.37620	5.28	0.28499	5.73	0.21322
4.39	0.48687	4.84	0.37394	5.29	0.28319	5.74	0.21182
4.40	0.48401	4.85	0.37169	5.30	0.28141	5.75	0.21043
4.41	0.48135	4.86	0.36945	5.31	0.27963	5.76	0.20905
4.42	0.47860	4.87	0.36722	5.32	0.27787	5.77	0.20767
4.43	0.47587	4.88	0.36501	5.33	0.27611	5.78	0.20631
4.44	0.47315	4.89	0.36280	5.34	0.27436	5.79	0.20495

II. NUMERICAL DATA

x	$C_{vib.}$	x	$C_{vib.}$	x	$C_{vib.}$	x	$C_{vib.}$
5.80	0.20360	6.25	0.15042	6.70	0.11006	7.15	0.07985
5.81	0.20226	6.26	0.14939	6.71	0.10929	7.16	0.07928
5.82	0.20092	6.27	0.14837	6.72	0.10852	7.17	0.07871
5.83	0.19960	6.28	0.14736	6.73	0.10774	7.18	0.07814
5.84	0.19828	6.29	0.14635	6.74	0.10700	7.19	0.07758
5.85	0.19697	6.30	0.14535	6.75	0.10625	7.20	0.07702
5.86	0.19566	6.31	0.14436	6.76	0.10550	7.21	0.07646
5.87	0.19437	6.32	0.14337	6.77	0.10476	7.22	0.07591
5.88	0.19308	6.33	0.14239	6.78	0.10402	7.23	0.07536
5.89	0.19180	6.34	0.14141	6.79	0.10329	7.24	0.07482
5.90	0.19052	6.35	0.14044	6.80	0.10256	7.25	0.07428
5.91	0.18926	6.36	0.13948	6.81	0.10184	7.26	0.07374
5.92	0.18800	6.37	0.13852	6.82	0.10112	7.27	0.07321
5.93	0.18675	6.38	0.13757	6.83	0.10040	7.28	0.07268
5.94	0.18550	6.39	0.13662	6.84	0.09969	7.29	0.07215
5.95	0.18427	6.40	0.13568	6.85	0.09899	7.30	0.07164
5.96	0.18304	6.41	0.13475	6.86	0.09829	7.31	0.07111
5.97	0.18181	6.42	0.13382	6.87	0.09759	7.32	0.07059
5.98	0.18060	6.43	0.13289	6.88	0.09690	7.33	0.07008
5.99	0.17939	6.44	0.13198	6.89	0.09621	7.34	0.06957
6.00	0.17819	6.45	0.13107	6.90	0.09533	7.35	0.06907
6.01	0.17700	6.46	0.13016	6.91	0.09485	7.36	0.06857
6.02	0.17581	6.47	0.12926	6.92	0.09418	7.37	0.06807
6.03	0.17463	6.48	0.12837	6.93	0.09351	7.38	0.06757
6.04	0.17346	6.49	0.12748	6.94	0.09284	7.39	0.06708
6.05	0.17230	6.50	0.12660	6.95	0.09218	7.40	0.06659
6.06	0.17114	6.51	0.21572	6.96	0.09153	7.41	0.06611
6.07	0.16999	6.52	0.21485	6.97	0.09088	7.42	0.06562
6.08	0.16884	6.53	0.12398	6.98	0.09023	7.43	0.06515
6.09	0.16771	6.54	0.12312	6.99	0.08958	7.44	0.06467
6.10	0.16658	6.55	0.12226	7.00	0.08895	7.45	0.06420
6.11	0.16545	6.56	0.12141	7.01	0.08831	7.46	0.06373
6.12	0.16433	6.57	0.12057	7.02	0.08768	7.47	0.06326
6.13	0.16322	6.58	0.11973	7.03	0.08705	7.48	0.06280
6.14	0.16212	6.59	0.11889	7.04	0.08643	7.49	0.06234
6.15	0.16102	6.60	0.11807	7.05	0.08581	7.50	0.06189
6.16	0.15993	6.61	0.11724	7.06	0.08520	7.51	0.06143
6.17	0.15885	6.62	0.11642	7.07	0.08459	7.52	0.06098
6.18	0.15779	6.63	0.11561	7.08	0.08398	7.53	0.06054
6.19	0.15670	6.64	0.11480	7.09	0.08338	7.54	0.06009
6.20	0.15564	6.65	0.11400	7.10	0.08278	7.55	0.05965
6.21	0.15458	6.66	0.11320	7.11	0.08219	7.56	0.05921
6.22	0.15353	6.67	0.11241	7.12	0.08160	7.57	0.05878
6.23	0.15249	6.68	0.11162	7.13	0.08101	7.58	0.05835
6.24	0.15145	6.69	0.11084	7.14	0.08043	7.59	0.05792

(Table continued)

TABLE 78 (*continued*)

x	$C_{vib.}$	x	$C_{vib.}$	x	$C_{vib.}$	x	$C_{vib.}$
7.60	0.05749	8.25	0.03535	10.50	0.00603	12.75	0.00094
7.61	0.05707	8.30	0.03404	10.55	0.00579	12.80	0.00090
7.62	0.05665	8.35	0.03277	10.60	0.00556	12.85	0.00086
7.63	0.05623	8.40	0.03154	10.65	0.00534	12.90	0.00083
7.64	0.05582	8.45	0.03036	10.70	0.00513	12.95	0.00079
7.65	0.05541	8.50	0.02922	10.75	0.00492	13.00	0.00076
7.66	0.05500	8.55	0.02812	10.80	0.00473	13.05	0.00073
7.67	0.05459	8.60	0.02706	10.85	0.00454	13.10	0.00070
7.68	0.05419	8.65	0.02605	10.90	0.00436	13.15	0.00067
7.69	0.05379	8.70	0.02506	10.95	0.00418	13.20	0.00064
7.70	0.05339	8.75	0.02411	11.00	0.00402	13.25	0.00061
7.71	0.05300	8.80	0.02320	11.05	0.00385	13.30	0.00059
7.72	0.05261	8.85	0.02232	11.10	0.00370	13.35	0.00056
7.73	0.05222	8.90	0.02147	11.15	0.00355	13.40	0.00054
7.74	0.05183	8.95	0.02065	11.20	0.00341	13.45	0.00052
7.75	0.05145	9.00	0.01987	11.25	0.00327	13.50	0.00050
7.76	0.05107	9.05	0.01911	11.30	0.00314	13.55	0.00047
7.77	0.05069	9.10	0.01838	11.35	0.00301	13.60	0.00045
7.78	0.05032	9.15	0.01767	11.40	0.00289	13.65	0.00044
7.79	0.04994	9.20	0.01700	11.45	0.00277	13.70	0.00042
7.80	0.04957	9.25	0.01634	11.50	0.00266	13.75	0.00040
7.81	0.04921	9.30	0.01572	11.55	0.00255	13.80	0.00038
7.82	0.04884	9.35	0.01511	11.60	0.00245	13.85	0.00037
7.83	0.04848	9.40	0.01453	11.65	0.00235	13.90	0.00035
7.84	0.04812	9.45	0.01396	11.70	0.00226	13.95	0.00034
7.85	0.04776	9.50	0.01342	11.75	0.00218	14.00	0.00032
7.86	0.04741	9.55	0.01291	11.80	0.00208	14.05	0.00031
7.87	0.04705	9.60	0.01240	11.85	0.00199	14.10	0.00030
7.88	0.04670	9.65	0.01192	11.90	0.00191	14.15	0.00028
7.89	0.04635	9.70	0.01146	11.95	0.00183	14.20	0.00027
7.90	0.04601	9.75	0.01101	12.00	0.00176	14.25	0.00026
7.91	0.04567	9.80	0.01058	12.05	0.00169	14.30	0.00025
7.92	0.04533	9.85	0.01017	12.10	0.00162	14.35	0.00024
7.93	0.04499	9.90	0.00977	12.15	0.00155	14.40	0.00023
7.94	0.04465	9.95	0.00939	12.20	0.00149	14.45	0.00022
7.95	0.04432	10.00	0.00902	12.25	0.00143	14.50	0.00021
7.96	0.04399	10.05	0.00867	12.30	0.00137	14.55	0.00020
7.97	0.04366	10.10	0.00833	12.35	0.00131	14.60	0.00019
7.98	0.04333	10.15	0.00800	12.40	0.00126	14.65	0.00018
7.99	0.04301	10.20	0.00768	12.45	0.00121	14.70	0.00018
8.00	0.04269	10.25	0.00738	12.50	0.00116	14.75	0.00017
8.05	0.04111	10.30	0.00709	12.55	0.00111	14.80	0.00016
8.10	0.03959	10.35	0.00681	12.60	0.00106	14.85	0.00015
8.15	0.03813	10.40	0.00654	12.65	0.00102	14.90	0.00015
8.20	0.03672	10.45	0.00628	12.70	0.00098	14.95	0.00014
						15.00	0.00014

Author Index

The numbers in parentheses are footnote numbers and are inserted to enable the reader to locate a cross reference when the author's name does not appear at the point of reference in the text.

A

Andersen, J. W., 57, 58, 61(33), 62, 67(33), 68, 70(33), 130, 179, 180, 181, 182, 183
Arnett, R. L., 6, 48(1), 57(1) 62(1), 71(1), 73(1), 74(1), 84(1), 87(1), 93(1), 94(1), 102(1), 106(1), 122(1), 124(1), 135(1), 137(1), 140(1)
Aston, J. G., 34, 64, 92, 137, 139(96), 140(96), 175

B

Barriol, J., 36
Barrow, G. M., 94
Beckett, C. W., 86, 91(48), 92(48), 133
Bellamy, L. J., 19
Bennewitz, K., 119
Benson, S. W., 76, 79(46a), 147, 204, 205, 206, 207
Bernstein, H. J., 175
Beyer, G. H., 57, 58, 61(33), 62(33), 67(33), 68(33), 70(33), 130(33), 179, 180, 181, 182, 183
Bichowsky, F. R., 61, 109
Billmeyer, F. W., Jr., 147
Blade, E., 34, 175
Bliss, R. H., 146
Braun, R. M., 6, 48(1), 57(1), 62(1), 71(1), 73(1), 74(1), 84(1), 87(1), 93(1), 94(1), 102(1), 106(1), 122(1), 124(1), 135(1), 137(1), 140(1)
Bremmer, J. G. M., 95, 96(54), 130
Brewer, L., 102
Brickwedde, F., 137, 139(96), 140(96)
Brown, J. M., 62, 179, 180, 181, 182, 183
Bryant, W. M. D., 79, 81(46b), 82(46b), 147, 208
Burrows, G. H., 135
Buss, J. H., 76, 79(46a), 147, 204, 205, 206, 207

C

Chermin, H. A. G., 58, 74, 125, 130, 147, 199, 200, 201, 203
Ciola, R., 83, 84(46e), 209, 210, 211, 212, 213, 214, 215
Cottrell, T. L., 100, 101(56), 151
Cross, P. C., 19

D

Dailey, B. P., 175
Dainton, F. S., 147
Davidson, F. G., 80
Dawson, J. P., 126
Dean, L. B., Jr., 30
Decius, J. C., 19
Deering, R. F., 136, 138(94), 139(94)
DeVries, T., 133
Dewar, M. J. S., 100
Dobratz, C. J., 62, 64(39), 120, 228
Dodge, B. F., 146, 147
Doriaswamy, L. K., 125, 126, 217, 218, 219, 220, 221, 222
Doumani, T. F., 136, 138(94), 139
Douslin, D. R., 126
Duncan, N. E., 31, 94(20), 136, 140, 144(100), 175

E

Eaton, M., 133
Egloff, G., 141
Epstein, M. B., 137
Evans, W. H., 6, 57(2), 70(2), 123(2), 124(2), 133(2)
Ewell, R. B., 21, 55, 56(31), 153
Eyring, H., 138

F

Farmer, J. B., 80

AUTHOR INDEX

Finke, H. L., 57, 75(32), 94, 137(46), 175
Fishtine, S. H., 113
Fitzgerald, W. E., 32, 95
Forsythe, W. R., 175
Franklin, J. L., 58, 67(34), 70, 106, 107, 108(62), 126, 130, 184, 185, 186, 187, 188, 203
Fraser, F. M., 111, 112(67), 113(67)
Freeman, N. K., 133
Frost, A. V., 135
Fugassi, P., 120

G

Gaydon, A. G., 102
George, A., 126
Giauque, F. W., 75, 175
Gilles, P., 102
Gittler, F. L., 175
Given, P. H., 147
Glasstone, S., 138
Goldbing, D. R. V., 175
Goldstein, J., 28
Good, W. D., 126
Gordon, D. G., 116, 118(76)
Gross, M. E., 57, 75(32), 94, 137(46), 175
Gurney, R. W., 13
Gwinn, W. D., 28, 169, 170, 171, 172, 173, 174

H

Halford, J. O., 34
Halverson, F., 140
Hammick, D. L. O., 147
Hawkins, P. J., 140
Henderson, I. H. S., 80
Herzberg, G., 13, 19(6)
Hildebrand, J. H., 115
Honig, R. E., 107
Hougen, O. A., 116, 118(76), 152
Hubbard, W. N., 175
Huffman, H. M., 11, 12, 50, 52(4), 57, 64, 75(32), 81, 91, 94, 130, 133, 137(46), 141, 175
Huggins, M. L., 45
Hurd, C. O., 58, 67(35), 72, 97(35), 126(35), 130(35), 189, 190, 191, 192, 193, 194, 195, 196, 197, 198

I

Ince, E. L., 28
Ivin, K. J., 147

J

Janz, G. J., 22, 31, 32, 33, 94(20), 95, 123, 133, 136, 140, 144, 145, 146, 147, 175
Jenkins, F. A., 102
Johnson, W. H., 104, 105(61)

K

Kemp, J. D., 27, 75
Kharasch, M. S., 61, 109, 110, 111(38), 224, 227
Kilpatrick, J. E., 70, 71(43), 74(43), 86, 88, 89(47), 90(47), 91(48), 92(48), 94(47), 130(43, 47)
Kimball, G. E., 34, 175
Kirkbride, F. W., 80
Kirkwood, J. G., 36
Kistiakowsky, G. B., 133, 136, 139(92), 140
Kistiakowsky, W., 115
Kuchler, L., 139

L

Lacher, J. R., 140
Lacina, J. L., 126
Laidler, J. C., 127
Laidler, K. J., 138, 223
Lassettre, E. L., 30
Lerner, R. G., 175
Levedahl, B. H., 175
Lewis, W. K., 115
Lielmezs, J., 85, 216
Lossing, F. P., 80
Lucarni, C., 135

M

McAdams, W. H., 115
McCulloch, W. J. G., 140, 145(105), 146
McCullough, J. P., 75, 137(46)
McKinnis, A. C., 136, 138(94), 139(94)
Marsden, D. G. H., 80

Matthews, C. S., 58, 67(35), 72, 97(35), 126(35), 130(35), 189, 190, 191, 192, 193, 194, 195, 196, 197, 198
Matthews, J. H., 133
Mayer, J. E., 26
Mayer, M. G., 26
Mayfield, F. D., 62, 121, 229, 230
Mecke, H., 119
Meissner, H. P., 116
Messerly, G. H., 175
Montgomery, J. B., 133
Morgan, J. P., 85, 216
Morrell, J. C., 115, 141
Morris, J. C., 175
Mortimer, C. T., 151
Moskow, M., 137, 139(96), 140(96)
Myers, R. E., 144

N

Nielsen, H. H., 28, 36
Nyholm, R. S., 147

O

Oliver, G. D., 91, 133, 175
Othmer, D. F., 115

P

Pace, E. L., 175
Pardee, W. A., 147
Parks, G. S., 11, 12, 50, 52(4), 64, 81, 130, 141
Pennington, R. E., 75, 137(46)
Pergiel, F. Y., 133
Person, W. B., 35, 39(26), 42(26), 45, 131, 176, 177, 178
Pimentel, G. C., 6, 35, 39(26), 42(26), 45, 48(1), 57(1), 62(1), 71(1), 73(1), 74(1), 84(1), 87(1), 93(1), 94(1), 102(1), 106(1), 122(1), 124(1), 131, 135(1), 137(1), 140(1), 176, 177, 178
Pitzer, K. S., 6, 12, 19, 27, 28, 34, 35, 40(5), 45(5), 46, 48(1), 57(1), 59, 62(1), 70, 71(1, 43), 73(1), 74(1, 43), 84(1), 86(44), 87(1), 88(47), 89(47), 90(47), 91(48), 92(48), 93(1), 94(1), 102(1), 106(1), 122(1), 124(1), 130, 133, 135(1), 137(1), 140(1), 151, 169, 170, 171, 172, 173, 174, 175, 178

Prosen, E. J., 6, 57(2), 70(2), 86, 88(47), 89(47), 90(47), 91(48), 92(48), 94(47), 104, 105(61), 111, 112(67), 113(67), 123(2), 124(2), 130(47), 133(2)

R

Ransom, W. W., 136, 139(92)
Redding, E. R., 116
Reid, R. C., 119, 128
Rossini, F. D., 6, 13, 14(7), 48(1), 57(1, 2), 61, 62, 70(2), 71(43), 73(1), 74(1, 43), 84, 86(44), 87(1), 88(47), 89(47), 90(47), 91(48), 92(48), 93, 94(1), 100, 102(1), 103, 104, 105(61), 106, 108, 109, 111, 113(7), 122(1), 123(2), 124(1, 2), 130, 133(2), 135(1), 137(1), 140(1), 147
Rossner, W., 119
Rowley, D., 136, 139(39), 140(93)
Rubin, T. R., 175
Rubin, W., 146
Rudy, C. E., Jr., 120
Ruhoff, J. R., 133
Russell, H., 175
Russell, K. E., 33, 175

S

Scott, D. W., 57, 75(32), 94, 126, 137(46), 175
Sher, B., 109
Sherman, J., 21, 153
Sherwood, T. K., 119, 128
Smith, H. A., 133
Souclers, M., 58, 67(35), 72, 97(35), 126, 130, 189, 190, 191, 192, 193, 194, 195, 196, 197, 198
Stamm, R., 140
Steiner, H., 136, 139(93), 140(93), 141, 143, 146
Stull, D. R., 62, 121, 229
Szasz, G. J., 92
Szwarc, M., 107

T

Taylor, H. S., 135, 141
Taylor, W. J., 71, 86(44), 130(44)
Thomas, C. T., 141
Thomas, G. D., 95, 96(54), 130
Thomas, L. H., 36

Thompson, H. W., 146
Timpane, N. E., 140, 145(105)
Todd, S. S., 91
Turkevich, J., 141

V

Van Krevelen, D. W., 58, 74, 125, 130, 199, 200, 201, 203
Vaughan, W. E., 133
Verma, K. K., 125, 126, 217, 218, 219, 220, 221, 222

W

Waddington, G., 57, 75(32), 94, 126, 137(46)
Wagman, D. D., 6, 57(2), 70(2), 71(43), 74(43), 86(44), 123(2), 124(2), 130(43, 44), 133(2)
Wait, S. C., 22

Wassermann, A., 138, 146
Watson, K. M., 57, 58, 61(33), 62(33), 67(33), 68(33), 70(33), 113, 115, 116, 118(76), 130(33), 152, 179, 180, 181, 182, 183
Weber, H., 115
Whalen, J., 140
Whitcomb, S. E., 36
Williams, M. G., 71, 86(44), 130(44)
Williamson, K. D., 75, 137(46)
Wilson, E. B., Jr., 19, 26
Wood, J. L., 175

Y

Yost, D. M., 175

Z

Zharkov, V. R., 135
Zolki, T., 175

Subject Index

A

Acetaldehyde, 67, 175
Acetals, heat of combustion equations, 224
Acetic acid, 67
Acetone, 67
  barrier, 175
Acetylene(s), 67, 123
  heat of combustion equations, 224
  dimethyl, barrier, 81, 175
  ethyl, 91
  isopropyl, 91
  n-propyl, 91
Acids, aliphatic, 54
Acid amides, heat of combustion equations, 224
Acid anhydrides, heat of combustion equations, 224
Acids, heat of combustion equations, 224
Acrylonitrile, 139–141
Activity, 8
  coefficient, 10, 11
Additivity, and structure; see Partition function, Vibrational frequencies, Statistical thermodynamic functions
Alcohols, 53
  heat of combustion equations, 224
  heat of formation, 103
  polyhydroxy; heat of combustion equation, 224
Aldehyde group, 52, 183, 188
Aldehydes, heat of combustion equations, 224
1-Alkenes, 104, 112
Alkylbenzenes (see also Benzene), 104
n-Alkylcyclohexanes, 104, 112
n-Alkylcyclopentanes, 104, 112
Alkyl halides, 85, 216
Allene, 92
Amines, heat of combustion equations, 224
Amino group, 52, 183, 188
Andersen, Beyer and Watson, method of, (see Group Increments, Construction of Correlations)
Anti-configuration, 32
Antoine vapor pressure equation, 113

Appearance potentials (see Ionization potentials)
Aromatic ring contributions, 184–187
Aromatization, 97, 143, 144
Atom, models, use of, 21, 23, 33, 45
  table of covalent radii, 152
Atomic properties, additivity of, 76–77
Avogadro's number, 17, 18, 46

B

Ball-like molecules, 45–48
Barrier, potential, and hindered rotation (see also Hindered rotation, Internal rotation, Potential barriers), 28–30
  long chain paraffins, 38, 41
  mono-olefins, 88–89
  table of values, 175
Bennewitz and Rossner, heat capacity equation, 119
Benson and Buss method of (see Group Increments, Construction of Correlations)
Benzene, 87, 123, 141–143
  alkyl, 92, 104, 112
  1,3-dimethyl, 92
  entropy, 179
  ethyl, 92
  group increments, entropy, 179
  heat capacity, 179
  heat of formation, 179
  heat capacity, 179
  heat of formation, 179
  isopropyl, 92
  1-methyl-3-ethyl, 71, 92
  thermodynamics of hydrogenation, 133–136, 143
  1,3,5-trimethyl, 71, 73
Benzonitrile, cyclization with butadiene, 144–145
Bingham equation, 114
Boltzmann constant, 18–19
Bond angles, table, 152
Bond properties, additivity of, 77–78
Bond energy, cyclic compounds, 108
  heat of formation and, 101

ring compounds, 107, 108
   table, 151
Bremner and Thomas, method of, 95–98
Bromine increment, 52, 183
1,2-Butadiene, 91, 92
   3-methyl, 92
1,3-Butadiene, 52, 94, 133, 139–141
   2-methyl, 92
   thermal dimerization, 136–139
Bryant method of (see Group Increments, Construction of Correlations)
Butane, 41, 91, 94, 95, 100, 101, 120, 123
   barrier, 175
   2,2-dimethyl, 93, 100, 102
   2,2,3,3-tetramethyl, 70, 71
   2,2,3-trimethyl, 74
1-Butene, 88, 90, 91, 92, 100, 133
   barrier, 175
   3,3-dimethyl, 93
   2-methyl, 89, 90, 94
   3-methyl, 89
2-Butene, 88, 90, 91, 94, 97, 107
   2,3-dimethyl, 90
   2-methyl, 91, 94
1-Butyne, 94
   3-methyl, 69, 71, 74, 89, 90
2-Butyne, 94

C

Carbohydrates, heat of combustion equations, 224
Carbon dioxide, 68
   free energy of formation, 203
Carbon disulfide, free energy of formation, 203
Carbon monoxide, free energy of formation, 203
Carbon oxi-sulfide, free energy of formation, 203
Carbonium ions, heat of formation of, 106–107, 203
Carboxylic group, 52, 183, 188
Carbylamines, heat of combustion equations, 224
Catalysis, 3, 139, 141, 144, 146
Chemical binding energy, 100–101
Chemical reactions, thermodynamic or kinetic control of, 3
Chermin, method of (see Construction of Correlations)
Chlorine increment, 52, 183

Ciola, method of (see Construction of Correlations)
Clapeyron equation, 113
Classical partition function, see Partition function, Long chain hydrocarbons, 36, 38
Clausius-Clapeyron equation, 114, 116
Collidine, 60
Combined thermodynamic functions, definition, 7
Comparison of methods, 129–132
Compressibility factor, 113, 114
Construction of correlation, Andersen, Benson and Buss, 76–79
   Beyer and Watson, 60–62
   Bremner and Thomas, 96
   Bryant, 79–83
   Chermin, 147–148
   Ciola, 83–84
   Franklin, 62–64
   heat capacity equations, 118–122
   heat of combustion, 109–112
   heat of formation, 103, 106–108
   heat of vaporization, 115–118
   Huggins, 45–48
   Laidler, 127–128
   Morgan and Lielmezs, 85
   Parks and Huffman, 50–56
   Pitzer, 36–39
   Reid and Sherwood, 128
   Rossini, Pitzer, and associates, 86–89
   Souders, Matthews, and Hurd, 64
   statistical thermodynamics, 15–19
   Van Krevelen and Chermin, 66
   Verma and Doriaswamy, 125–126
   Waddington *et al.*, 126–127
Cottrell binding energies, 100
Cracking reactions, 143
Critical pressure, Meissner and Redding equation, 116
Critical temperature, 115, 116
   Meissner and Redding equation, 116
Critical volume, Meissner and Redding equation, 117
Cyanogen, 68
   free energy of formation, 203
Cyanogen chloride, free energy of formation, 203
Cyanogen iodide, free energy of formation, 203
Cyclic addition reactions, 144–146
Cyclic hydrocarbons, 62

SUBJECT INDEX

Cyclobutane group, 184–187
Cyclohexadiene, 133, 134
Cyclohexane, 95, 96, 97, 107, 133, 134, 143
  alkyl, 104
Cyclohexane group, 52, 184–187
Cyclohexene, 95, 96, 97, 133, 134, 143
  4-cyano, 139–141
  vinyl, 136–139
Cycloparaffins, thermodynamic properties from paraffins and deviations from addivity, 63
Cyclopentane, 143
  alkyl, 104
  entropy, 179
  heat capacity, 179
  heat of formation, 179
  group increments, entropy, 179
  heat capacity, 179
  heat of formation, 179
Cyclopentane group, 52, 179, 184–187
Cyclopropane group, 184–187

D

Decane, 43, 44
Degeneracy of energy levels, 15
Degrees of freedom, polyatomic system, 14–15
  rotational, 15
  vibrational, 18, 19
Depolymerization, 136
Deviations from additivity, 62, 63, 64, 101, 102
Dewar equation, 100
Diatomic molecule, moment of inertia, 21
Dibromoethane, 57, 70
Dichloroethane, barrier, 175
Diels-Alder Reaction, 139–141, 144, 146
Dienophilic group, 144
Dimerization, 136
Dimethylamine, barrier, 175
  entropy, 179
  heat capacity, 179
  heat of formation, 179
  group increments, entropy, 179
  heat capacity, 179
  heat of formation, 179
Dimethylether, 58
  barrier, 175
  entropy, 179
  heat capacity, 179
  heat of formation, 179

Dimethylhydrazine, barrier, 175
Dimethylsulfide, 67
  barrier, 175
Direction cosines, 26–27, 31
Disaccharides, heat of combustion equation, 224
Dobratz, heat capacity equation, 120
Dodecane, 49

E

n-Eicosane, 41
Einstein functions, 19–21, 228
  table, 153–168
Energy, and partition function, 15
  of polyatomic system (see also Molecular energy, Partition function), 13
  separations of levels, 15
Energy difference, rotational isomers, 31
Energy of activation, 3, 136, 138, 139, 140, 141
Entropy, of activation, 140
  changes associated with resonance effects, 97–98
  correction for optical isomers, 33, 65
  examples of calculations, 22, 26, 30–31, 32–33
  of formation, 64, 72
  of internal randomness, 48
  of mixing, 32, 33
  and partition function, 17
  rotational isomers, 33
  of translation and ring closure, 96
Equilibrium constant, 1–5
  and additivity of, 83–85
  and conversions, 8–11
  and pressure, 9
  from kinetic rate equations, 136
  temperature dependent equation, 5
Equilibrium conversions, 8–11
  effect of pressure, 9–11
  examples of calculation of, 8, 9, 10, 11
Ester group, 52, 188
Esters, heat of combustion equation, 224
Estimation methods (see also Structural similarity, Group equations, Group increments), 12
  comparison of, 129–133
  general classification, 11–12
Ethane, 28, 48, 68, 97, 102, 106, 123
  barrier, 175
  1,2-dichloro, 29, 31

SUBJECT INDEX

free energy of formation, 203
hindered rotation, 27–28
trifluoro, 67
Ether linkage, 52, 181, 182, 188
Ethers, heat of combustion equation, 224
Ethylene, 26, 92
Ethylene oxide, 106
Ethyl cyanide, 30–31
   barrier, see also Propionitrile, 175
Ethyl cyclopentane, 111–113
Ethyl group, 52, 184–187
Eugenol acetate, 111
Ewell, method of, 55

F

Fluorine increment, 79–83, 183, 202, 204, 206, 208, 212, 126
Formamide, entropy, 179
   group increments, entropy, 179
   heat capacity, 179
   heat of formation, 179
Franklin, method of, 70–72, 103, 106
Free energy of formation, 62
   temperature dependence, 66
Free energy change (see Standard free energy change, Statistical thermodynamic functions), 4
   criteria for reaction feasibility, 3
   and equilibrium conversions, 7–11
   prediction from statistical thermodynamic functions, 6
   from thermodynamic functions, 7
   by van 't Hoff Isochore, 5–6
Free energy function, and partition function, 17
   long chain hydrocarbons, 41
Free energy increments of Parks and Huffman, 52, 55
Free internal rotation (see Partition function, Internal rotation), 26
Free radicals, group increments, heats of formation, 82–83, 203
Frequency factor, 139
Fugacity, 10
Fugassi and Rudi, method of, 120
Fundamental vibrational frequencies (see also Vibrational frequencies), 14
Furane, 67

G

Gauche configuration, 32, 76–79, 100

Generalized bond frequencies (see Heat capacity), 228–229
Graphite, heat of sublimation, 11, 65, 102
Group contributions, method of
   Andersen, Beyer, and Watson, 58, 60–62, 68–70, 130, 132
   examples of estimates by, 69–71, 73–75, 145
   Benson and Buss, 76–79
   Bryant, 79–83
   Chermin, 147–148
   Ciola, 83–84
   Franklin, 59, 62–64, 70–71, 130, 132
   Laidler, 127–128
   Morgan and Lielmezs, 85
   Souders, Matthews and Hurd, 59, 64–66, 68–72, 130, 131
   Van Krevelen and Chermin, 60, 66–68, 74–75, 130, 132
   Verma and Doriaswamy, 125–126
   Waddington et al., 126–127
Group equations, method of Rossini, Pitzer, and associates, 86–95; 130, 131
   examples of, 87, 88, 89, 90, 91–95, 97
   extension of aliphatic data to aromatics, 95–98
Group increment tables,
   Andersen, Beyer, and Watson, 179–183
   Benson and Buss, 204–206
   Bryant, 208
   Chermin, 147
   Ciola, 209–215
   Franklin, 184–188, 203
   long chain paraffins, 176–178
   Laidler, 223
   Morgan and Lielmezs, 216
   Souders, Matthews and Hurd, 187–198
   Van Krevelen and Chermin, 202–203
   Verma and Doriaswamy, 217–222

H

Halogen, compounds containing, 54, 67, 85, 216
Heat capacity, equations, 64, 123–124
   examples of estimates, 121, 123, 124
   generalized bond frequency, contributions, 229
   and Einstein functions, 228
   method of, 119–123
   long chain hydrocarbons, 41
   and partition function, 17
   solutions to Einstein functions, 230–236

Heat of combustion, 108–113, 127, 223
    alkyl paraffin increments, 111
    by group increments, 108–113, 127, 223
    empirical equations, 224
    equation of Prosen and co-workers, 111–112
    of Kharasch and Sher, 109–110
    examples of estimates, 110, 111, 117, 128
    structural correlations, 223, 225–227
Heat content function, and partition function, 17
    long chain hydrocarbons, 41
Heat of formation, 62, 99, 100–102
    additivity and ring closure, 96
    from bond energies, 100–102, 126, 127
    from group increments, 72, 103–106, 125–126, 127
    examples of estimations, 73, 102, 105, 117, 125, 126, 127
    gaseous free radicals and carbonium ions, 106–108
    increments by Ewell, 55, 56
Heat of reaction, 5, 137
Heat of vaporization, 113–117
    graphical methods, 116
    Othmer equation, 115–116
    Watson equation, 115
Heptane, 39, 42, 43, 95
    thermodynamic properties, table, 177
1-Hexadecene, 111–113
Hexafluoroethane, barrier, 175
Hexane, 41, 42, 95, 100, 106, 141–143
    various reactions, 142–143
Hexene, 90, 106, 141–143
Hildebrand principle, 115
Hindered internal rotation (see also Internal rotation), 34
    changes in cyclization, 96, 97
    example of calculation, 30
Huggins, method of, 45
Hydrazine, barrier, 175
Hydrocarbons, 35–49
    acetylene, 91
    branching increments, free energy of formation, 200
    conjugation and adjacency increments, entropy of formation, 194
    cyclic, 51
    diolefins, 91–92
    entropy of formation, 193
    entropy of kinking, 45, 46
    free energy, 186
    free energy of formation, 187, 199
    group increments, entropy, 179–182
        heat of formation, 194
        heat capacity, 179–182
        heat content, 184
        heat of formation, 179–182, 185, 193
        heat of combustion equations, 224
    internal rotational increments,
        entropy of formation, 196
        heat capacity, 192
        heat content, 190
        heat of formation, 198
    mono-olefins, 90, 104
    polycyclic, heat of combustion equations, 224
    ring formation increments, 200
    unsaturated (see also Long chain hydrocarbons), 51
    vibrational group increments, entropy of formation, 195
        heat capacity, 191
        heat content, 189
        heat of formation, 197
Hydrogen, 11, 65
Hydroxyl group, 52, 183, 188

I

Infinite chain model, examples of estimates, 43
    long chain hydrocarbons, 36
Imides, heat of combustion equations, 224
Internal rotation (see also Hindered internal rotation), 26–34, 88–90
    free, 26–27
    long chain hydrocarbons, 38
    rotational isomers, 32
Iodine increment, 52, 183, 216
Ionization potentials, and heat of formation, 106
Isobutane, 44, 92, 94, 100, 123
    barrier, 175
Isobutene, 90, 91, 92, 100
Isonitrile increment, 188

K

Kharasch and Sher, method of, 109–111, 224–227
Ketone group, 52, 183, 188
Ketones, heat of combustion equation, 224

SUBJECT INDEX

Kinetics (see Reaction rates)
Kistiakowsky equation, 115

L

Lactones, heat of combustion equation, 224
Laidler, method of (see, Construction of Correlations)
Linear molecule, moment of inertia, 21
   vibrational degrees of freedom, 19
Long chain hydrocarbons, 35–49
   additive increments, 176–178
   branch chain, 44
   infinite chain molecular model, 36
   partition function, 37–39
   random kinked and ball-like models, 45–49
   steric factor, 38
   thermodynamic properties equation, 39–41
   vibrational frequencies, 36
   unsaturated, 44

M

Maxwell-Boltzmann distribution law, 15
Mecke's generalized vibrational modes, 119
Meissner and Redding equations, 116
Mercaptan, barrier, 175
Methane, 123, 124
   entropy, 179
   group increments, entropy, 179
   heat capacity, 179
   heat of formation, 179
   heat capacity, 179
   heat of formation, 179
Methanol, barrier, 175
   free energy of formation, 203
Method of group equations, 9
   of group increments, examples of estimates, 137
   of structural similarity, example of estimates, 127, 133
Methyl alcohol (see also Methanol), 31, 34, 68
Methyl cyanide, 61
Methyl group, 52, 179, 181, 182, 184–187
Methylamine, barrier, 175
   entropy, 179
   group increments, entropy, 179
   heat capacity, 179
   heat of formation, 179
   heat capacity, 179
   heat of formation, 179
Methylene group, 52, 184–187
Methylcyclopentane, 142
Methylhydrazine, 32
   barrier, 175
Molecular energy, electronic, 14
   external rotational, 14
   internal rotational, 14
   nuclear spin, 15
   translational, 13–14
Moment of inertia, 21–22
   example of calculation, 23–24
   product of principal, 22
   randomly kinked and ball-like molecules, 47
   reduced, 26–27
Monosaccharides, heat of combustion equation, 224
Morgan and Lielmezs, method of (see Construction of Correlations)
Multiple bond contributions, 181, 182, 184–187

N

Naphthalene, entropy, 179
   group increments, entropy, 179
   heat capacity, 179
   heat of formation, 179
   heat capacity, 179
   heat of formation, 179
Neopentane, 44, 100
   barrier, 175
Nitrate increment, 188
Nitric acid, barrier, 175
Nitric oxide, free energy of formation, 203
Nitriles, heat of combustion equations, 224
Nitro compounds, heat of combustion equations, 224
Nitro group, 52, 183, 188
Nitrogen, compounds containing, 54, 67
Nitrogen, increments, 52, 183, 188
Nitrogen dioxide, 68
   free energy of formation, 203
Nitrogen tetroxide, free energy of formation, 203
Nitromethane, 60

SUBJECT INDEX

Nonhydrocarbons, 63, 67, 85, 94
  group increments, entropy, 183
    free energy of formation, 188, 201–202
    heat capacity, 183
    heat of formation, 183, 188
  group substitutions, 52
Nonlinear molecule, moments of inertia, 21–22
  number of vibrational modes, 25
  vibrational degrees of freedom, 14, 19

O

Octane, 11, 42
Olefins, heat of combustion equations, 224
  hydrogenation, 84, 209
Optical isomerism, contribution to entropy, 33, 65
Othmer equation, 115
Oxygen compounds, 53–54, 67
  increments, 52, 181, 182, 183, 188

P

Parachor, 117
  atomic and structural increments, 228
Paraffins, 36, 45, 51, 104
  additive increments for infinite chain model, 176–178
  branch chain, 44, 51
  unsaturated, 44, 51, 90–92
Parent group, 58, 60, 68
Parks and Huffman, method of, 50
Partition function, 15–19
  additive properties, 18
  approximations for evaluation for polyatomic molecules, 16
  evaluation at high temperatures, 26
  free rotation, 26
  high temperatures, 35
  hindered rotation, 28–30
  long chain hydrocarbons, 36–39
  rigid rotator-simple vibrator, 17–18
  thermodynamic functions, 16–17
Pentane, 28, 41, 90, 95, 100
  2-methyl, 102
1,2-Pentadiene, 91
1,3-Pentadiene, 94, 133
1,4-Pentadiene, 92
2,3-Pentadiene, 91
Pentanethiol, 56–57

1-Pentene, 3-methyl, 94
cis-2-Pentene, 88–90, 133
Person and Pimentel, method of, 39
Phenol(s), 52, 183, 188
  heat of combustion equation, 224
Phenyl group, 52, 183
Phenylpyridine, 145
Phosgene, free energy of formation, 203
Picoline, 60
Pitzer, method of, 36
Planar zig-zag molecules, 35–44
Planck constant, 17–18
Potential barrier, 28–30
  long chain hydrocarbons, 37
  non-equivalent minima, 31–32
Polyenes, 63
Pressure, effect of, on equilibrium conversions, 9–11
Principal axes, 22
  direction cosines, 27
Products of principal moments of inertia, 22
  example of calculation, 24
Propadiene, 81
Propane, 90, 91, 92, 93, 94, 106, 120, 121, 142
  barrier, 175
Propionitrile (see also Ethyl cyanide), 94
Propylene, 88, 90, 91, 92, 93, 94, 106, 142
  barrier, 84, 175
Propyne, 94
Prosen and co-workers, method of, 104, 111
Pyridine, 60, 67

Q

Quadrupole forces and hindered internal rotation, 30
Quinoline, 60
Quinones, heat of combustion equation, 224

R

Randomly kinked molecules, 45–49
Reaction rate, 146
  butadiene dimerization, 136
  catalysis, 141
  frequency factor and entropy of activation, 138

relative, predicted from kinetic and thermodynamic estimates, 140
and yields, 3
Reduced moment of inertia (see also moment of inertia)
example of calculation, 31
long chain paraffins, 41
Reduced pressure, 116
Reduced temperature, 116
Reid and Sherwood, method of (see Construction of Correlations)
Resonance energy, method of group equations, 97
ring stability, 144
Retenequinone, 111
Rigid rotator-simple vibrator, 17–26
Ring formation, 144
group equations, 95–98
Rossini effect, 100
Rossini, Pitzer, and associates, method of, 86
Rotational isomers (see Statistical thermodynamic functions, Potential barrier)

S

Saccharinic acid lactone, 110
Sakur-Tetrode equation of translational entropy, 96
Safrole, 111
Skew configuration, 32–33
Souders, Matthews, and Hurd, method of, 64
Stability (see Thermodynamic stability)
Standard free energy change, 4
effect of pressure on, 9
estimation by modified van 't Hoff Isochore, 5
change, procedure for obtaining, 4
temperature dependent equation, 5
and thermodynamic equilibrium, 8
Standard free energy of formation, 4
Standard states, 4, 9
and thermodynamic activity, 8
State sum (see also Partition function), 16
Statistical thermodynamic functions, 6
entropy of ball-like and randomly kinked molecules, 48
equations for calculation of, 19, 20
examples of calculations of, 22–26, 30–31, 32–33

free internal rotation, 27
group equations and increments, 70, 72, 90
n-heptane, table of, 177
hindered internal rotational contributions, tables, 169–174
hindered rotational contribution, calculations of, 30–31
(see also Ethyl cyanide)
methylene increment, equation, 42
long chain hydrocarbon calculations, 42
equations, 39–41
rigid rotator-simple vibrator calculations, 22–26
rotational isomers, calculations, 32–33
Statistical thermodynamic methods, 11–45, 129–131
of Huggins for long chain hydrocarbons, 45–49
Steric factor, 45
Strain, in rings, 107
Structural similarity, method of, 50, 54, 55, 130, 132
example of estimates, 50–51, 140
increments by Ewell, 56
increments by Parks and Huffman, 52
Stull and Mayfield, heat capacity equation, 121
Sub-molecule, of a polyatomic system, 46
Succinonitrile, 32, 94
Sulfur compounds, 54, 67, 75
increment, 52, 183, 188
Sulfur dioxide, 68
free energy of formation, 203
Sulphur trioxide, free energy of formation, 203
Symmetrical top, 27, 29
internal rotation, 28, 30
long chain hydrocarbons, 38
total symmetry number, 65, 87, 93
Symmetry correction, 65, 66, 70, 87, 90, 93, 94, 97, 131, 133
Symmetry number, 18–19, 21, 23, 32

T

Temperature, units of, 12
Terpenes, heat of combustion equations, 224
Tetramethylsilane, barrier, 175

SUBJECT INDEX

Thermodynamics, acrylonitrile butadiene cyclization, 139–141
    dimerization of butadiene, 136–139
    hydrogenation of benzene, 133
    ring closure of paraffins, 141–143
Thermodynamic criteria for reactions, 8, 138–140, 141–143
Thermodynamic properties and molecular energy levels, 13–15
    and molecular structure, 17
    and partition function, 15–17
Thermodynamic stability, branch chain paraffins, 51, 101
    of various rings, 53
Thiocyclopropane, 106
Thiols, 85, 202, 204, 207
o-Toluic acid, 110
Toluene, 92, 125
p-Toluidene, 110
Torsion-oscillator-rotator, 28
Trans configuration, internal rotational isomers, 32–33
Translational contributions (see also Molecular energy Partition function), 17, 18
    example of calculation, 22
Trichloroethane, barrier, 175
Trifluoroacetonitrile, statistical thermodynamic properties, 22–25
Trifluoroethane barrier, 175
Trimethylamine, barrier, 175
    entropy, 179
    heat capacity, 179
    heat of formation, 179
    group increments, entropy, 179
        heat capacity, 179
        heat of formation, 179, 224
Triethylamine, 118
Trouton's rule, and modifications 114–115

V

Van Krevelen and Chermin, method of, 66
Van't Hoff equations, 5, 8, 134, 137, 145
Verma and Doriaswamy, method of (see Construction of Correlations)
Vibrational contribution, 19–21
    example of calculation, 25
Vibrational frequencies (see also Ethyl cyanide, Ethylene, Long chain hydrocarbons, Trifluoroacetonitrile)
    characteristic frequencies for paraffins 46
    and degrees of freedom, 14
    force constants and, 18
    fundamental skeletal-modes, 36
    generalized bond assignments and heat capacity, 119–123
    valence and deformation modes, 119
Vinyl chloride, estimation of thermodynamic properties for, 85, 204
2-Vinylpyridine, 3,6-dihydro-, 139–141

W

Waddington *et al.*, method of (see Construction of Correlations)
Watson equation, 113

X

Xylene, barrier, 175

Z

Zero point energy, 6, 15, 100